Foundations of Engineering Mechanics

Series editors

V.I. Babitsky, Loughborough, Leicestershire, UK
Jens Wittenburg, Karlsruhe, Germany

More information about this series at http://www.springer.com/series/3582

Ranjan Ganguli

Finite Element Analysis of Rotating Beams

Physics Based Interpolation

Springer

Ranjan Ganguli
Department of Aerospace Engineering
Indian Institute of Science
Bengaluru, Karnataka
India

ISSN 1612-1384 ISSN 1860-6237 (electronic)
Foundations of Engineering Mechanics
ISBN 978-981-10-1901-2 ISBN 978-981-10-1902-9 (eBook)
DOI 10.1007/978-981-10-1902-9

Library of Congress Control Number: 2016947015

Printed on acid-free paper

This Springer imprint is published by Springer Nature
The registered company is Springer Science+Business Media Singapore Pte Ltd.

Preface

Rotating beams are important mathematical models for structures such as helicopter rotor blades, wind turbine rotor blades, propellers, turbine blades and robotic manipulators. Therefore, the modeling and analysis of rotating beams is an important practical problem. The natural frequencies of the rotating beam should be kept away from multiples of the rotor speed. Therefore, accurate frequency prediction is important. Control of rotating beams requires the development of low-order models. This motivates the development of efficient mathematical models. Since the rotating beam equation does not have a simple exact solution, approximate methods such as Rayleigh–Ritz, Galerkin, and the finite element methods are widely used for the vibration analysis of rotating beams. This book provides an introduction to the finite element for rotating beams. A background on the Rayleigh–Ritz and Galerkin method is also provided.

The first chapter gives a detailed introduction to the rotating beam equation and illustrates the Rayleigh–Ritz, Galerkin, and finite element methods for its solution. Several example problems are given to illustrate the methods. A MATLAB finite element code for the rotating beam is also provided. The following chapters give adaptations of the basis functions which can accelerate the convergence of rotating beam finite elements, thus allowing for efficient low-order models. Some of these basis functions are based on analogies between the piano string and rotating Euler–Bernoulli and between the violin string and rotating Timoshenko beams. A major theme here is to choose finite element interpolation functions which are closer to the problem physics.

This book should be useful to engineers, graduate students and researchers working on rotating beam problems. It is also useful for people in the area of computational mechanics and applied mathematics.

The author is grateful to his advisors Dr. P.K. Datta and Dr. Indrajit Chopra for introducing him to the rotating beam problem. He is also grateful to his students

Jagadish Babu Gunda, Parampal Chhabria, A.S. Vinod Kumar, K.G. Vinod, D. Sushma, and R. Ganesh for helping in the research. He is grateful to his research collaborators Dr. R.K. Gupta and Dr. S. Gopalkrishnan. He is grateful to Mr. K. Kalaiselvam for helping in the preparation of the manuscript.

Bengaluru, India
2016

Ranjan Ganguli

Contents

About the Author

Prof. Ranjan Ganguli obtained his M.S. and Ph.D. in Aerospace Engineering from the University of Maryland, College Park, in 1991 and 1994, respectively, and his B.Tech degree in Aerospace Engineering from the Indian Institute of Technology, Kharagpur, in 1989.

Following his Ph.D., he worked at the Alfred Gessow Rotorcraft Center of the University of Maryland as Assistant Research Scientist until 1997 on projects on rotorcraft health monitoring and vibratory load validation for the Naval Surface Warfare Center and United Technology Research Center, respectively. He also worked at the GE Research Lab in Schenectady, New York, and at Pratt and Whitney, East Hartford, Connecticut, from 1997 to 2000. He joined the Aerospace Engineering department of the Indian Institute of Science, Bangalore, as Assistant Professor in July 2000. He was promoted to Associate Professor in 2005 and to Full Professor in 2009. He is currently the Satish Dhawan Chair Professor at the Indian Institute of Science, Bangalore. He has held visiting positions at TU Braunschweig, University of Ulm and Max Planck Institute of Metal Research, Stuttgart, in Germany; University Paul Sabatier and Institute of Mathematics, in Toulouse, France; Konkuk University in South Korea, the University of Michigan, Ann Arbor, in USA, and the Nanyang Technological University, Singapore.

Professor Ganguli's research interests are in helicopter aeromechanics, aeroelasticity, structural dynamics, composite and smart structures, design optimization, finite element methods and health monitoring. He has published 178 articles in refereed journals and over 100 conference papers. He was awarded the ASME Best Paper Award in 2001, the Golden Jubilee Award of the Aeronautical Society of India in 2002, the Alexander von Humboldt fellowship in 2007 and the Fulbright Senior Research fellowship in 2010. Professor Ganguli is a Fellow of the American Society of Mechanical Engineers, a Fellow of the Royal Aeronautical Society, UK, a Fellow of the Indian National Academy of Engineering, a Fellow of the Aeronautical Society of India, an Associate Fellow of the American Institute of Aeronautics and Astronautics, and a Senior Member of IEEE.

He has taught courses on flight and space mechanics, engineering optimization, helicopter dynamics, aircraft structures, structural mechanics, aeroelasticity, and navigation. He has supervised the thesis of 15 Ph.D. and 35 Master's degree students. He has written books "Engineering Optimization" and "Gas Turbine Diagnostics," both published by CRC Press, New York, and books titled "Structural Damage Detection using Genetic Fuzzy Systems" and "Smart Helicopter Rotors," published by Springer.

Chapter 1
Introduction

1.1 Introduction

Rotating beams are useful mathematical models for helicopter rotor blades, wind turbine blades, turbine blades, robotic manipulators and propeller blades. Therefore, the rotating beam equation is a topic of great practical interest. In this chapter, we derive the equation of motion of the rotating beam and present some approximate methods for its solution. Approximate methods are very useful for obtaining frequencies of rotating beams. The Galerkin method and Rayleigh–Ritz method can be used for this purpose.

1.1.1 Elastic Blade

A good model of a rotor blade is an elastic beam restrained at the root, as shown in Fig. 1.1. Typically, helicopter rotor blades are treated as long slender beams. Treating the blade as a slender beam is appropriate since cross-sectional dimensions are much smaller than the length. At first, we will consider only out-of-plane or flap bending. Also, we make the Bernoulli–Euler assumption: a plane section normal to the neutral axis (beam axis) remains plane after deformation. It is also assumed that shear deflection is negligible and rotary inertia is neglected. This implies that the effect of rotation of element is small compared to vertical displacement. In addition, structural damping is neglected. With these assumptions, we are ready to derive the elastic blade governing differential equation.

Consider a blade section as shown in Fig. 1.2. Here $f_z(r, t)$ is the vertical load per unit length and $w(r, t)$ is the vertical deflection at station r. Several forces act on this blade section: $m\ddot{w}dr$ is the inertia force, f_z dr is the external force, $f_H dr$ is the axial force parallel to r, T is the axial force (Positive for tension), S is the shear force at the cut section (Positive as shown), M is the bending moment (Positive when top fiber is under compression).

R. Ganguli, *Finite Element Analysis of Rotating Beams*,
Foundations of Engineering Mechanics, DOI 10.1007/978-981-10-1902-9_1

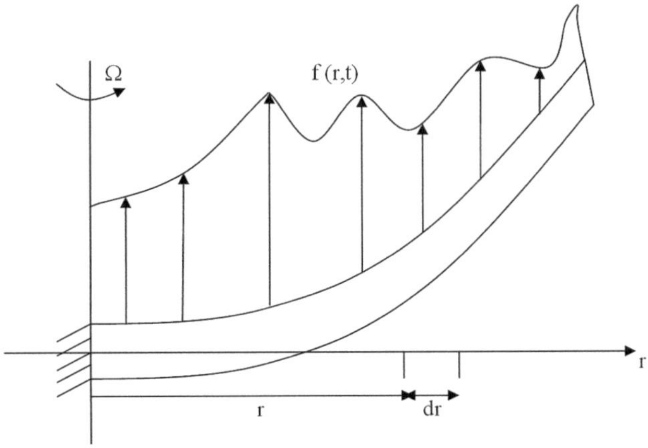

Fig. 1.1 Elastic blade with distributed load

Fig. 1.2 Elastic blade section

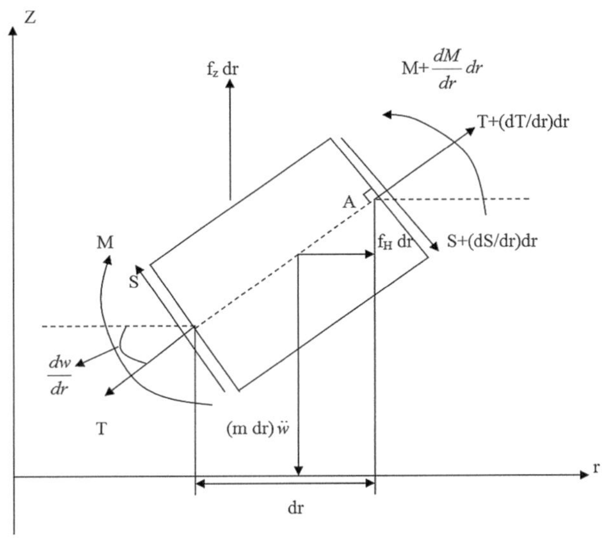

Also, $\frac{dw}{dr}$ is the bending deflection slope, assumed to be small. Therefore,

$$\cos\left(\frac{dw}{dr}\right) \approx 1 \qquad \sin\left(\frac{dw}{dr}\right) = \frac{dw}{dr} \tag{1.1}$$

Equilibrium of forces on element, in the r-direction yields

$$\sum F_r = -T + T + \frac{dT}{dr}dr + f_H dr - S\frac{dw}{dr} + \left(S + \frac{dS}{dr}dr\right)\left(\frac{dw}{dr} + \frac{d^2w}{dr^2}dr\right) = 0$$

$$\Rightarrow \frac{dT}{dr} + f_H + \frac{d}{dr}\left(S\frac{dw}{dr}\right) = 0 \tag{1.2}$$

Forces in the Z-direction yield,

$$\sum F_Z = f_z dr + S - S - \frac{dS}{dr}dr - m\ddot{w}dr - T\frac{dw}{dr} + \left(T + \frac{dT}{dr}\right)\left(\frac{dw}{dr} + \frac{d^2w}{dr^2}dr\right) = 0$$

$$\Rightarrow f_z - \frac{dS}{dr} - m\ddot{w} + T\frac{d^2w}{dr^2} + \frac{dT}{dr}\frac{dw}{dr} = 0 \tag{1.3}$$

Or

$$\frac{dS}{dr} = f_z - m\ddot{w} + \frac{d}{dr}\left(T\frac{dw}{dr}\right)$$

The above equation tells that differential of shear is loading distribution.

Now, we take moment about the point A in Fig. 1.2,

$$\sum M = M + Sdr - M - \frac{dM}{dr}dr = 0 \tag{1.4}$$

$$\frac{dM}{dr} = S$$

Thus, differential of moment is shear distribution. From beam theory,

$$M = EI\frac{d^2w}{dr^2} \tag{1.5}$$

Also,

$$S = \frac{dM}{dr} = \frac{d}{dr}EI\frac{d^2w}{dr^2} \tag{1.6}$$

$$\frac{\partial^2}{\partial r^2}\left(EI\frac{\partial^2 w}{\partial r^2}\right) + m\frac{\partial^2 w}{\partial t^2} - \frac{\partial}{\partial r}\left(T\frac{\partial w}{\partial r}\right) = f_z(r,t) \tag{1.7}$$

The above equation is the PDE for beam bending, Force equilibrium in horizontal direction gives,

$$\frac{dT}{dr} + f_H + \frac{d}{dr}\left(S\frac{dw}{dr}\right) = 0 \tag{1.8}$$

$$\frac{d}{dr}\left(S\frac{dw}{dr}\right) \rightarrow Second\quad Order\quad term \approx 0$$

$$\frac{dT}{dr} + f_H = 0. \tag{1.9}$$

To get a physical feel of the problem, we consider a rectangular cross-section as shown in Fig. 1.3.

$$I = \frac{bh^3}{12}$$

m = mass/unit length and E = Young's Modulus. The unit of I is m^4, unit of m is Kg/m and the unit of E is GPa.

1.1.2 Horizontal Force Equilibrium

Consider a case with uniform mass and a small mass element from Fig. 1.4. From Eq. (1.9) we get,

$$f_H = m\Omega^2 r \tag{1.10}$$

$$\frac{dT}{dr} = -m\Omega^2 r \tag{1.11}$$

Integrating,

$$T = -m\Omega^2 \frac{r^2}{2} + C \tag{1.12}$$

At tip, $r = R, T = 0, \Rightarrow C = m\Omega^2 \frac{R^2}{2}$

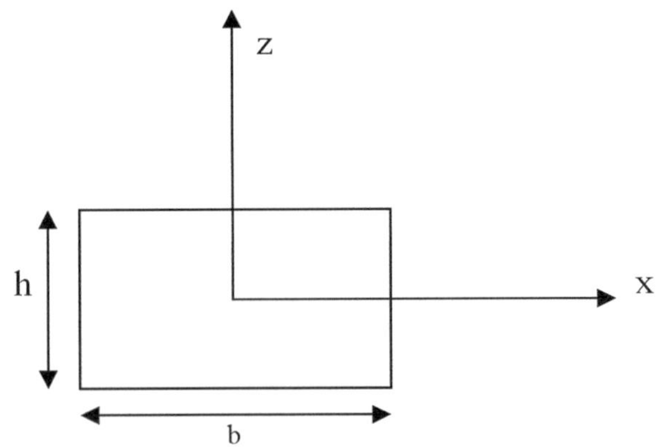

Fig. 1.3 Rectangular cross-section

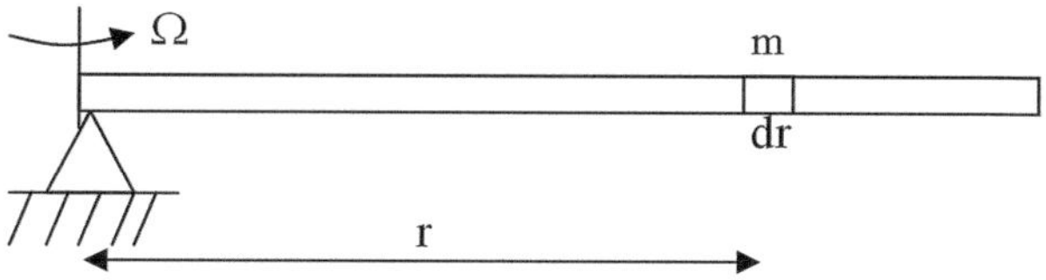

Fig. 1.4 Horizontal force equilibrium

$$T = \frac{m\Omega^2}{2}\left(R^2 - r^2\right) \tag{1.13}$$

In general,

$$T = \int_r^R m\Omega^2 r dr = \int_r^R mr\Omega^2 dr \tag{1.14}$$

1.1.3 Boundary Conditions

The governing equation is a fourth order partial differential equation (PDE) in space and needs 4 boundary conditions. These boundary conditions depend on the physics of the problem (Figs. 1.5, 1.6, 1.7, 1.8 and 1.9). We look at some typical boundary conditions.

(a) Pin end or simply supported (Fig. 1.5)

$$w = 0; M = EI\frac{d^2w}{dr^2} = 0$$

(b) Cantilevered or fixed end (Fig. 1.6)

$$w = 0; \frac{dw}{dr} = 0$$

(c) Free end (Fig. 1.7)

$$M = EI\frac{d^2w}{dr^2} = 0; S = \frac{d}{dr}\left(EI\frac{d^2w}{dr^2}\right) = 0$$

Fig. 1.5 Pinned end

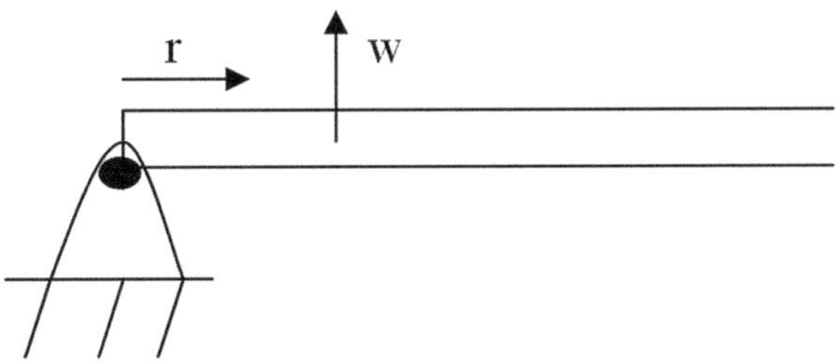

Fig. 1.6 Fixed end

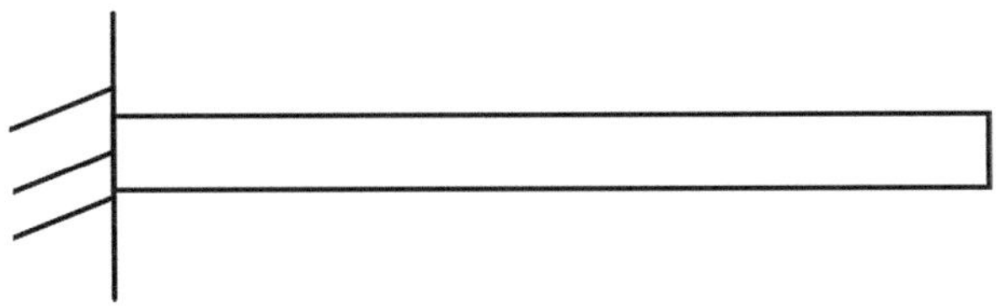

Fig. 1.7 Free end

(d) Vertical spring (Fig. 1.8)

$$M = EI\frac{d^2w}{dr^2} = 0;\ S = \frac{d}{dr}\left(EI\frac{d^2w}{dr^2}\right) = -Kw$$

(e) Leaf spring (Fig. 1.9)

$$M = EI\frac{d^2w}{dr^2} = K_\theta\frac{dw}{dr};\ S = \frac{d}{dr}\left(EI\frac{d^2w}{dr^2}\right) = 0$$

1.1.4 Initial Conditions

The beam PDE is second order in time and so needs two initial conditions: at $t = 0$, $\frac{\partial w}{\partial t}$ and w are prescribed.

1.1.5 Cantilever Beam Vibrations (Non-rotating)

Recall the PDE of the rotating beam,

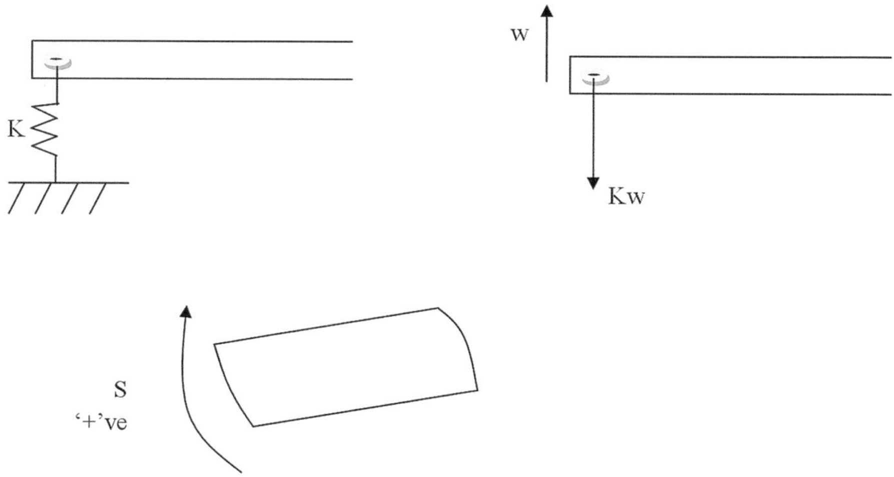

Fig. 1.8 Vertical spring

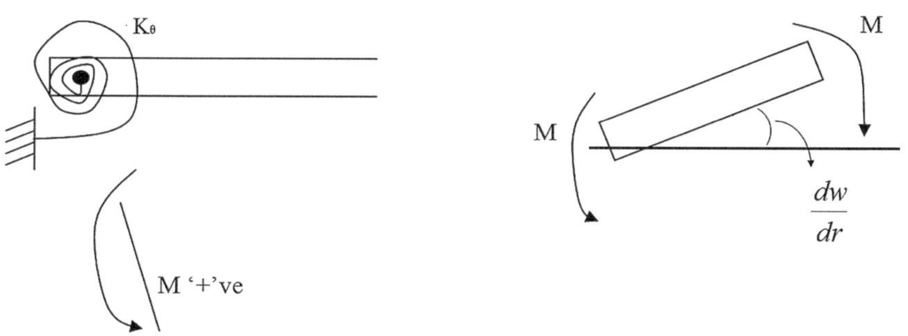

Fig. 1.9 Leaf spring

$$\frac{\partial^2}{\partial r^2}\left(EI\frac{\partial^2 w}{\partial r^2}\right)+m\frac{\partial^2 w}{\partial t^2}-\frac{\partial}{\partial r}\left(T\frac{\partial w}{\partial r}\right)=f_z(r,t) \tag{1.15}$$

For no rotation speed, $T = 0$. If we assume that the mass and stiffness are constant or uniform along the beam,

$$EIw'''' + m\ddot{w} = f_z(r,t) \tag{1.16}$$

If we consider the cantilever boundary conditions:

$$r = 0, w = 0, w' = 0$$

$$r = R, M = EIw'' = 0, w'' = 0; S = EIw''' = 0, w''' = 0$$

To get natural frequencies, set $f_z(r, t) = 0$ (free vibration so force is zero). Then put,

$$w(r, t) = W(r)e^{i\omega t}$$

Here we assume harmonic motion,

$$\frac{d^4W}{dr^4} - \frac{m\omega^2}{EI}W = 0$$

Let, $W = e^{Pr}$

$$P^4 = \frac{m\omega^2}{EI} \Rightarrow P = \left(\frac{m\omega^2}{EI}\right)^{\frac{1}{4}} \tag{1.17}$$

$$P = \pm\lambda, \pm i\lambda$$

where,

$$\lambda = \left(\frac{m\omega^2}{EI}\right)^{\frac{1}{4}}$$

The solution of the equation is then given by

$$w(r) = C\sinh(\lambda r) + D\cosh(\lambda r) + E\sin(\lambda r) + F\cos(\lambda r) \tag{1.18}$$

Boundary conditions are now used to determine the constants C, D, E and F. At $r = 0$,

$$w(0) = C\sinh(0) + D\cosh(0) + E\sin(0) + F\cos(0)$$

$$0 = C.0 + D.1 + E.0 + F.1 = D + F = 0$$

where we use the fact that $\sinh(0) = 0$, $\cosh(0) = 1$.

$$w'(r) = C\lambda\cosh(\lambda r) - D\lambda\sinh(\lambda r) + E\lambda\cos(\lambda r) - F\lambda\sin(\lambda r) \tag{1.19}$$

$$w'(0) = C\lambda + E\lambda = 0$$

Similarly, the boundary conditions for the free end can be enforced.
Collecting all the four equations for the boundary conditions,

$$\begin{bmatrix} 0 & 1 & 0 & 1 \\ 1 & 0 & 1 & 0 \\ \sinh(\lambda R) & \cosh(\lambda R) & -\sin(\lambda R) & -\cos(\lambda R) \\ \cosh(\lambda r) & \sinh(\lambda R) & -\cos(\lambda R) & \sin(\lambda R) \end{bmatrix} \begin{Bmatrix} C \\ D \\ E \\ F \end{Bmatrix} = 0$$

Note the following is used for the hyperbolic functions.

$$\cosh(x) = \frac{e^x - e^{-x}}{2}, \quad \sinh(x) = \frac{e^x + e^{-x}}{2}$$

$$\frac{d(\sinh(x))}{dx} = \cosh(x) = \frac{e^x - e^{-x}}{2}, \quad \frac{d(\cosh(x))}{dx} = \sinh(x) = \frac{e^x + e^{-x}}{2}$$

For non-trivial solution of the four equations for the boundary conditions, we get,

$$\cos(\lambda R)\cosh(\lambda R) = -1$$

$$\cos(\lambda R) = \frac{-1}{\cosh(\lambda R)} \qquad (Transcendental \;\; equation)$$

Such an equation can be solved numerically or graphically. The graphical solution is shown in Fig. 1.10.

$$\lambda R = 0.597\pi, 1.49\pi, \frac{5\pi}{2}, \frac{7\pi}{2}, \ldots$$

There are an infinite number of solutions and therefore, an infinite number of eigenvalues

$$\lambda = \left(\frac{m\omega^2}{EI}\right)^{\frac{1}{4}} \Rightarrow \omega_j = (\lambda R)_j^2 \sqrt{\frac{EI}{mR^4}}$$

Mode shape corresponding to a particular eigenvalue can be obtained by substituting R in any three equations, say the first three equations.

$$\begin{bmatrix} 0 & 1 & 0 \\ 1 & 0 & 1 \\ \sinh(\lambda R) & \cosh(\lambda R) & -\sin(\lambda R) \end{bmatrix} \begin{Bmatrix} C \\ D \\ E \end{Bmatrix} = \begin{bmatrix} 1 \\ 0 \\ -\cos(\lambda R) \end{bmatrix}$$

Set F = 1, Calculate C, D and E.
Then,

$$\phi(r) = \frac{\cosh(\lambda R) + \cos(\lambda R)}{\sinh(\lambda R) + \sin(\lambda R)} \{(\sinh(\lambda r) - \sin(\lambda r)) - \cosh(\lambda r) + \cos(\lambda r)\}$$

For each $(\lambda r)_j$, we have $\phi_j(r)$ as the mode shape. These are free vibration modes (Fig. 1.11). Natural mode shapes represent the unique ways in which a beam can vibrate when it is excited. If an impulsive force is applied at the tip, almost all modes get excited. Typically, most of the response contribution comes from the first three to four modes.

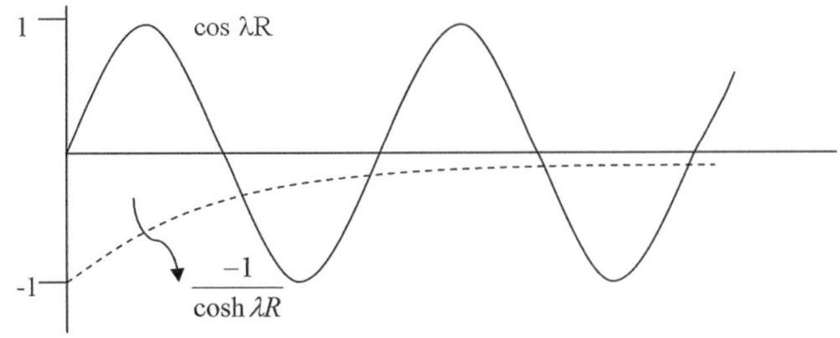

Fig. 1.10 Graphical solution of beam transcendental equation

Fig. 1.11 Mode shapes of cantilever beam

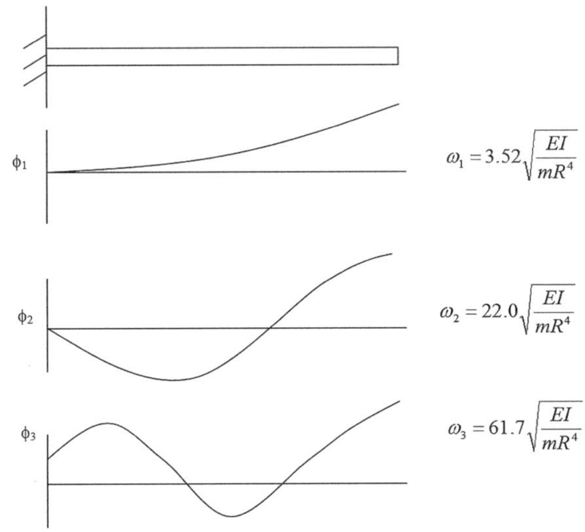

1.1.6 *Beam Functions*

The mode shapes of the cantilever beam are used in many situations and are called beam functions,

$$\phi_j(x) = \cosh(\lambda_j x) - \cos(\lambda_j x) - \alpha_j\left(\sinh(\lambda_j x) - \sin(\lambda_j x)\right) \tag{1.20}$$

where,

$$\alpha_j = \frac{\cosh(\lambda_j R) + \cos(\lambda_j R)}{\sinh(\lambda_j R) + \sin(\lambda_j R)}$$

They satisfy all the boundary conditions.

$$\omega_j = \left(\lambda_j R\right)^2 \sqrt{\frac{EI}{mR^4}}$$

i	1	2	3	4	j
$\lambda_j R$	1.8751	4.6941	7.8548	10.9955	$(2j-1)\frac{\pi}{2}$
α_j	0.7341	1.0185	0.9992	1.000	1

For a cantilever beam, mode shapes are orthogonal,

$$\int_0^R m\phi_i(r)\phi_j(r)dr = 0 \qquad i \neq r \tag{1.21}$$
$$= \delta_{ij} M_i$$

$$\int_0^R \phi_i \frac{d^2}{dr^2}\left(EI\frac{d^2\phi_j}{dr^2}\right) dr = 0 \qquad i \neq j \tag{1.22}$$
$$= \delta_{ij}\omega_i^2 M_i$$

Here δ_{ij} is the Kronecker delta. Also M_i = Generalized mass = $\int_0^R m\phi_i^2 dr$.

Beam functions are of great value for calculating an approximate solution as they satisfy all the boundary conditions for the problem. We can assume the deflection in terms of a series of beam functions. Because of orthogonality, the beam functions have good convergence characteristics. The beam functions for uniform beams can be used as basis functions for approximate methods used for non-uniform and rotating beams.

1.1.7 Rotating Beam Vibration

The PDE for free vibration is given by

$$\frac{\partial^2}{\partial r^2}\left(EI\frac{\partial^2 w}{\partial r^2}\right) + m\frac{\partial^2 w}{\partial t^2} - \frac{\partial}{\partial r}\left(T\frac{\partial w}{\partial r}\right) = 0 \tag{1.23}$$

where

$$T(r) = \int_r^R mr\Omega^2 dr$$

Let

$$w(r,t) = W(r)e^{i\omega t}$$

$$\frac{d^2}{dr^2}\left(EI\frac{d^2W}{dr^2}\right) - \frac{d}{dr}\left(\int_r^R mr\Omega^2 dr \frac{dW}{dr}\right) - m\omega^2 W = 0 \tag{1.24}$$

This equation has no simple solution. We can also write this equation in non-dimensional form. Let,

$$\eta = \frac{r}{R}, \quad w = \frac{W}{R}$$

Also, m_0 is the reference mass per unit length.

$$\frac{d^2}{dx^2}\left(\frac{EI}{m_0\Omega^2R^4}\frac{d^2w}{dx^2}\right) - \frac{d}{dx}\left(\int_\eta^1 \frac{m}{m_0}\eta d\eta \frac{dw}{d\eta}\right) - \frac{m}{m_0}\nu^2 w = 0 \tag{1.25}$$

Here, $\frac{EI}{m_0\Omega^2R^4}$ is the non-dimensional stiffness, $\frac{m}{m_0}$ is the non-dimensional mass per unit length and ν is the non-dimensional bending frequency. We now consider the Galerkin method for solving this equation.

1.2 Galerkin Method

The Galerkin method is based on the equilibrium equation of the structure. We assume the deflection as a function which must satisfy all the boundary conditions. We substitute this assumed deflection into the differential equation, and obtain the error. Then we minimize the error by integrating the weighted error over the complete domain.

Consider the governing partial differential equation (PDE) of the rotating beam,

$$\frac{\partial^2}{\partial r^2}(EI\frac{\partial^2 w}{\partial r^2}) - \frac{\partial}{\partial r}(T\frac{\partial w}{\partial r}) + m\ddot{w} = f_z(r,t) \tag{1.26}$$

We assume a series solution,

$$w_{app} = \sum_{i=1}^{N} Y_i(r)q_i(t) \tag{1.27}$$

Here $Y_i(r)$ is the assumed deflection shape which must satisfy all boundary conditions. If w_{app} is the exact solution, then PDE is fully satisfied. However, this is a hypothetical case. In general, we will get an error term of the form,

$$\varepsilon(r,t) = m\ddot{w}_{app} + \frac{\partial^2}{\partial r^2}(EI\frac{\partial^2 w_{app}}{\partial r^2}) - \frac{\partial}{\partial r}(T\frac{\partial w_{app}}{\partial r}) - f_z \tag{1.28}$$

$$\varepsilon(r,t) = \sum_{i=1}^{N}[mY_i\ddot{q}_i + (EIY_i^{''})^{''}q_i - (TY_i^{'})^{'}q_i - f_z] \tag{1.29}$$

where $()^{'} = \frac{\partial}{\partial r}$. We want to find q_i that will reduce this error. To do so, we orthogonalize the error using the weighing function Y_j

$$\int_0^R \varepsilon(r,t)Y_j dr = 0 \qquad \text{j} = 1, 2, \ldots \text{N}$$

$$\sum_{i=1}^{N}\{(\int_0^R Y_i m(r)Y_j dr)\ddot{q}_i + ((\int_0^R Y_j(EI(r)Y_i^{''})^{''})dr - (\int_0^R Y_j(T(r)Y_i^{'})^{'})dr)q_i$$
$$-(\int_0^R Y_j f_z(r,t)dr)\} = 0 \tag{1.30}$$

The above equation can be written as,

$$\sum_{i=1}^{N}[m_{ij}\ddot{q}_i + K_{ij}q_i] = Q_j \qquad \text{j} = 1, 2, \ldots \text{N}$$

where

$$m_{ij} = (\int_0^R m(r)Y_iY_j dr) \tag{1.31}$$

$$K_{ij} = (\int_0^R Y_j(EI(r)Y_i^{''})^{''}dr) - (\int_0^R Y_j(T(r)Y_i^{'})^{'}dr) \tag{1.32}$$

$$Q_j = (\int_0^R Y_j f_z(r,t)dr) \tag{1.33}$$

Putting in matrix form

$$M\ddot{q} + Kq = Q \tag{1.34}$$

which is a set of coupled differential equations. For free vibration, set $Q = 0$, and assume harmonic motion

$$q(t) = qe^{i\omega t} \tag{1.35}$$

$$\omega^2 Mq = Kq \tag{1.36}$$

This is a general algebraic eigenvalue problem. For beam vibration problems, it is convenient to express displacement in terms of beam functions.

$$w_{app}(r,t) = \sum_{i=1}^{N} \phi_i(r) q_i(t) \tag{1.37}$$

Here $\phi_i(r)$ is the ith mode shape. An advantage of this approach is that since the beam functions are orthogonal, the matrices are simple in structure. Moreover, beam functions satisfy all the boundary conditions for the problem. Beam functions are available for any given mode. To summarize, we see that the procedure for the Galerkin method is simple, but it is quite difficult to choose a function which satisfies all the boundary conditions. The Galerkin method over estimates the solution, i.e., the calculated natural frequencies are higher than the correct value. Furthermore, estimation error increases for higher modes. We need a large number of terms to accurately estimate the higher modes.

Example 1.1 For a cantilevered uniform rotating beam, calculate the fundamental out-of-plane frequency using the Galerkin method.

Solution: Assume one term deflection as,

$$w(r,t) = Y_1(r) q_1(t)$$

where

$$Y_1(r) = 3(\frac{r}{R})^2 - 2(\frac{r}{R})^3 + \frac{1}{2}(\frac{r}{R})^4$$

This function can be obtained by assuming a fourth order polynomial and then ensuring that it satisfies the boundary conditions. We now check to see if the cantilever boundary conditions are satisfied,

$$\frac{dY_1}{dr} = Y_1' = 6(\frac{r}{R^2}) - 6(\frac{r^2}{R^3}) + \frac{2}{R}(\frac{r}{R})^3$$

$$Y_1(0) = 0,\ Y_1'(0) = 0$$

$$\frac{d^2Y_1}{dr^2} = Y_1'' = (\frac{6}{R^2}) - (\frac{6}{R^2})(2\frac{r}{R}) + (\frac{6}{R^2})(\frac{r}{R})^2$$

$$\frac{d^2Y_1}{dr^2}|_{r=R} = (\frac{6}{R^2}) - (\frac{12}{R^2}) + (\frac{6}{R^2}) = 0$$

$$\frac{d^3Y_1}{dr^3} = Y_1''' = (\frac{-12}{R^2})\frac{1}{R} + (\frac{6}{R^2})(2\frac{r}{R})(\frac{1}{R}) = \frac{-12}{R^3} + \frac{12r}{R^4}$$

$$\frac{d^3Y_1}{dr^3}\,|_{r=R} = \frac{-12}{R^3} + \frac{12}{R^3} = 0$$

$$\frac{d^4Y_1}{dr^4} = Y_1^{''''} = \frac{12}{R^4} = 0$$

Therefore, we see that all four boundary conditions are satisfied.
$N = 1$, So

$$m_{11} = \int_0^R mY_1^2 dr$$

Let, $\frac{r}{R} = \eta \Rightarrow dr = Rd\eta$; when $r = 0 \rightarrow \eta = 0$; $r = R \rightarrow \eta = 1$.

$$m_{11} = \int_0^1 mY_1^2(\eta)Rd\eta = mR\int_0^1 \left(3\eta^2 - 2\eta^3 + \frac{\eta^4}{2}\right)^2 d\eta = \frac{26}{45}mR$$

$$T = \frac{m\Omega^2}{2}(R^2 - r^2)$$

$$T^{'} = -mr\Omega^2$$

$$K_{11} = \int_0^R EIY_1Y_1^{''''}dr - \int_0^R TY_1Y_1^{''}dr - \int_0^R T^{'}Y_1Y_1^{'}dr$$

$$K_{11} = \int_0^R EI\left(3(\frac{r}{R})^2 - 2(\frac{r}{R})^3 + \frac{1}{2}(\frac{r}{R})^4\right)\left(\frac{12}{R^4}\right)dr$$

$$- \int_0^R T\left(3(\frac{r}{R})^2 - 2(\frac{r}{R})^3 + \frac{1}{2}(\frac{r}{R})^4\right)\left((\frac{6}{R^2}) - (\frac{6}{R^2})(2\frac{r}{R}) + (\frac{6}{R^2})(\frac{r}{R})^2\right)dr$$

$$- \int_0^R T^{'}\left(3(\frac{r}{R})^2 - 2(\frac{r}{R})^3 + \frac{1}{2}(\frac{r}{R})^4\right)\left(6(\frac{r}{R^2}) - 6(\frac{r^2}{R^3}) + \frac{2}{R}(\frac{r}{R})^3\right)dr$$

simplifying we get,

$$= \frac{36}{5}\frac{EI}{R^3} + \frac{61}{90}m\Omega^2 R$$

$$\omega_1^2 = \frac{K_{11}}{m_{11}} = 12.46\frac{EI}{mR^4} + 1.173\Omega^2$$

$$\nu_1^2 = \frac{\omega_1^2}{\Omega^2} = 12.46\frac{EI}{mR^4\Omega^2} + 1.173$$

For $\Omega = 0$, $\omega_1 = 3.53\sqrt{\frac{EI}{mR^4}}$.

Recall the exact solution for the first frequency of a non-rotating uniform cantilever beam,

$$\omega_1 = 3.53\sqrt{\frac{EI}{mR^4}}$$

For rotation, if $\frac{EI}{m\Omega^2 R^4} = 0.1$

$$\nu_1^2 = 12.46 \times 0.1 + 1.173 = 2.419$$
$$\nu_1 = 1.56/\text{rev}$$

In many situations, the frequency of rotating beams is expressed in this non-dimensional form of '/rev' called 'per rev'. To convert this to dimensional form, multiply by Ω. Thus 1.56/rev = 1.56 Ω. If $\Omega = 5\,\text{rad/s}$, $\omega = 1.56 \times 5 = 7.8\,\text{rad/s}$.

Special Case: Let,

$$\frac{EI}{m\Omega^2 R^4} = 0.1$$

Put, $EI = 100000\ \text{Nm}^2$, $m = 5\ \text{Kg/m}$ and $R = 10\,\text{m}$

$$\frac{10^5}{5\Omega^2 10^4} = 0.1$$

$$\Omega = 4.4721\ \text{rad/s}$$

Substitute the value of "Ω" in ω_1^2 equation we get,

$$\omega_1^2 = 12.46\frac{10^5}{5 \times 10^4} + 1.173\left(4.4721^2\right)$$

$$\omega_1 = 6.956\,\text{rad/s} \qquad @\Omega = 4.4721\ \text{rad/s}$$

Similarly,

$$\omega_1 = 4.991\,\text{rad/s} \qquad @\Omega = 0$$

From converged finite element method (to be discussed later) we get, for $\Omega = 0$, $\omega = 4.9724$ and $\Omega = 4.4721, \omega = 6.9551$. So, obtained solution is closer to the exact solution and this occurs because the function considered for this problem satisfies all the boundary conditions.

Example 1.2 For a cantilevered uniform rotating beam, calculate the frequency using the Galerkin method with beam function. Check the results for non-rotating and rotating beam with the exact solution and finite element method, respectively. Consider $EI = 100000\ \text{Nm}^2$, $m = 5\ \text{Kg/m}$, $R = 10\,\text{m}$, $\Omega = 0$ rad/s and $\Omega = 4.4721$ rad/s.

Solution: Thus beam function is,

$$\phi_j(x) = \cosh(\lambda_j x) - \cos(\lambda_j x) - \alpha_j \left(\sinh(\lambda_j x) - \sin(\lambda_j x)\right)$$

Since the function contains trigonometric terms, manual calculation is cumbersome. Therefore, a Matlab code is written to solve the problem. Calculations are done for a non-rotating and rotating beam.

```
% Code for Galerkin method using Uniform cantilever beam function
clc;
clear all;
syms x R O;
m=5; % mass per unit length of the beam in kg/m
R=10; % Radius of the beam in m
EI=100000; % Flexural Rigidity of beam in N-m^2
O=0; % Non-Rotating Beam (For simplification)in rad/sec
l=1.8751/R; % Using Standard Table value of lambda for 1st mode
a=0.7341; % Corresponding alpha value
T=(m*O^2)*((R^2)-(x^2))/2;
T1=diff(T,x);
Y =(cosh(l*x)-cos(l*x)-(a*(sinh(l*x)-sin(l*x)))); % Beam function
m1 = int((m*(Y^2)),x,0,R); % Mass
Y1 = diff(Y,x);
Y2 = diff(Y1,x);
Y3 = diff(Y2,x);
Y4 = diff(Y3,x);
K11=int((Y*diff(diff(EI*Y2))),x,0,R);
K12=int((Y*diff(T*Y1)),x,0,R);
K=K11-K12;
K1=vpa(simplify(K),4);
omsq = (K1)/(m1); % Frequency obtained by (omega^2)= K1/m1
omega =vpa(sqrt(omsq),4)% Actual Frequency of the beam in rad/sec
frequency = vpa(omega/(sqrt((EI)/(m*(R^4)))),4);
% Frequency in Non Dimensional form

Output:frequency
4.972
```

In the code l and a are the values of λ and α respectively, the values can be obtained from the table as below, where i is the mode shape.

i	1	2	3	4	j
λ_j	1.8751	4.6941	7.8548	10.9955	$(2j-1)\frac{\pi}{2}$
α_j	0.7341	1.0185	0.9992	1.000	1

Exact solution of the problem is obtained by substituting the values of $EI = 100000\ \text{Nm}^2$, $m = 5\ \text{Kg/m}$ and $R = 10\,\text{m}$ in the equation:

$$\omega = 3.515\sqrt{\frac{EI}{mR^4}}$$

$$\omega = 4.97\,\text{rad/s}$$

Therefore, the output of the Matlab code for Galerkin method is 4.972 rad/s and exact solution for the problem is 4.97 rad/s, the frequency prediction is very good. If we put $\Omega = 4.4721$ in the code, we get $\omega = 6.971$. So the Galerkin method results are satisfactory.

1.3 Rayleigh–Ritz Method

In contrast to the Galerkin method which is based on the governing differential equation, the Rayleigh–Ritz method is based on energy principles. We assume the deflection as a series of functions which satisfy only geometric boundary conditions.

$$w(r,t) = \sum_{i=1}^{N} Y_i(r) q_i(t) \tag{1.38}$$

For cantilever beam, geometric boundary conditions are: $r = 0w = 0w' = 0$. Thus, $Y_i(0) = 0 \quad Y_i'(0) = 0$.

We can calculate the kinetic energy (KE) as,

$$KE = T = \frac{1}{2}\int_0^R m\dot{w}^2 dr$$

$$= \frac{1}{2}\int_0^R m\left[\sum_{i=1}^{N} Y_i(r)\dot{q}_i(t)\right]\left[\sum_{j=1}^{N} Y_j(r)\dot{q}_j(t)\right] dr = \frac{1}{2}\sum_{i=1}^{N}\sum_{j=1}^{N} m_{ij}\dot{q}_i\dot{q}_j \tag{1.39}$$

where

$$m_{ij} = \int_0^R mY_iY_j dr$$

Potential or strain energy (V) = Strain energy due to flexure + strain energy due to centrifugal force

$$V = \frac{1}{2}\int_0^R EI\left(\frac{d^2w}{dr^2}\right)^2 dr + \frac{1}{2}\int_0^R T\left(\frac{dw}{dr}\right)^2 dr \tag{1.40}$$

$$V = \frac{1}{2}\int_0^R EI\left[\sum_{i=1}^{N} Y_i'' q_i\right]\left[\sum_{j=1}^{N} Y_j'' q_j\right] dr + \frac{1}{2}\int_0^R T\left[\sum_{i=1}^{N} Y_i' q_i\right]\left[\sum_{j=1}^{N} Y_j' q_j\right] dr$$

$$V = \frac{1}{2}\sum_{i=1}^{N}\sum_{j=1}^{N} K_{ij}q_iq_j$$

$$K_{ij} = \int_0^R EIY_i''Y_j''dr + \int_0^R Y_i'TY_j'dr \tag{1.41}$$

and $T = \int_0^R m\Omega^2 rdr$

We also need to calculate the virtual work done due to virtual displacement,

$$\delta W = \int_0^R f_Z(r,t)\delta w dr = \int_0^R f_Z \sum_{i=1}^N Y_i \delta q_i dr = \sum_{i=1}^N Q_i \delta q_i \tag{1.42}$$

where $Q_i = \int_0^R f_Z Y_i dr$

The minimization of energy is expressed in terms of Lagrange's equations,

$$\frac{\partial}{\partial t}\left(\frac{\partial T}{\partial \dot{q}_i}\right) - \frac{\partial T}{\partial q_i} + \frac{\partial V}{\partial q_i} = Q_i \qquad i = 1, 2 \ldots N \tag{1.43}$$

or

$$\sum_{j=1}^N (m_{ij}\ddot{q}_j + K_{ij}q_j) = Q_i \qquad i = 1, 2 \ldots N \tag{1.44}$$

Or in matrix form,

$$\underset{N \times N}{m} \quad \underset{N \times 1}{\ddot{q}} \quad + \quad \underset{N \times N}{K} \quad \underset{N \times 1}{q} \quad = \quad \underset{N \times 1}{Q}$$

Sometimes, a one term use of the Rayleigh–Ritz method is called the Rayleigh solution. For N = 1,

$$w(r,t) = Y_1(r)q_1(t) \tag{1.45}$$

This leads to one equation for a single dof system,

$$m_{11}\ddot{q}_1 + K_{11}q_1 = 0 \tag{1.46}$$

The frequency is then obtained as,

$$\omega^2 = \frac{K_{11}}{m_{11}} = \frac{\int_0^R EI\left(Y_i''\right)^2 dr + \int_0^R T\left(Y_i'\right)^2 dr}{\int_0^R mY_1^2 dr} \tag{1.47}$$

Example 1.3 Calculate Rayleigh solution for a uniform rotating blade with cantilever condition. Assume,

$$w(r,t) = \left(\frac{r}{R}\right)^2 q_1(t)$$

Solution: We first check if the assumed solution satisfies the geometric boundary condition.

$$@r = 0 \qquad w = w' = 0$$

$$Y_1 = \frac{dY_1}{dr} = 0$$

We can now calculate,

$$m_{11} = \int_0^R mY_1^2 dr = \int_0^R m\frac{r^4}{R^4}dr = \frac{mR}{5}$$

$$T = \frac{m}{2}\left(R^2 - r^2\right)\Omega^2$$

$$K_{11} = \int_0^R EI\left(Y_1''\right)^2 dr + \int_0^R T\left(Y_1'\right)^2 dr$$

$$= \frac{4EI}{R^3} + \frac{4}{15}m\Omega^2 R$$

Thus, the frequency predicted by the one term approximation is,

$$\omega^2 = \frac{K_{11}}{m_{11}} = \frac{\frac{4EI}{R^3} + \frac{4}{15}m\Omega^2 R}{\frac{mR}{5}} = 20\frac{EI}{mR^4} + \frac{4}{3}\Omega^2$$

Let us check the special cases,

For $\Omega = 0$, $\omega_1 = 4.47\sqrt{\frac{EI}{mR^4}}$

Exact, $\omega_1 = 3.515\sqrt{\frac{EI}{mR^4}}$

Galerkin, $\omega_1 = 3.53\sqrt{\frac{EI}{mR^4}}$

We should note that if a function satisfying all boundary conditions is used with Rayleigh−Ritz method, a better solution is obtained.

Example 1.4 Calculate Rayleigh–Ritz solution for a rotating uniform cantilever beam using the beam function. Find the frequency for a beam with $m = 5\,\text{Kg/m}$, $EI = 100000\,\text{N} - \text{m}^2$, $R = 10\,\text{m}$, $\Omega = 0\,\text{rad/s}$ and $\Omega = 4.4721\,\text{rad/s}$.

Solution: Thus beam function is,

$$\phi_j(x) = \cosh(\lambda_j x) - \cos(\lambda_j x) - \alpha_j\left(\sinh(\lambda_j x) - \sin(\lambda_j x)\right)$$

The Matlab code is given below,

```
% Code for Rayleigh Ritz method using Uniform cantilever beam function
clc;
clear all;
syms x R;
EI=100000; % Flexural Rigidity of beam in N-m^2
m=5; % mass per unit length of the beam in kg/m
```

```
R=10; % Radius of the beam in m
O=0; % Non-Rotating Beam (For simplification)
l=1.8751/R; % Using Standard Table value of lambda for 1st mode
a=0.7341; % Corresponding alpha value
T=(m*O^2)*((R^2)-(x^2))/2;
Y =(cosh(l*x)-cos(l*x)-(a*(sinh(l*x)-sin(l*x)))); % Beam function
m1 = int((m*Y^2),x,0,R); % Mass
Y1 = diff(Y,x);
Y2 = diff(Y1,x);
Yi=int((EI*(Y2^2)),x,0,R);
Yt=int((T*(Y1^2)),x,0,R);
K = Yi+Yt;
K1=vpa(simplify(K),4);
omsq = K1/(m1); % Frequency obtained by (omega^2)= K1/m1
omega =vpa(sqrt(omsq),4)% Actual Frequency of the beam in rad/sec
frequency = vpa(omega/(sqrt((EI)/(m*(R^4)))),4);
% Frequency in Non Dimensional form
```

```
Output:
frequency  4.972
```

If we put $\Omega = 4.4721$ in the code, we get $\omega = 6.971$.

Example 1.5 Calculate Rayleigh solution for a rotating blade with cantilever condition and the following properties: $EI(r) = a_0 + a_1 r$, $m(r) = b_0 + b_1 r$. This is typical for a rotating wind turbine blade.

$$w(r,t) = \left(\frac{r}{R}\right)^2 q_1(t)$$

Solution:
We can now calculate,

$$m_{11} = \int_0^R mY_1^2 dr = \int_0^R (b_0 + b_1 r)\left(\frac{r^4}{R^4}\right) dr$$

$$m_{11} = \int_0^R \left[b_0\left(\frac{r^4}{R^4}\right) + b_1\left(\frac{r^5}{R^4}\right)\right] dr$$

$$m_{11} = \left[\frac{b_0}{R^4}\left(\frac{r^5}{5}\right) + \frac{b_1}{R^4}\left(\frac{r^6}{6}\right)\right]_0^R$$

$$m_{11} = \frac{Rb_0}{5} + \frac{R^2 b_1}{6}$$

$$T = \frac{m}{2}\left(R^2 - r^2\right)\Omega^2$$

$$K_{11} = \int_0^R EI\left(Y_1^{''}\right)^2 dr + \int_0^R T\left(Y_1^{'}\right)^2 dr$$

$$Y_1 = \left(\frac{r}{R}\right)^2, \quad Y_1^{'} = \frac{2r}{R^2}, \quad Y_1^{''} = \frac{2}{R^2}$$

$$K_{11} = \int_0^R (a_0 + a_1 r)\left(\frac{2}{R^2}\right)^2 dr + \int_0^R \left(\frac{b_0 + b_1 r}{2}\left(R^2 - r^2\right)\Omega^2\right)\left(\frac{2r}{R^2}\right)^2 dr$$

$$K_{11} = \int_0^R \left(\frac{4a_0}{R^4} + \frac{4a_1 r}{R^4}\right) dr + \int_0^R \frac{\Omega^2}{2}(b_0 + b_1 r)\left(\frac{4r^2}{R^2}\right) dr$$

$$- \int_0^R \frac{\Omega^2}{2}(b_0 + b_1 r)\left(\frac{4r^4}{R^4}\right) dr$$

$$K_{11} = \frac{4a_0}{R^3} + \frac{2a_1}{R^2} + \frac{2b_0\Omega^2 R}{3} + \frac{b_1\Omega^2 R^2}{2} - \frac{2b_0\Omega^2 R}{5} - \frac{b_1\Omega^2 R^2}{3}$$

$$K_{11} = \frac{4a_0}{R^3} + \frac{2a_1}{R^2} + \frac{4b_0\Omega^2 R}{15} + \frac{b_1\Omega^2 R^2}{6}$$

Thus, at $\Omega = 0$ the frequency can be determined by,

$$\omega^2 = \frac{K_{11}}{m_{11}} = \frac{\frac{4a_0}{R^3} + \frac{2a_1}{R^2}}{\frac{Rb_0}{5} + \frac{R^2 b_1}{6}}$$

For previous problem the solution obtained for Rayleigh method is,

$$\omega = 4.47\sqrt{\frac{EI}{mR^4}}$$

To verify the obtained solution, put $a_1 = 0$ and $b_1 = 0$ in $EI(r)$ and $m(r)$ equation respectively and simplify the equation. We get,

$$EI(r) = a_0$$

$$m(r) = b_0$$

$$\omega = 4.47\sqrt{\frac{a_0}{b_0 R^4}} = 4.47\sqrt{\frac{EI}{mR^4}}$$

Thus obtained Rayleigh solution is verified in the limiting case. We also observe that the Rayleigh–Ritz method can be easily used for tapered beams.

Example 1.6 Calculate Rayleigh solution for a rotating blade with cantilever condition and the following properties: $EI(r) = a_0 + a_1 r + a_2 r^2 + a_3 r^3$, $m(r) = b_0 + b_1 r$. This is typical for a rotating helicopter blade.

$$w(r, t) = \left(\frac{r}{R}\right)^2 q_1(t)$$

Solution:
We can now calculate,

$$m_{11} = \int_0^R mY_1^2 dr = \int_0^R (b_0 + b_1 r)\left(\frac{r^4}{R^4}\right) dr$$

This integral is the same as considered in the previous problem. So,

$$m_{11} = \frac{Rb_0}{5} + \frac{R^2 b_1}{6}$$

To evaluate stiffness,

$$K_{11} = \int_0^R EI\left(Y_1''\right)^2 dr + \int_0^R T\left(Y_1'\right)^2 dr$$

$$Y_1 = \left(\frac{r}{R}\right)^2, \quad Y_1' = \frac{2r}{R^2}, \quad Y_1'' = \frac{2}{R^2}$$

$$K_{11} = \int_0^R \left(a_0 + a_1 r + a_2 r^2 + a_3 r^3\right)\left(\frac{2}{R^2}\right)^2 dr + \int_0^R \left(\frac{b_0 + b_1 r}{2}\left(R^2 - r^2\right)\Omega^2\right)\left(2\left(\frac{r}{R^2}\right)\right)^2 dr$$

$$K_{11} = \int_0^R \left(\frac{4a_0}{R^4}\right) dr + \int_0^R \left(\frac{4a_1 r}{R^4}\right) dr + \int_0^R \left(\frac{4a_2 r^2}{R^4}\right) dr$$

$$+ \int_0^R \left(\frac{4a_3 r^3}{R^4}\right) dr + \int_0^R \left(\frac{2(b_0 + b_1 r)\Omega^2 r^2}{R^2}\right) dr - \int_0^R \left(\frac{2(b_0 + b_1 r)\Omega^2 r^4}{R^4}\right) dr$$

$$K_{11} = \frac{4a_0}{R^3} + \frac{2a_1}{R^2} + \frac{4a_2}{3R} + a_3 + \frac{2b_0\Omega^2 R}{3} + \frac{b_1\Omega^2 R^2}{2} - \frac{2b_0\Omega^2 R}{5} - \frac{b_1\Omega^2 R^2}{3}$$

$$K_{11} = \frac{4a_0}{R^3} + \frac{2a_1}{R^2} + \frac{4a_2}{3R} + a_3 + 0.2666\left(b_0\Omega^2 R\right) + 0.1666\left(b_1\Omega^2 R^2\right)$$

Thus, at $\Omega = 0$ the frequency can be determined by,

$$\omega^2 = \frac{K_{11}}{m_{11}} = \frac{\frac{4a_0}{R^3} + \frac{2a_1}{R^2} + \frac{4a_2}{3R} + a_3}{\frac{Rb_0}{5} + \frac{R^2 b_1}{6}}$$

For previous problem the solution obtained for Rayleigh method is,

$$\omega = 4.47\sqrt{\frac{EI}{mR^4}}$$

To verify the obtained solution, put $a_1 = 0, a_2 = 0, a_3 = 0$ and $b_1 = 0$ in $EI(r)$ and $m(r)$ equation respectively and simplify the equation. We get,

$$EI(r) = a_0$$

$$m(r) = b_0$$

$$\omega = 4.47\sqrt{\frac{a_0}{b_0 R^4}} = 4.47\sqrt{\frac{EI}{mR^4}}$$

Thus obtained Rayleigh solution is verified is the limiting case.

Example 1.7 Calculate Rayleigh solution for the admissible function.

$$w(r,t) = Y_1(r)q_1(t)$$

where, $Y_i(r)$ is the admissible function,

$$Y_i(r) = \frac{\left(\frac{r}{R}\right)^{i+1}\left[2+i-i\left(\frac{r}{R}\right)\right]}{i(i+1)(i+2)}$$

Solution: In Rayleigh solution, we need one term of the admissible function. For first term, $i = 1$,

$$Y_1(r) = \left[\frac{r^2}{2R^2} - \frac{r^3}{6R^3}\right]$$

We first check if the assumed solution satisfies the geometric boundary condition.

$$@r = 0 \qquad w = w' = 0$$

$$Y_1 = \frac{dY_1}{dr} = 0$$

We can now calculate,

$$m_{11} = \int_0^R mY_1^2 dr = \int_0^R m\left[\frac{r^2}{2R^2} - \frac{r^3}{6R^3}\right]^2 dr$$

$$m_{11} = m\left[\frac{R}{20} - \frac{R}{72} - \frac{R}{72} + \frac{R}{252}\right]$$

$$m_{11} = 0.02619mR$$

$$T = \frac{m}{2}\left(R^2 - r^2\right)\Omega^2$$

$$K_{11} = \int_0^R EI\left(Y_1''\right)^2 dr + \int_0^R T\left(Y_1'\right)^2 dr$$

$$Y_1 = \left(\frac{r^2}{2R^2} - \frac{r^3}{6R^3}\right), \quad Y_1' = \left(\frac{r}{R^2} - \frac{r^2}{2R^3}\right), \quad Y_1'' = \left(\frac{1}{R^2} - \frac{r}{R^3}\right)$$

$$K_{11} = \int_0^R EI\left(\frac{1}{R^2} - \frac{r}{R^3}\right)^2 dr + \int_0^R \left(\frac{m\left(R^2 - r^2\right)\Omega^2}{2}\right)\left(\frac{r}{R^2} - \frac{r^2}{2R^3}\right)^2 dr$$

$$K_{11} = EI\left[\frac{1}{R^3} - \frac{1}{R^3} + \frac{1}{3R^3}\right]_0^R + \frac{m\Omega^2}{2}\left[\frac{r^3}{3R^2} - \frac{r^5}{5R^4} - \frac{r^4}{4R^3} + \frac{r^6}{6R^5} + \frac{r^5}{20R^4} - \frac{r^7}{28R^6}\right]_0^R$$

$$K_{11} = \frac{EI}{3R^3} + 0.06428\frac{m\Omega^2}{2}$$

Thus, at $\Omega = 0$ the frequency can be determined by,

$$\omega^2 = \frac{K_{11}}{m_{11}} = \frac{\frac{EI}{3R^3}}{0.02619mR} = 12.7275\frac{EI}{mR^4}$$

$$\omega = 3.567\sqrt{\frac{EI}{mR^4}}$$

For previous Example 1.3 the solution obtained for Rayleigh method is,

$$\omega = 4.47\sqrt{\frac{EI}{mR^4}}$$

The exact solution can be given by,

$$\omega = 3.515\sqrt{\frac{EI}{mR^4}}$$

Comparing the solution obtained with previous Rayleigh problem solution and exact solution, our present solution is closer to the exact solution. This occurs because the admissible function considered for this problem satisfies all the boundary conditions.

Example 1.8 Try to solve the rotating uniform beam equation using the symbolic processing software Mathematica. Consider $EI = 1, m = 1, \Omega = 1, R = 1$. Check for the non-rotating uniform beam. Also try special cases for $EI = x^2, m = 1$ and $EI = x^2, m = \frac{1}{x}$.

Solution:
Case 1: For rotating beam equation, Mathematica code is given below. We can see that the equation cannot be solved by this approach.

```
EI = 1;
m = 1;
R = 1;
om2 = 1;
T = Integrate[m om2 x, { x, x, R}];
DSolve[ D[D[EI w''[x], x], x] - m f^2 w[x] - D[ T w'[x], x] == 0,w[x], x]
```

Output:

```
{{w[x] → DifferentialRoot[
    Function[{ẏ, ẋ}, {-2 f² ẏ[ẋ] + 2 ẋ ẏ'[ẋ] + (-1 + ẋ²) ẏ''[ẋ] + 2 ẏ⁽⁴⁾ [ẋ] == 0,
      ẏ[0] == C[1], ẏ'[0] == C[2], ẏ''[0] == C[3], ẏ⁽³⁾ [0] == C[4]}]] [x]}}
```

Case 2: For non-rotating beam, Mathematica code is given below in which Ω value is zero.

```
EI = 1;
m = 1;
R = 1;
om2 = 0;
T = Integrate[m om2 x, { x, x, R}];
DSolve[ D[D[EI w''[x], x], x] - m f^2 w[x] - D[ T w'[x], x] == 0,w[x], x]
```

Output:

$$\{\{w[x] \to e^{-\sqrt{f}x}C[2] + e^{\sqrt{f}x}C[4] + C[1]\mathrm{Cos}[\sqrt{f}x] + C[3]\mathrm{Sin}[\sqrt{f}x]\}\}$$

This is the well known solution with hyperbolic sine and cosine replaced by exponentials.

Case 3: For rotating beam special case at $E = x^2$ and $m = 1$, Mathematica code is:

```
EI = x^2;
m = 1;
R = 1;
om2 = 1;
T = Integrate[m om2 x, { x, x, R}];
DSolve[ D[D[EI w''[x], x], x] - m f^2 w[x] - D[ T w'[x], x] == 0,w[x], x]
```

Output:

$$\{\{w[x] \to C[1]\mathrm{HypergeometricPFQ}[\{\frac{1}{4} - \frac{1}{4}\sqrt{1+8f^2}, \frac{1}{4} + \frac{1}{4}\sqrt{1+8f^2}\}, \{\frac{1}{2}, \frac{3}{4} - \frac{\sqrt{3}}{4}, \frac{3}{4} + \frac{\sqrt{3}}{4}\}, -\frac{x^2}{8}] + \frac{xC[2]\mathrm{HypergeometricPFQ}[\{\frac{3}{4} - \frac{1}{4}\sqrt{1+8f^2}, \frac{3}{4} + \frac{1}{4}\sqrt{1+8f^2}\}, \{\frac{3}{2}, \frac{5}{4} - \frac{\sqrt{3}}{4}, \frac{5}{4} + \frac{\sqrt{3}}{4}\}, -\frac{x^2}{8}]}{2\sqrt{2}} + 2^{-\frac{3}{4}(1-\sqrt{3})}x^{\frac{1}{2}(1-\sqrt{3})}C[3]\mathrm{HypergeometricPFQ}[\{\frac{1}{2} - \frac{\sqrt{3}}{4} - \frac{1}{4}\sqrt{1+8f^2}, \frac{1}{2} - \frac{\sqrt{3}}{4} + \frac{1}{4}\sqrt{1+8f^2}\}, \{1 - \frac{\sqrt{3}}{2}, \frac{3}{4} - \frac{\sqrt{3}}{4}, \frac{5}{4} - \frac{\sqrt{3}}{4}\}, -\frac{x^2}{8}] + 2^{-\frac{3}{4}(1+\sqrt{3})}x^{\frac{1}{2}(1+\sqrt{3})}C[4]\mathrm{HypergeometricPFQ}[\{\frac{1}{2} + \frac{\sqrt{3}}{4} - \frac{1}{4}\sqrt{1+8f^2}, \frac{1}{2} + \frac{\sqrt{3}}{4} + \frac{1}{4}\sqrt{1+8f^2}\}, \{\frac{3}{4} + \frac{\sqrt{3}}{4}, \frac{5}{4} + \frac{\sqrt{3}}{4}, 1 + \frac{\sqrt{3}}{2}\}, -\frac{x^2}{8}]\}\}$$

We see that some solutions exist for a rotating beam in the form of special functions.

Case 4: For rotating beam special case at $E = x^2$ and $m = \frac{1}{x}$, Mathematica code is:

```
EI = x^2;
m = 1/x;
R = 1;
om2 = 1;
T = Integrate[m om2 x, { x, x, R}];
DSolve[ D[D[EI w''[x], x], x] - m f^2 w[x] - D[ T w'[x], x] == 0,w[x], x]
```

Output:

$$\{\{w[x] \to -xC[3]\text{HypergeometricPFQ}[\{1-f,1+f\},\{2,\frac{3}{2}-\frac{\sqrt{5}}{2},\frac{3}{2}+\frac{\sqrt{5}}{2}\},-x] + x^{\frac{1}{2}(1-\sqrt{5})}C[2]\text{HypergeometricPFQ}[\{\frac{1}{2}-\frac{\sqrt{5}}{2}-f,\frac{1}{2}-\frac{\sqrt{5}}{2}+f\},\{1-\sqrt{5},\frac{1}{2}-\frac{\sqrt{5}}{2},\frac{3}{2}-\frac{\sqrt{5}}{2}\},-x] + x^{\frac{1}{2}(1+\sqrt{5})}C[1]\text{HypergeometricPFQ}[\{\frac{1}{2}+\frac{\sqrt{5}}{2}-f,\frac{1}{2}+\frac{\sqrt{5}}{2}+f\},\{\frac{1}{2}+\frac{\sqrt{5}}{2},\frac{3}{2}+\frac{\sqrt{5}}{2},1+\sqrt{5}\},-x] + C[4]\text{MeijerG}[\{\{1-f,1+f\},\{\}\},\{\{0,1\},\{\frac{1}{2}(1-\sqrt{5}),\frac{1}{2}(1+\sqrt{5})\}\},-x]\}\}$$

Again, we see a beam with a closed form solution in terms of special functions.

Example 1.9 Find the first two frequencies of a uniform rotating beam using the Rayleigh–Ritz method. Use $\left(\frac{r}{R}\right)^2$ and $\left(\frac{r}{R}\right)^3$ as the two admissible functions.

Solution:
For N = 2,

$$w(r,t) = \left(\frac{r}{R}\right)^2 q_1(t) + \left(\frac{r}{R}\right)^3 q_2(t)$$

For, cantilever beam, geometric BC are

$$@r = 0 \qquad w_1 = 0, w_2 = 0, w_1' = 0, w_2' = 0$$

Thus, $Y_1(0) = 0$, $Y_2(0) = 0$, $Y_1'(0) = 0$, $Y_2'(0) = 0$

We can now calculate,

$$m_{11} = \int_0^R mY_1Y_1dr = \int_0^R m\left(\frac{r}{R}\right)^4 dr = \frac{m}{R^4}\left[\frac{r^5}{5}\right]_0^R = \frac{mR}{5}$$

$$m_{12} = \int_0^R mY_1Y_2dr$$

$$m_{12} = \int_0^R m\left(\frac{r}{R}\right)^2\left(\frac{r}{R}\right)^3 dr = \frac{m}{R^5}\left[\frac{r^6}{6}\right]_0^R = \frac{mR}{6} = m_{21}$$

$$m_{22} = \int_0^R mY_2Y_2dr$$

$$m_{22} = \int_0^R m\left(\frac{r}{R}\right)^3\left(\frac{r}{R}\right)^3 dr = \frac{m}{R^6}\left[\frac{r^7}{7}\right]_0^R = \frac{mR}{7}$$

$$K_{ij} = \int_0^R EI \quad Y_i'' Y_j'' \quad dr + \int_0^R Y_i' Y_j' \quad Tdr$$

$$T = \int_0^R m\Omega^2 \, rdr = m\Omega^2 \left[\frac{r^2}{2}\right]_r^R = \frac{m\Omega^2}{2}\left[R^2 - r^2\right]$$

Now,

$$Y_1 = \left(\frac{r}{R}\right)^2 = \frac{r^2}{R^2} \qquad Y_2 = \left(\frac{r}{R}\right)^3 = \frac{r^3}{R^3}$$

$$Y_1' = \frac{2r}{R^2} \qquad Y_2' = \frac{3r^2}{R^3}$$

$$Y_1'' = \frac{2}{R^2} \qquad Y_2'' = \frac{6r}{R^3}$$

$$K_{11} = \int_0^R EI \quad \left(\frac{2}{R^2}\right)\left(\frac{2}{R^2}\right) dr + \int_0^R \left(\frac{2r}{R^2}\right)\left(\frac{2r}{R^2}\right) \; \frac{m\Omega^2}{2}\left[R^2 - r^2\right] dr$$

$$K_{11} = \frac{4EI}{R^4}\left[r\right]_0^R + \frac{2m\Omega^2}{R^4}\int_0^R \left[R^2 r^2 - r^4\right] dr$$

$$K_{11} = \frac{4EI}{R^3} + \frac{2m\Omega^2}{R^4}\left[R^2\left(\frac{r^3}{3}\right) - \frac{r^5}{5}\right]_0^R$$

$$K_{11} = \frac{4EI}{R^3} + \frac{4m\Omega^2 R}{15}$$

$$K_{12} = K_{21} = \int_0^R EI \quad Y_1'' Y_2'' dr + \int_0^R Y_1' Y_2' \, Tdr$$

$$K_{12} = \int_0^R EI\left(\frac{2}{R^2}\right)\left(\frac{6r}{R^3}\right) dr + \int_0^R \left(\frac{2r}{R^2}\right)\left(\frac{3r^2}{R^3}\right)\left(\frac{m\Omega^2}{2}(R^2 - r^2)\right) dr$$

$$K_{12} = \frac{12EI}{R^5}\int_0^R rdr + \frac{3m\Omega^2}{R^5}\int_0^R r^3(R^2 - r^2)dr$$

$$K_{12} = \frac{12EI}{R^5}\left[\frac{r^2}{2}\right]_0^R + \frac{3m\Omega^2}{R^5}\left[R^2\left(\frac{r^4}{4}\right) - \left(\frac{r^6}{6}\right)\right]_0^R$$

$$K_{12} = \frac{6EI}{R^3} + \frac{m\Omega^2 R}{4} = K_{21}$$

$$K_{22} = \int_0^R EI \quad Y_2'' Y_2'' dr + \int_0^R Y_2' Y_2' \, T dr$$

$$K_{22} = \int_0^R EI \left(\frac{6r}{R^3}\right)\left(\frac{6r}{R^3}\right) dr + \int_0^R \left(\frac{3r^2}{R^3}\right)\left(\frac{3r^2}{R^3}\right) \frac{m\Omega^2}{2}(R^2 - r^2) dr$$

$$K_{22} = \frac{36EI}{R^6} \int_0^R r^2 dr + \frac{9m\Omega^2}{2R^6} \int_0^R r^4 (R^2 - r^2) dr$$

$$K_{22} = \frac{36EI}{R^6} \left[\frac{r^3}{3}\right]_0^R + \frac{9m\Omega^2}{2R^6} \left[R^2\left(\frac{r^5}{5}\right) - \left(\frac{r^7}{7}\right)\right]_0^R$$

$$K_{22} = \frac{12EI}{R^3} + \frac{9m\Omega^2 R}{35}$$

Summarizing, the mass and stiffness matrices are,

$$m = \begin{bmatrix} \frac{mR}{5} & \frac{mR}{6} \\ \frac{mR}{6} & \frac{mR}{7} \end{bmatrix}$$

$$K = \begin{bmatrix} \frac{4EI}{R^3} + \frac{4m\Omega^2 R}{15} & \frac{6EI}{R^3} + \frac{m\Omega^2 R}{4} \\ \frac{6EI}{R^3} + \frac{m\Omega^2 R}{4} & \frac{12EI}{R^3} + \frac{9m\Omega^2 R}{35} \end{bmatrix}$$

Using m and K, we can determine eigenvalues using the characteristic equation,

$|K - Dm| = 0$

Take, $A = \frac{EI}{R^3}$, $B = m\Omega^2 R$, $C = mR$ for simplification, D are eigenvalues

$$\left| \begin{bmatrix} 4A + \frac{4B}{15} & 6A + \frac{B}{4} \\ 6A + \frac{B}{4} & 12A + \frac{9B}{35} \end{bmatrix} - D \begin{bmatrix} \frac{C}{5} & \frac{C}{6} \\ \frac{C}{6} & \frac{C}{7} \end{bmatrix} \right| = 0$$

$$\begin{vmatrix} 4A + \frac{4B}{15} - \frac{DC}{5} & 6A + \frac{B}{4} - \frac{DC}{6} \\ 6A + \frac{B}{4} - \frac{DC}{6} & 12A + \frac{9B}{35} - \frac{DC}{7} \end{vmatrix} = 0$$

$$\left(4A + \frac{4B}{15} - \frac{DC}{5}\right) \times \left(12A + \frac{9B}{35} - \frac{DC}{7}\right) - \left(6A + \frac{B}{4} - \frac{DC}{6}\right)^2 = 0$$

Substituting the values of A, B, C we get, $@\Omega = 0$,

$$12\left(\frac{EI}{R^3}\right)^2 - 0.9714\left(\frac{EI}{R^3}\right)DmR + 0.000793\left(D^2m^2R^2\right) = 0$$

By solving equation for D we get,
Eigenvalues,

$$D_1 = 12.48\frac{EI}{mR^4}$$

$$D_2 = 1212\frac{EI}{mR^4}$$

Natural frequency are given by,

$$f_1 = \sqrt{D_1} \qquad f_2 = \sqrt{D_2}$$

$$f_1 = 3.5327\sqrt{\frac{EI}{R^4m}} \qquad f_2 = 34.8138\sqrt{\frac{EI}{R^4m}}$$

Here, the first frequency is well predicted.

Example 1.10 Find $Y_2(r)$ for use in the Rayleigh–Ritz method. Assume $Y_1(r)$ used in Example 1.1. Using this as admissible function for Rayleigh–Ritz method, find the first two frequencies of a cantilever beam.

Solution:

$$w(r,t) = Y_1(r)q_1t + Y_2(r)q_2(t)$$

where,

$$Y_1(r) = 3\left(\frac{r}{R}\right)^2 - 2\left(\frac{r}{R}\right)^3 + \frac{1}{2}\left(\frac{r}{R}\right)^4$$

We now assume that

$$Y_2(r) = a\left(\frac{r}{R}\right)^3 + b\left(\frac{r}{R}\right)^4 + c\left(\frac{r}{R}\right)^5$$

This function already satisfies the geometric boundary conditions $Y_2(0) = 0$ and $Y_2^{'}(0) = 0$. Now, $Y_2(r)$ must also satisfy the force boundary conditions.

$$Y_2^{''}(R) = \frac{6a}{R^2} + \frac{12b}{R^2} + \frac{20c}{R^2} = 0$$

$$Y_2^{'''}(R) = \frac{6a}{R^3} + \frac{24b}{R^3} + \frac{60c}{R^3} = 0$$

By solving these two equations we get,

$$a = -b$$

Substitute b in terms of a in one of the equations we get,

$$\frac{6a}{R^2} - \frac{12a}{R^2} + \frac{20c}{R^2} = 0 \implies a = \frac{10c}{3}$$

Substitute values of b and c in terms of a in main equation $Y_2(r)$ we get,

$$Y_2(r) = \left(\frac{r}{R}\right)^3 - \left(\frac{r}{R}\right)^4 + \frac{3}{10}\left(\frac{r}{R}\right)^5$$

Therefore, we see that all boundary conditions are satisfied, so for $N = 2$,

$$m_{11} = \int_0^R mY_1Y_1dr$$

$$m_{11} = \int_0^R m\left(3\left(\frac{r}{R}\right)^2 - 2\left(\frac{r}{R}\right)^3 + \frac{1}{2}\left(\frac{r}{R}\right)^4\right)^2 dr$$

$$m_{11} = \int_0^R \left[\frac{1}{4}\left(\frac{r}{R}\right)^8 - 2\left(\frac{r}{R}\right)^7 + 7\left(\frac{r}{R}\right)^6 - 12\left(\frac{r}{R}\right)^5 + 9\left(\frac{r}{R}\right)^4\right] dr$$

$$m_{11} = \left[\frac{1}{4R^8}\left(\frac{r^9}{9}\right) - \frac{2}{R^7}\left(\frac{r^8}{8}\right) + \frac{7}{R^6}\left(\frac{r^7}{7}\right) - \frac{12}{R^5}\left(\frac{r^6}{6}\right) + \frac{9}{R^4}\left(\frac{r^5}{5}\right)\right]_0^R$$

$$m_{11} = \frac{26mR}{45} = 0.5777mR$$

Similarly,

$$m_{12} = \int_0^R mY_1Y_2dr$$

$$m_{12} = \int_0^R m\left(3\left(\frac{r}{R}\right)^2 - 2\left(\frac{r}{R}\right)^3 - \frac{1}{2}\left(\frac{r}{R}\right)^4\right)\left(\left(\frac{r}{R}\right)^3 - \left(\frac{r}{R}\right)^4 + \frac{3}{10}\left(\frac{r}{R}\right)^5\right) dr$$

$$m_{12} = mR\,[0.103492] = 0.103492\ \ mR = m_{21}$$

$$m_{22} = \int_0^R mY_2^2dr$$

$$m_{22} = m\int_0^R \left(\left(\frac{r}{R}\right)^3 - \left(\frac{r}{R}\right)^4 - \frac{3}{10}\left(\frac{r}{R}\right)^5\right)\left(\left(\frac{r}{R}\right)^3 - \left(\frac{r}{R}\right)^4 - \frac{3}{10}\left(\frac{r}{R}\right)^5\right) dr$$

$$m_{22} = m\,[0.018816] = 0.018816mR$$

Mass matrix,

$$m = \begin{bmatrix} 0.5777mR & 0.103492mR \\ 0.103492mR & 0.018816mR \end{bmatrix}$$

Solve for stiffness matrix (K) let,

$$K_{ij} = \int_0^R EIY_i''Y_j''dr + \int_0^R Y_i'Y_j'Tdr$$

$$T = \int_r^R m\Omega^2 r dr$$

$$T = m\Omega^2 \left(\frac{r^2}{2}\right)_r^R$$

$$T = \frac{m\Omega^2}{2}\left(R^2 - r^2\right)$$

$$K_{11} = \int_0^R EIY_1''Y_1''dr + \int_0^R TY_1'Y_1'dr$$

Let,

$$Y_1(r) = 3\left(\frac{r}{R}\right)^2 - 2\left(\frac{r}{R}\right)^3 + \frac{1}{2}\left(\frac{r}{R}\right)^4$$

Differentiate with respect to (r)

$$Y_1'(r) = \frac{6r}{R^2} - \frac{6r^2}{R^3} + \frac{2r^3}{R^4}$$

Differentiate with respect to (r)

$$Y_1''(r) = \frac{6}{R^2} - \frac{12r}{R^3} + \frac{6r^2}{R^4}$$

Let,

$$Y_2(r) = \left(\frac{r}{R}\right)^3 - \left(\frac{r}{R}\right)^4 + \frac{3}{10}\left(\frac{r}{R}\right)^5$$

Differentiate with respect to (r)

$$Y_2'(r) = \frac{3r^2}{R^3} - \frac{4r^3}{R^4} + \frac{3r^4}{2R^5}$$

Differentiate with respect to r

$$Y_2''(r) = \frac{6r}{R^3} - \frac{12r^2}{R^4} + \frac{6r^3}{R^5}$$

$$K_{11} = EI\int_0^R \left(\frac{6}{R^2} - \frac{12r}{R^3} + \frac{6r^2}{R^4}\right)\left(\frac{6}{R^2} - \frac{12r}{R^3} + \frac{6r^2}{R^4}\right)dr$$
$$+ \int_0^R \left(\frac{6r}{R^2} - \frac{6r^2}{R^3} + \frac{2r^3}{R^4}\right)\left(\frac{6r}{R^2} - \frac{6r^2}{R^3} + \frac{2r^3}{R^4}\right)\left(\frac{m\Omega^2}{2}(R^2 - r^2)\right)dr$$

Simplifying we get,

$$K_{11} = EI\int_0^R \frac{36}{R^8}(R-r)^4 dr + \int_0^R \frac{2m\Omega^2 r^2}{R^8}(R^2 - r^2)(3R^2 - 3Rr + r^2)^2 dr$$

$$K_{11} = \frac{36EI}{R^8}\left[\int_0^R (R-r)^4 dr\right] + \frac{2m\Omega^2}{R^8}\int_0^R \left[\left(R^2 r^2 - r^4\right)(3R^2 - 3Rr + r^2)^2\right]dr$$

$$K_{11} = \frac{36EI}{5R^3} + \frac{2m\Omega^2}{R^8}\left[3R^9 - \frac{9}{2}R^9 + \frac{6}{5}R^9 + 2R^9 - 2R^9 + \frac{3}{4}R^9 - \frac{R^9}{9}\right]$$

$$K_{11} = \frac{36EI}{5R^3} + \frac{2m\Omega^2}{R^8}\left(R^9\right)\left[\frac{1080 - 1620 + 432 + 270 - 40}{360}\right]$$

$$K_{11} = \frac{36EI}{5R^3} + 2m\Omega^2 R\left[\frac{122}{360}\right] = \frac{36EI}{5R^3} + \frac{61m\Omega^2 R}{90}$$

$$K_{11} = \frac{7.2EI}{R^3} + 0.6777m\Omega^2 R$$

Similarly,

$$K_{12} = K_{21} = \int_0^R EI \;\; Y_1'' Y_2'' dr + \int_0^R \;\; Y_1' Y_2' T dr$$

$$K_{12} = \int_0^R EI\left(\frac{6}{R^2} - \frac{12r}{R^3} + \frac{6r}{R^3}\right)\left(\frac{6r}{R^3} - \frac{12r^2}{R^4} + \frac{6r^3}{R^5}\right) dr$$

$$+ \int_0^R \left(\frac{6r}{R^2} - \frac{6r^2}{R^3} + \frac{2r^3}{R^4}\right)\left(\frac{3r^2}{R^3} - \frac{4r^3}{R^4} + \frac{3r^4}{2R^5}\right)\frac{m\Omega^2}{2}\left(R^2 - r^2\right) dr$$

$$K_{12} = \frac{1.19988EI}{R^3} + \left(0.13015m\Omega^2 R\right) = K_{21}$$

$$K_{22} = \int_0^R EI \;\; Y_2'' Y_2'' dr + \int_0^R \;\; Y_2' Y_2' T dr$$

$$K_{22} = EI\int_0^R \left(\frac{6r}{R^3} - \frac{12r}{R^4} + \frac{6r}{R^5}\right)^2 dr$$

$$+ \int_0^R \left(\frac{3r^2}{R^3} - \frac{4r^3}{R^4} + \frac{3r^4}{2R^5}\right)^2 \left(\frac{m\Omega^2}{2}\left(R^2 - r^2\right)\right) dr$$

$$K_{22} = \frac{(0.34285)EI}{R^3} + (0.026695)\, m\Omega^2 R$$

Stiffness matrix,

$$K = \begin{bmatrix} \frac{7.2EI}{R^3} + 0.6777m\Omega^2 R & \frac{1.19988EI}{R^3} + 0.13015m\Omega^2 R \\ \frac{1.19988EI}{R^3} + 0.13015m\Omega^2 R & \frac{0.34285EI}{R^3} + 0.026695m\Omega^2 R \end{bmatrix}$$

To find eigenvalues use characteristics equation, $|K - Dm| = 0$

$$\begin{vmatrix} \frac{7.2EI}{R^3} + 0.6777m\Omega^2 R - 0.5777mRD & \frac{1.19988EI}{R^3} + 0.13015m\Omega^2 R - 0.103492mRD \\ \frac{1.19988EI}{R^3} + 0.13015m\Omega^2 R - 0.103492mRD & \frac{0.34285EI}{R^3} + 0.026695m\Omega^2 R - 0.018816mRD \end{vmatrix} = 0$$

Solving the equation at $\Omega = 0$ we get,

$$D_1 = 12.36\frac{EI}{mR^4}$$

$$D_2 = 515.9\frac{EI}{mR^4}$$

$$F_1 = \sqrt{D_1} = 3.515\sqrt{\frac{EI}{mR^4}}$$

$$F_1 = \sqrt{D_1} = 22.713\sqrt{\frac{EI}{mR^4}}$$

Here, both the first and second frequencies are well predicted.

Example 1.11 Using $Y_1(r)$ and $Y_2(r)$ from Example 1.10 as admissible functions for Galerkin method, find the first two frequencies of a cantilever beam.

Solution:

$$w(r,t) = Y_1(r)q_1t + Y_2(r)q_2(t)$$

where,

$$Y_1(r) = 3\left(\frac{r}{R}\right)^2 - 2\left(\frac{r}{R}\right)^3 + \frac{1}{2}\left(\frac{r}{R}\right)^4$$

and

$$Y_2(r) = a\left(\frac{r}{R}\right)^3 + b\left(\frac{r}{R}\right)^4 + c\left(\frac{r}{R}\right)^5$$

The mass matrix is same as for the rayleigh ritz method in Example 1.10.

Mass matrix,

$$m = \begin{bmatrix} 0.5777mR & 0.103492mR \\ 0.103492mR & 0.018816mR \end{bmatrix}$$

Solve for stiffness matrix (K)

$$K_{ij} = \int_0^R EIY_jY_i^{''''}dr - \int_0^R TY_jY_i^{''}dr - \int_0^R T^{'}Y_jY_i^{'}dr$$

$$T = \int_r^R m\Omega^2 r dr = \frac{m\Omega^2}{2}\left(R^2 - r^2\right)$$

$$T^{'} = -mr\Omega^2$$

$$K_{11} = \int_0^R EIY_1Y_1^{''''}dr - \int_0^R TY_1Y_1^{''}dr - \int_0^R T^{'}Y_1Y_1^{'}dr$$

Let,

$$Y_1(r) = 3\left(\frac{r}{R}\right)^2 - 2\left(\frac{r}{R}\right)^3 + \frac{1}{2}\left(\frac{r}{R}\right)^4$$

Differentiate with respect to (r)

$$Y_1'(r) = \frac{6r}{R^2} - \frac{6r^2}{R^3} + \frac{2r^3}{R^4}$$

Differentiate with respect to (r)

$$Y_1''(r) = \frac{6}{R^2} - \frac{12r}{R^3} + \frac{6r^2}{R^4}$$

Differentiate with respect to (r)

$$Y_1'''(r) = \frac{12r}{R^4} - \frac{12}{R^3}$$

Differentiate with respect to (r)

$$Y_1''''(r) = \frac{12}{R^4}$$

Let,

$$Y_2(r) = \left(\frac{r}{R}\right)^3 - \left(\frac{r}{R}\right)^4 + \frac{3}{10}\left(\frac{r}{R}\right)^5$$

Differentiate with respect to (r)

$$Y_2'(r) = \frac{3r^2}{R^3} - \frac{4r^3}{R^4} + \frac{3r^4}{2R^5}$$

Differentiate with respect to (r)

$$Y_2''(r) = \frac{6r}{R^3} - \frac{12r^2}{R^4} + \frac{6r^3}{R^5}$$

Differentiate with respect to (r)

$$Y_2'''(r) = \frac{6}{R^3} - \frac{24r}{R^4} + \frac{18r^2}{R^5}$$

Differentiate with respect to (r)

$$Y_2''''(r) = \frac{36}{R^5} - \frac{24}{R^4}$$

$$K_{11} = \int_0^R EI\left(3\left(\frac{r}{R}\right)^2 - 2\left(\frac{r}{R}\right)^3 + \frac{1}{2}\left(\frac{r}{R}\right)^4\right)\left(\frac{12}{R^4}\right)dr$$

$$-\int_0^R T\left(3\left(\frac{r}{R}\right)^2 - 2\left(\frac{r}{R}\right)^3 + \frac{1}{2}\left(\frac{r}{R}\right)^4\right)\left(\frac{6}{R^2} - \frac{12r}{R^3} + \frac{6r^2}{R^4}\right)dr$$

$$-\int_0^R T'\left(3\left(\frac{r}{R}\right)^2-2\left(\frac{r}{R}\right)^3+\frac{1}{2}\left(\frac{r}{R}\right)^4\right)\left(\frac{6r}{R^2}-\frac{6r^2}{R^3}+\frac{2r^3}{R^4}\right)dr$$

Simplifying we get,

$$K_{11}=\frac{7.2EI}{R^3}+\left(0.67m\Omega^2R\right)$$

Similarly,

$$K_{12}=\int_0^R EIY_2Y_1''''\,dr-\int_0^R TY_2Y_1''dr-\int_0^R T'Y_2Y_1'dr$$

$$K_{12}=\int_0^R EI\left(\left(\frac{r}{R}\right)^3-\left(\frac{r}{R}\right)^4+\frac{3}{10}\left(\frac{r}{R}\right)^5\right)\left(\frac{12}{R^4}\right)dr$$

$$-\int_0^R T\left(\left(\frac{r}{R}\right)^3-\left(\frac{r}{R}\right)^4+\frac{3}{10}\left(\frac{r}{R}\right)^5\right)\left(\frac{6}{R^2}-\frac{12r}{R^3}+\frac{6r^2}{R^4}\right)dr$$

$$-\int_0^R T'\left(\left(\frac{r}{R}\right)^3-\left(\frac{r}{R}\right)^4+\frac{3}{10}\left(\frac{r}{R}\right)^5\right)\left(\frac{6r}{R^2}-\frac{6r^2}{R^3}+\frac{2r^3}{R^4}\right)dr$$

Simplifying we get,

$$K_{12}=\frac{1.2EI}{R^3}+\left(0.1302m\Omega^2R\right)=K_{21}$$

$$K_{22}=\int_0^R EIY_2Y_2''''\,dr-\int_0^R TY_2Y_2''dr-\int_0^R T'Y_2Y_2'dr$$

$$K_{22}=\int_0^R EI\left(\left(\frac{r}{R}\right)^3-\left(\frac{r}{R}\right)^4+\frac{3}{10}\left(\frac{r}{R}\right)^5\right)\left(\frac{36}{R^5}-\frac{24}{R^4}\right)dr$$

$$-\int_0^R T\left(\left(\frac{r}{R}\right)^3-\left(\frac{r}{R}\right)^4+\frac{3}{10}\left(\frac{r}{R}\right)^5\right)\left(\frac{6r}{R^3}-\frac{12r^2}{R^4}+\frac{6r^3}{R^5}\right)dr$$

$$-\int_0^R T'\left(\left(\frac{r}{R}\right)^3-\left(\frac{r}{R}\right)^4+\frac{3}{10}\left(\frac{r}{R}\right)^5\right)\left(\frac{3r^2}{R^3}-\frac{4r^3}{R^4}+\frac{3r^4}{2R^5}\right)dr$$

Simplifying we get,

$$K_{22}=\frac{(0.3429)EI}{R^3}+(0.0267)\,m\Omega^2R$$

Stiffness matrix,

$$K = \begin{bmatrix} \frac{7.2EI}{R^3} + \left(0.67m\Omega^2R\right) & \frac{1.2EI}{R^3} + \left(0.1302m\Omega^2R\right) \\ \frac{1.2EI}{R^3} + \left(0.1302m\Omega^2R\right) & \frac{(0.3429)EI}{R^3} + (0.0267)\, m\Omega^2R \end{bmatrix}$$

To find eigenvalues use characteristics equation, $|K - Dm| = 0$

$$\begin{vmatrix} \frac{7.2EI}{R^3} + \left(0.67m\Omega^2R\right) - 0.5777mRD & \frac{1.2EI}{R^3} + \left(0.1302m\Omega^2R\right) - 0.103492mRD \\ \frac{1.2EI}{R^3} + \left(0.1302m\Omega^2R\right) - 0.103492mRD & \frac{(0.3429)EI}{R^3} + (0.0267)\, m\Omega^2R - 0.018816mRD \end{vmatrix} = 0$$

Solving the equation at $\Omega = 0$ we get,

$$D_1 = 12.36\frac{EI}{mR^4}$$

$$D_2 = 519.9\frac{EI}{mR^4}$$

$$F_1 = \sqrt{D_1} = 3.515\sqrt{\frac{EI}{mR^4}}$$

$$F_1 = \sqrt{D_1} = 22.801\sqrt{\frac{EI}{mR^4}}$$

Example 1.12 Calculate Rayleigh solution for a uniform rotating beam with the polynomial satisfying all boundary conditions as used in the Galerkin method.

Solution:

$$w(r,t) = Y_1(r)q_1(t)$$

where

$$Y_1(r) = 3\left(\frac{r}{R}\right)^2 - 2\left(\frac{r}{R}\right)^3 + \frac{1}{2}\left(\frac{r}{R}\right)^4$$

This function is a 4th order polynomial. Now we can check if it satisfies the cantilever boundary condition,

$$@r = 0,\ w = w' = 0$$

$$\frac{dY_1}{dr} = 6\frac{r}{R^2} - 6\frac{r^2}{R^3} + \frac{2r^3}{R^4} = Y_1'$$

$$@r = 0,\ Y_1(0) = 0,\ Y_1' = 0$$

$$\frac{d^2Y_1}{dr^2} = \frac{6}{R^2} - \frac{6}{R^2}\left(\frac{2r}{R}\right) + \frac{6}{R^2}\left(\frac{r}{R}\right)^2 = Y_1''$$

$$@r = R,\ Y_1'' = 0$$

$$\frac{d^3Y_1}{dr^3} = \frac{-12}{R^3} + \frac{6}{R^2}\left(\frac{2r}{R^2}\right) = Y_1^{'''}$$

$$@r = R,\ Y_1^{'''} = 0$$

Therefore, we see that all boundary conditions are satisfied.
We can now calculate,

$$m_{11} = \int_0^R mY_1Y_1dr$$

$$m_{11} = \int_0^R m\left(3\left(\frac{r}{R}\right)^2 - 2\left(\frac{r}{R}\right)^3 + \frac{1}{2}\left(\frac{r}{R}\right)^4\right)^2 dr$$

$$m_{11} = \int_0^R m\left[\left(\frac{r^8}{4R^8}\right) - 2\left(\frac{r^7}{R^7}\right) + 7\frac{r^6}{R^6} - 12\left(\frac{r^5}{R^5}\right) + 9\frac{r^4}{R^4}\right] dr$$

$$m_{11} = m\left[\frac{R}{36} - \frac{R}{4} + R - 2R + \frac{9R}{5}\right] = \frac{26Rm}{45}$$

$$K_{ij} = \int_0^R EI\quad Y''_iY''_j\ \ dr + \int_0^R Y_i'Y_j'\ \ Tdr$$

$$T = \frac{m}{2}\left(R^2 - r^2\right)\Omega^2$$

$$K_{11} = \int_0^R EI\quad Y_1^{''}Y_1^{''}dr + \int_0^R Y_1'Y_1'Tdr$$

Let,

$$\frac{dY_1}{dr} = Y_1^{'} = 6\frac{r}{R^2} - 6\frac{r^2}{R^3} + \frac{2r^3}{R^4}$$

$$\frac{d^2Y_1}{dr^2} = Y_1^{''} = \frac{6}{R^2} - \frac{6}{R^2}\left(\frac{2r}{R}\right) + \frac{6}{R^2}\left(\frac{r}{R}\right)^2$$

$$K_{11} = \int_0^R EI\left(\frac{6}{R^2} - \frac{6}{R^2}\left(\frac{2r}{R}\right) + \frac{6}{R^2}\left(\frac{r}{R}\right)^2\right)^2 dr$$

$$+ \int_0^R \left(6\frac{r}{R^2} - 6\frac{r^2}{R^3} + \frac{2r^3}{R^4}\right)^2 Tdr$$

$$K_{11} = \int_0^R EI\left(\frac{36}{R^4} - \frac{144r}{R^5} + \frac{216r^2}{R^6} - \frac{144r^3}{R^7} + \frac{36r^4}{R^8}\right) dr$$

$$+\frac{m\Omega^2}{2}\int_0^R \left(\frac{36r^2}{R^2} - \frac{36r^4}{R^4} - \frac{72r^3}{R^3} + \frac{72r^5}{R^5} + \frac{60r^4}{R^4} - \frac{60r^6}{R^6} - \frac{24r^5}{R^5} + \frac{24r^7}{R^7} + \frac{4r^6}{R^6} - \frac{4r^8}{R^8}\right) dr$$

$$K_{11} = \left(\frac{36EI}{5R^3}\right) + \frac{m\Omega^2}{2}\left[\frac{36R}{3} - \frac{36R}{5} - \frac{72R}{4} + \frac{72R}{6} + \frac{60R}{5} - \frac{60R}{7} - \frac{24R}{6} + \frac{24R}{8} + \frac{4R}{7} - \frac{4R}{9}\right]$$

$$K_{11} = \frac{36EI}{5R^3} + 0.6777m\Omega^2 R$$

Thus, at $\Omega = 0$ the frequency can be determined by,

$$\omega^2 = \frac{K_{11}}{m_{11}} = \frac{\frac{36EI}{5R^3}}{\frac{26Rm}{45}}$$

$$\omega^2 = 12.4615\frac{EI}{mR^4}$$

$$\omega = 3.53\sqrt{\frac{EI}{mR^4}}$$

For previous Example 1.3 the solution obtained for Rayleigh method is,

$$\omega = 4.47\sqrt{\frac{EI}{mR^4}}$$

The Galerkin solution from Example 1.1 is,

$$\omega = 3.53\sqrt{\frac{EI}{mR^4}}$$

We see that Rayleigh–Ritz and Galerkin methods give the same solution in this case.

1.4 Finite Element Method

The finite element method is a very powerful method in structural mechanics. It can take care of complex geometries and non-uniformities. We will use a simple displacement based finite element analysis for the rotating beam. Consider a rotating free-free beam. Let us divide the beam into 3 finite elements, as shown in Fig. 1.12. We see from this figure that there are 4 joints, called nodes. Each node undergoes two degrees of motion; vertical displacement and rotation, thus, in Fig. 1.12, there are a total of eight nodal degree of freedom (dof):

Vertical displacements: q_1 q_2 q_3 q_4

Slopes: q_2 q_4 q_6 q_8

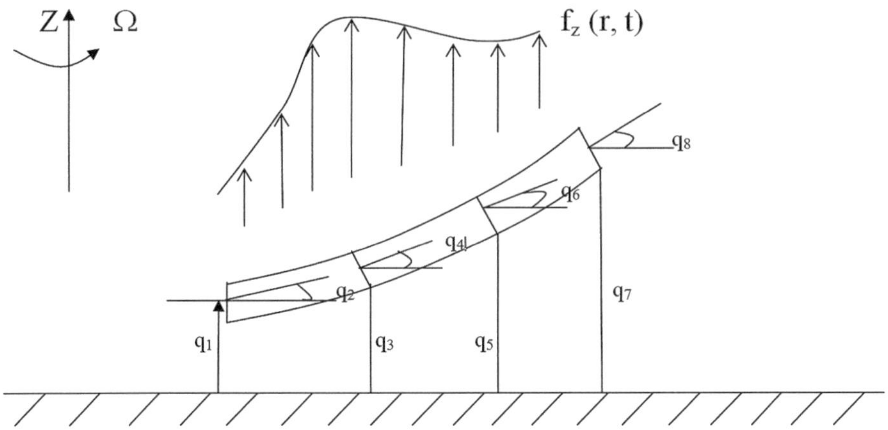

Fig. 1.12 Finite element model of rotating beam

We will now illustrate the two major steps in finite element analysis

1. Development of elemental properties
2. Assembly of elements

1.4.1 Element Properties

The expanded view of a beam finite element is shown in Fig. 1.13. Here, l = Length of element and x = Local coordinate of element.

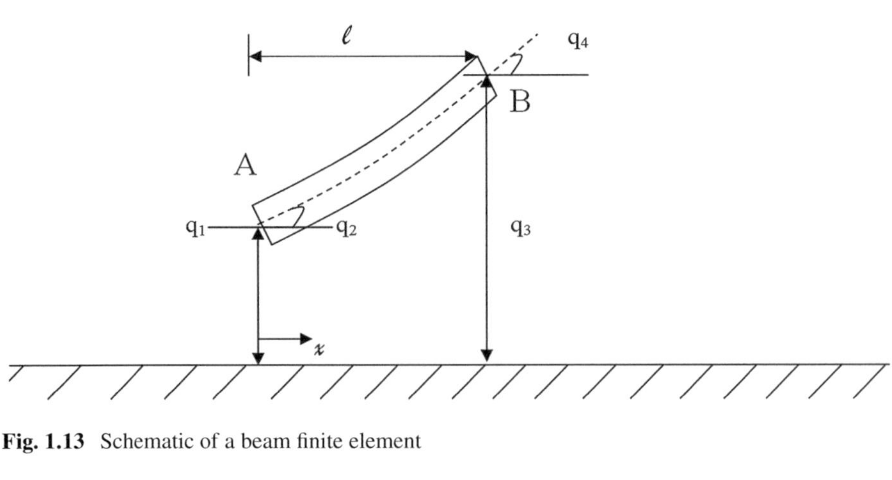

Fig. 1.13 Schematic of a beam finite element

Assume polynomial distribution of displacement $w(x)$ within the element,

$$w(x) = \alpha_1 + \alpha_2 x + \alpha_3 x^2 + \alpha_4 x^3 \tag{1.48}$$

Here, we have selected a cubic polynomial with four coefficients as there are four end conditions for the elements. The displacement must satisfy end conditions,

$$\begin{aligned}
@ \quad x &= 0, \qquad w = q_1 = \alpha_1 \\
@ \quad x &= 0, \qquad w' = q_2 = \alpha_2 \\
@ \quad x &= l, \qquad w = q_3 = \alpha_1 + \alpha_2 l + \alpha_3 l^2 + \alpha_4 l^3 \\
@ \quad x &= l, \qquad w' = q_4 = \alpha_2 + 2\alpha_3 l + 3\alpha_4 l^2
\end{aligned}$$

We can use these element end conditions to write,

$$\begin{Bmatrix} q_1 \\ q_2 \\ q_3 \\ q_4 \end{Bmatrix} = \begin{bmatrix} 1 & 0 & 0 & 0 \\ 0 & 1 & 0 & 0 \\ \frac{-3}{l^2} & \frac{-2}{l} & \frac{3}{l^2} & \frac{-1}{l} \\ \frac{2}{\left(\frac{2}{l^3}\right)} & \frac{1}{l^2} & \frac{-2}{l^3} & \frac{1}{l^2} \end{bmatrix} \begin{Bmatrix} \alpha_1 \\ \alpha_2 \\ \alpha_3 \\ \alpha_4 \end{Bmatrix}$$

Inverting the above matrix yields,

$$\begin{Bmatrix} \alpha_1 \\ \alpha_2 \\ \alpha_3 \\ \alpha_4 \end{Bmatrix} = \begin{bmatrix} 1 & 0 & 0 & 0 \\ 0 & 1 & 0 & 0 \\ \frac{-3}{l^2} & \frac{-2}{l} & \frac{3}{l^2} & \frac{-1}{l} \\ \frac{2}{\left(\frac{2}{l^3}\right)} & \frac{1}{l^2} & \frac{-2}{l^3} & \frac{1}{l^2} \end{bmatrix} \begin{Bmatrix} q_1 \\ q_2 \\ q_3 \\ q_4 \end{Bmatrix}$$

Putting the values of α_i back into the assumed displacement $w(x)$ yields,

$$w(x, t) = H_1 q_1 + H_2 q_2 + H_3 q_3 + H_4 q_4 = \sum_{i=1}^{4} H_i q_i \tag{1.49}$$

where H_1, H_2, H_3 and H_4 are known as shape functions. Sometimes, they are also called interpolation functions or basis functions.

$$H_1 = 2\left(\frac{x}{l}\right)^3 - 3\left(\frac{x}{l}\right)^2 + 1$$

$$H_2 = \left[\left(\frac{x}{l}\right)^3 - 2\left(\frac{x}{l}\right)^2 + \frac{x}{l}\right] l$$

$$H_3 = -2\left(\frac{x}{l}\right)^3 + 3\left(\frac{x}{l}\right)^2$$

$$H_4 = \left[\left(\frac{x}{l}\right)^3 - \left(\frac{x}{l}\right)^2\right] l$$

To derive the finite element equations for the rotating beam, we make use of the energy method.

1.4.2 Energy Expressions

The kinetic energy of the element is,

$$T = \frac{1}{2}\int_0^l m\dot{w}^2 dx = \frac{1}{2}\int_0^l \left(\sum_{i=1}^{4} H_i \dot{q}_i\right)\left(\sum_{j=1}^{4} H_j \dot{q}_j\right) dx$$

$$= \frac{1}{2}\sum_{i=1}^{4}\sum_{j=1}^{4} m_{ij}\dot{q}_i\dot{q}_j \tag{1.50}$$

where

$$m_{ij} = \int_0^l mH_iH_j dx$$

We see that this mass matrix is a local form of the earlier Rayleigh–Ritz and Galerkin formulation. In those methods, the integration was over the whole domain. The potential or strain energy of the element is given by,

$$V = \frac{1}{2}\int_0^l EI\left(\frac{d^2w}{dx^2}\right)^2 dx + \frac{1}{2}\int_0^l T(x)\left(\frac{dw}{dx}\right)^2 dx \tag{1.51}$$

$$= \frac{1}{2}\int_0^l EI\left[\sum_{i=1}^{4}\frac{d^2H_i}{dx^2}q_i\right]\left[\sum_{j=1}^{4}\frac{d^2H_j}{dx^2}q_j\right] dx +$$

$$\frac{1}{2}\int_0^l T(x)\left[\sum_{i=1}^{4}\frac{dH_i}{dx}q_i\right]\left[\sum_{j=1}^{4}\frac{dH_j}{dx}q_j\right] dx$$

$$= \frac{1}{2}\sum_{i=1}^{4}\sum_{j=1}^{4} K_{ij}q_iq_j$$

$$K_{ij} = \int_0^l EI\frac{d^2H_i}{dx^2}\frac{d^2H_j}{dx^2}dx + \int_0^l T(x)\frac{dH_i}{dx}\frac{dH_j}{dx}dx \tag{1.52}$$

Virtual work done,

$$\delta W = \int_0^l f_z \delta w dx = \int_0^l f_z \sum_{i=1}^{4} H_i \delta q_i dx = \sum_{i=1}^{4} Q_i \delta q_i \tag{1.53}$$

where

$$Q_i = \int_0^l f_z H_i dx$$

Consider a beam element with uniform mass and stiffness properties. Assume that the left edge of an element is at a distance x_i from rotational axis and l is the element length, as shown in Fig. 1.14.

The mass matrix is given by,

$$m_{ij} = m \begin{bmatrix} \frac{13}{35}l & \frac{11}{210}l^2 & \frac{9}{70}l & \frac{-13}{420}l^2 \\ \frac{11}{210}l^2 & \frac{1}{105}l^3 & \frac{13}{420}l^2 & \frac{-1}{140}l^3 \\ \frac{9}{70}l & \frac{13}{420}l^2 & \frac{13}{35}l & \frac{-11}{210}l^2 \\ \frac{-13}{420}l^2 & \frac{-1}{140}l^3 & \frac{-11}{210}l^2 & \frac{l^3}{105} \end{bmatrix}$$

For example,

$$m_{13} = \int_0^l m H_1 H_3 dx$$

$$= \int_0^l m\left(2\left(\frac{x}{l}\right)^3 - 3\left(\frac{x}{l}\right)^2 + 1\right)\left(-2\left(\frac{x}{l}\right)^3 + 3\left(\frac{x}{l}\right)^2\right) dx$$

$$= m\left[\frac{9}{70}l\right]$$

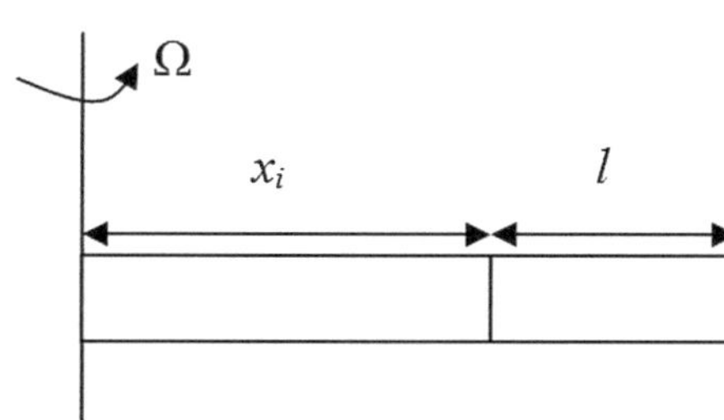

Fig. 1.14 Element on the blade

The stiffness matrix is given by,

$$k_{ij} = EI\begin{bmatrix} \frac{12}{l^3} & \frac{6}{l^2} & \frac{-12}{l^3} & \frac{6}{l^2} \\ \frac{6}{l^2} & \frac{4}{l} & \frac{-6}{l^2} & \frac{2}{l} \\ \frac{-12}{l^3} & \frac{-6}{l^2} & \frac{12}{l^3} & \frac{-6}{l^2} \\ \frac{6}{l^2} & \frac{2}{l} & \frac{-6}{l^2} & \frac{4}{l} \end{bmatrix} + \frac{\Omega^2 A_i}{2}\begin{bmatrix} \frac{6}{5l} & \frac{1}{10} & \frac{-6}{5l} & \frac{1}{10} \\ \frac{1}{10} & \frac{2l}{15} & \frac{-1}{10} & \frac{-l}{30} \\ \frac{-6}{5l} & \frac{-1}{10} & \frac{6}{5l} & \frac{-1}{10} \\ \frac{1}{10} & \frac{-l}{30} & \frac{-1}{10} & \frac{2l}{15} \end{bmatrix}$$

$$-m_i\Omega^2\begin{bmatrix} \frac{3}{5}x_i + \frac{6l}{35} & \frac{l}{10}x_i + \frac{l^2}{28} & \frac{-3}{5}x_i - \frac{6l}{35} & \frac{-l^2}{70} \\ \frac{l}{10}x_i + \frac{l^2}{28} & \frac{l^2}{30}x_i + \frac{l^3}{105} & \frac{-l}{10}x_i - \frac{l^2}{28} & \frac{-l^2}{60}x_i - \frac{l^3}{140} \\ \frac{-3}{5}x_i - \frac{6l}{35} & \frac{-l}{10}x_i - \frac{l^2}{28} & \frac{3}{5}x_i + \frac{6l}{35} & \frac{l^2}{70} \\ \frac{-l^2}{70} & \frac{-l^2}{60}x_i - \frac{l^3}{140} & \frac{l^2}{70} & \frac{l^2}{30}x_i + \frac{l^3}{105} \end{bmatrix}$$

$$A_i = \sum_{j=i}^{N} m_j \left(x_{j+1}^2 - x_j^2\right)$$

To understand the origin of the A_i term, consider, (Fig. 1.15)

$$\begin{aligned} T(r) &= \int_r^R m\Omega^2 r dr = \int_{x_i}^R m\Omega^2 r dr - \int_{x_i}^{x_i+x} m\Omega^2 x dx \\ &= \sum_{j=i}^{N} \int_{x_j}^{x_j+x} m_j\Omega^2 x dx - \int_{x_i}^{x_i+x} m_i\Omega^2 x dx \\ &= \sum_{j=i}^{N} m_j\Omega^2 \frac{\left(x_{j+1}^2 - x_j^2\right)}{2} - m_i\Omega^2 \frac{(x_i+x)^2 - x_i^2}{2} \end{aligned}$$

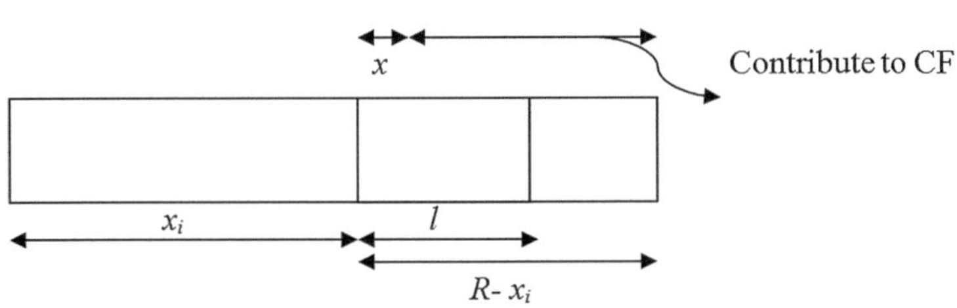

Fig. 1.15 Contribution to blade centrifugal force

$$T(x) = \frac{\Omega^2 A_i}{2} - m_i \Omega^2 \frac{(x_i + x)^2 - x_i^2}{2} \tag{1.54}$$

where

$$A_i = \sum_{j=i}^{N} m_j \left(x_{j+1}^2 - x_j^2\right)$$

For example,

$$K_{ij} = \int_0^l EI \frac{d^2 H_i}{dx^2} \frac{d^2 H_j}{dx^2} dx + \int_0^l T(x) \frac{dH_i}{dx} \frac{dH_j}{dx} dx$$

Substituting T(x) value in above equation we get,

$$K_{ij} = \int_0^l EI \frac{d^2 H_i}{dx^2} \frac{d^2 H_j}{dx^2} dx + \int_0^l \frac{\Omega^2 A_i}{2} - m_i \Omega^2 \frac{(x_i + x)^2 - x_i^2}{2} \frac{dH_i}{dx} \frac{dH_j}{dx} dx$$

Let solve for i = 1, j = 1

$$K_{11} = \int_0^l EI \frac{d^2 H_1}{dx^2} \frac{d^2 H_1}{dx^2} dx + \int_0^l \frac{\Omega^2 A_1}{2} - m_1 \Omega^2 \frac{(x_1 + x)^2 - x_1^2}{2} \frac{dH_1}{dx} \frac{dH_1}{dx} dx$$

$$H_1 = 2\left(\frac{x}{l}\right)^3 - 3\left(\frac{x}{l}\right)^2 + 1$$

$$\frac{dH_1}{dx} = 6\left(\frac{x}{l}\right)^2 \left(\frac{1}{l}\right) - 6\left(\frac{x}{l}\right)\left(\frac{1}{l}\right)$$

$$\frac{d^2 H_1}{dx^2} = 12\left(\frac{x}{l}\right)\left(\frac{1}{l^2}\right) - 6\left(\frac{1}{l^2}\right)$$

Substitute in above equation and solve, we get,

$$K_{11} = \frac{12EI}{l^3} + \frac{\Omega^2 A_i}{2}\left[\frac{6}{5l}\right] - m_i \Omega^2 \left[\frac{3}{5} x_i + \frac{6l}{35}\right]$$

The other elements can be obtained in a similar manner. Appendix A gives the derivation the stiffness and mass matrix terms.

1.4.3 Assembly of Elements

We have T, V and δW for each element. These are the energy values for each element. To apply the physical laws, we need to find the total energy. We assemble the element energies to get T_{TOT}, V_{TOT} and δW_{TOT}. Figure 1.16 shows the three finite elements and

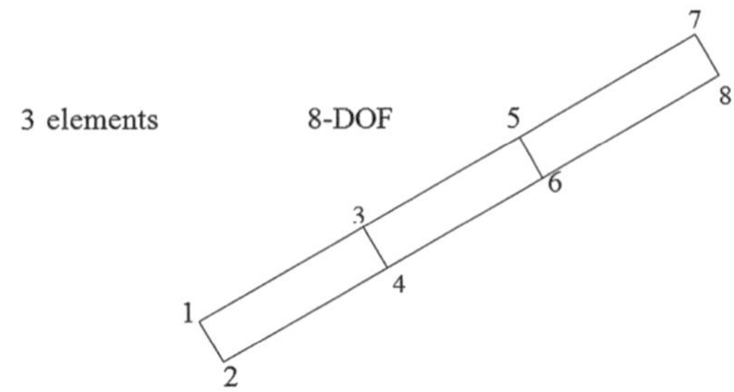

Fig. 1.16 Assembly of 3 elements

the corresponding degrees of freedom.

The total strain energy is given as, $V_{TOT} = V_1 + V_2 + V_3$

$$V_1 = \frac{1}{2}\begin{bmatrix} q_1 & q_2 & q_3 & q_4 & q_5 & q_6 & q_7 & q_8 \end{bmatrix} \begin{matrix} 1\\2\\3\\4\\5\\6\\7\\8 \end{matrix} \begin{bmatrix} \times & \times & \times & \times & & & & \\ \times & \times & \times & \times & & & & \\ \times & \times & \times & \times & & & & \\ \times & \times & \times & \times & & & & \\ & & & & & & & \\ & & & & & & & \\ & & & & & & & \\ & & & & & & & \end{bmatrix} \begin{Bmatrix} q_1\\q_2\\q_3\\q_4\\q_5\\q_6\\q_7\\q_8 \end{Bmatrix}$$

$$\downarrow$$

$$K_{1234}$$

$$V_2 = \frac{1}{2}\begin{bmatrix} q_1 & q_2 & q_3 & q_4 & q_5 & q_6 & q_7 & q_8 \end{bmatrix} \begin{matrix} 1\\2\\3\\4\\5\\6\\7\\8 \end{matrix} \begin{bmatrix} & & & & & & & \\ & & & & & & & \\ & & 0 & 0 & 0 & 0 & & \\ & & 0 & 0 & 0 & 0 & & \\ & & 0 & 0 & 0 & 0 & & \\ & & 0 & 0 & 0 & 0 & & \\ & & & & & & & \\ & & & & & & & \end{bmatrix} \begin{Bmatrix} q_1\\q_2\\q_3\\q_4\\q_5\\q_6\\q_7\\q_8 \end{Bmatrix}$$

$$\downarrow$$

$$K_{3456}$$

$$V_3 = \frac{1}{2}\begin{bmatrix} q_1 & q_2 & q_3 & q_4 & q_5 & q_6 & q_7 & q_8 \end{bmatrix} \begin{matrix} 1\\2\\3\\4\\5\\6\\7\\8 \end{matrix} \begin{bmatrix} & & & & & & & \\ & & & & & & & \\ & & & & & & & \\ & & & & & & & \\ & & & & + & + & + & + \\ & & & & + & + & + & + \\ & & & & + & + & + & + \\ & & & & + & + & + & + \end{bmatrix} \begin{Bmatrix} q_1\\q_2\\q_3\\q_4\\q_5\\q_6\\q_7\\q_8 \end{Bmatrix}$$

$$\downarrow$$

$$K_{5678}$$

To get the total energy, we combine the three matrices,

$$K_{TOT} = K_{1234} + K_{3456} + K_{5678}$$

$$K_{TOT} = \begin{array}{c} 1 \\ 2 \\ 3 \\ 4 \\ 5 \\ 6 \\ 7 \\ 8 \end{array} \begin{bmatrix} \times & \times & \times & \times & & & & \\ \times & \times & \times & \times & & & & \\ \times & \times & \otimes & \otimes & 0 & 0 & & \\ \times & \times & \otimes & \otimes & 0 & 0 & & \\ & & 0 & 0 & \oplus & \oplus & + & + \\ & & 0 & 0 & \oplus & \oplus & + & + \\ & & & & + & + & + & + \\ & & & & + & + & + & + \end{bmatrix} \begin{Bmatrix} q_1 \\ q_2 \\ q_3 \\ q_4 \\ q_5 \\ q_6 \\ q_7 \\ q_8 \end{Bmatrix} \quad \begin{array}{l} \times \rightarrow Element1 \\ 0 \rightarrow Element2 \\ + \rightarrow Element3 \end{array}$$

We can see that the assembly process is easy and can be programmed by computer.

Similarly

$$T_{TOT} = \frac{1}{2}\left[\dot{q}_1 \quad \dot{q}_2 \quad \dot{q}_3 \quad . \quad . \quad . \quad . \quad \dot{q}_8\right] m_{TOT} \begin{Bmatrix} q_1 \\ q_2 \\ q_3 \\ . \\ . \\ . \\ . \\ q_8 \end{Bmatrix}$$

$$m_{TOT} = m_{1234} + m_{3456} + m_{5678}$$

$$\begin{bmatrix} Q_1 \\ Q_2 \\ Q_3 \\ . \\ . \\ . \\ . \\ Q_8 \end{bmatrix} = Q_{1234} + Q_{3456} + Q_{5678}$$

The energy expressions have been expressed in terms of degrees of freedom.

Using Lagrange's equation,

$$\frac{\partial}{\partial T}\left(\frac{\partial T_{TOT}}{\partial \dot{q}_i}\right) - \left(\frac{\partial T_{TOT}}{\partial q_i}\right) + \left(\frac{\partial V_{TOT}}{\partial q_i}\right) = Q_i \qquad i = 1, 2, 3, \ldots N$$

Equation of motion becomes,

$$\begin{bmatrix} m_{11} & m_{12} & \cdot & \cdot & \cdot & \cdot & \cdot & m_{18} \\ m_{21} & m_{22} & \cdot & \cdot & \cdot & \cdot & \cdot & m_{28} \\ \cdot & \cdot & \cdot & \cdot & \cdot & \cdot & \cdot & \cdot \\ \cdot & \cdot & \cdot & \cdot & \cdot & \cdot & \cdot & \cdot \\ \cdot & \cdot & \cdot & \cdot & \cdot & \cdot & \cdot & \cdot \\ \cdot & \cdot & \cdot & \cdot & \cdot & \cdot & \cdot & \cdot \\ \cdot & \cdot & \cdot & \cdot & \cdot & \cdot & \cdot & \cdot \\ m_{81} & m_{82} & \cdot & \cdot & \cdot & \cdot & \cdot & m_{88} \end{bmatrix} \begin{Bmatrix} \ddot{q}_1 \\ \ddot{q}_2 \\ \cdot \\ \cdot \\ \cdot \\ \cdot \\ \cdot \\ \ddot{q}_8 \end{Bmatrix} +$$

$$\begin{bmatrix} K_{11} & K_{12} & \cdot & \cdot & \cdot & \cdot & \cdot & K_{18} \\ K_{21} & K_{22} & \cdot & \cdot & \cdot & \cdot & \cdot & K_{28} \\ \cdot & \cdot & \cdot & \cdot & \cdot & \cdot & \cdot & \cdot \\ \cdot & \cdot & \cdot & \cdot & \cdot & \cdot & \cdot & \cdot \\ \cdot & \cdot & \cdot & \cdot & \cdot & \cdot & \cdot & \cdot \\ \cdot & \cdot & \cdot & \cdot & \cdot & \cdot & \cdot & \cdot \\ \cdot & \cdot & \cdot & \cdot & \cdot & \cdot & \cdot & \cdot \\ K_{81} & K_{82} & \cdot & \cdot & \cdot & \cdot & \cdot & K_{88} \end{bmatrix} \begin{Bmatrix} q_1 \\ q_2 \\ \cdot \\ \cdot \\ \cdot \\ \cdot \\ \cdot \\ q_8 \end{Bmatrix} = \begin{Bmatrix} Q_1 \\ Q_2 \\ \cdot \\ \cdot \\ \cdot \\ \cdot \\ \cdot \\ Q_8 \end{Bmatrix}$$

Or in matrix form,

Here, K_{TOT} is singular because rigid body modes are present. These are zero frequency modes and prevent the matrix from being inverted. However, this equation is valid for a free−free rotating beam. For a typical rotor blade, there are boundary conditions imposed by the physics of the problem.

A Matlab code for the finite element method is given in the Appendix B.

1.4.4 Cantilever

$$at \quad r = 0, w = \frac{dw}{dr} = 0$$

When the beam is fixed at one end, we get the cantilever boundary condition, as shown in Fig. 1.17. In this case, we don't need to know q_1 and q_2, as they are zero. We can drop the first two rows and columns corresponding to q_1 and q_2.

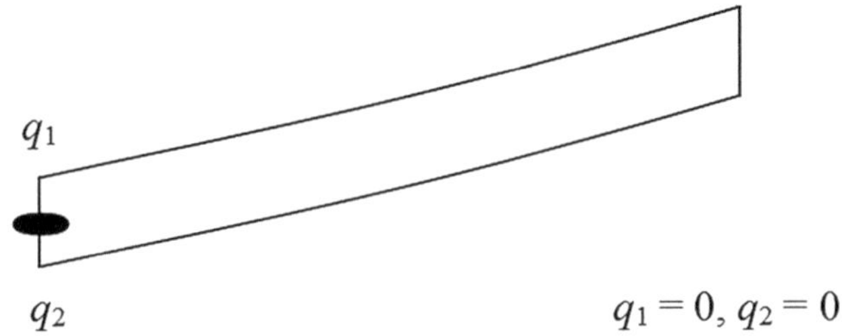

Fig. 1.17 Cantilever boundary condition

Equation of motion becomes,

$$\begin{bmatrix} m_{33} & m_{34} & . & . & . & m_{38} \\ m_{43} & m_{44} & . & . & . & m_{48} \\ . & . & . & . & . & . \\ . & . & . & . & . & . \\ . & . & . & . & . & . \\ m_{83} & m_{84} & . & . & . & m_{88} \end{bmatrix} \begin{Bmatrix} \ddot{q}_3 \\ \ddot{q}_4 \\ . \\ . \\ . \\ \ddot{q}_8 \end{Bmatrix} +$$

$$\begin{bmatrix} K_{33} & K_{34} & . & . & . & K_{38} \\ K_{43} & K_{44} & . & . & . & K_{48} \\ . & . & . & . & . & . \\ . & . & . & . & . & . \\ . & . & . & . & . & . \\ K_{83} & K_{84} & . & . & . & K_{88} \end{bmatrix} \begin{Bmatrix} q_3 \\ q_4 \\ . \\ . \\ . \\ q_8 \end{Bmatrix} = \begin{Bmatrix} Q_3 \\ Q_4 \\ . \\ . \\ . \\ Q_8 \end{Bmatrix}$$

The corresponding eigenvalue problem can be easily solved to obtain the frequency and mode shapes.

Example 1.13 To find the rotating and non-rotating natural frequencies of the cantilever uniform beam for 1 element with 2 degree of freedom. Assume EI (flexural rigidity) = 100000 Nm^2, m (mass per unit length) = 5 Kg/m and R (radius of beam) = 10 m.

Solution: Given Data,
$N = 1$
$EI = 100000\,\text{Nm}^2$
$DOF = 2$
$m = 5\,\text{Kg/m}$
$R = 10\,\text{m}$

Let, length of each element is given by,

$$l = \frac{R}{N} = \frac{10}{1} = 10\,\text{m}$$

The mass matrix is given by,

$$m_{ij} = m \begin{bmatrix} \frac{13}{35}l & \frac{11}{210}l^2 & \frac{9}{70}l & \frac{-13}{420}l^2 \\ \frac{11}{210}l^2 & \frac{1}{105}l^3 & \frac{13}{420}l^2 & \frac{-1}{140}l^3 \\ \frac{9}{70}l & \frac{13}{420}l^2 & \frac{13}{35}l & \frac{-11}{210}l^2 \\ \frac{-13}{420}l^2 & \frac{-1}{140}l^3 & \frac{-11}{210}l^2 & \frac{l^3}{105} \end{bmatrix}$$

By substituting the values of l and m in above matrix we get,

$$m_{ij} = \begin{bmatrix} 18.5714 & 26.1905 & 6.42857 & -15.4762 \\ 26.1905 & 47.619 & 15.4762 & -35.7143 \\ 6.42857 & 15.4762 & 18.5714 & -26.1905 \\ -15.4762 & -35.7143 & -26.1905 & 47.619 \end{bmatrix}$$

Case 1: $\lambda = 1$ (Non dimensional rotating speed)
The stiffness matrix is given by,

$$K_{ij} = EI \begin{bmatrix} \frac{12}{l^3} & \frac{6}{l^2} & \frac{-12}{l^3} & \frac{6}{l^2} \\ \frac{6}{l^2} & \frac{4}{l} & \frac{-6}{l^2} & \frac{2}{l} \\ \frac{-12}{l^3} & \frac{-6}{l^2} & \frac{12}{l^3} & \frac{-6}{l^2} \\ \frac{6}{l^2} & \frac{2}{l} & \frac{-6}{l^2} & \frac{4}{l} \end{bmatrix} + \frac{\Omega^2 A_i}{2} \begin{bmatrix} \frac{6}{5l} & \frac{1}{10} & \frac{-6}{5l} & \frac{1}{10} \\ \frac{1}{10} & \frac{2l}{15} & \frac{-1}{10} & \frac{-l}{30} \\ \frac{-6}{5l} & \frac{-1}{10} & \frac{6}{5l} & \frac{-1}{10} \\ \frac{1}{10} & \frac{-l}{30} & \frac{-1}{10} & \frac{2l}{15} \end{bmatrix}$$

$$-m_i\Omega^2 \begin{bmatrix} \frac{3}{5}x_i + \frac{6l}{35} & \frac{l}{10}x_i + \frac{l^2}{28} & \frac{-3}{5}x_i - \frac{6l}{35} & \frac{-l^2}{70} \\ \frac{l}{10}x_i + \frac{l^2}{28} & \frac{l^2}{30}x_i + \frac{l^3}{105} & \frac{-l}{10}x_i - \frac{l^2}{28} & \frac{-l^2}{60}x_i - \frac{l^3}{140} \\ \frac{-3}{5}x_i - \frac{6l}{35} & \frac{-l}{10}x_i - \frac{l^2}{28} & \frac{3}{5}x_i + \frac{6l}{35} & \frac{l^2}{70} \\ \frac{-l^2}{70} & \frac{-l^2}{60}x_i - \frac{l^3}{140} & \frac{l^2}{70} & \frac{l^2}{30}x_i + \frac{l^3}{105} \end{bmatrix}$$

Rotating speed (Ω) can be calculated by,

$$\Omega = \left(\lambda\sqrt{\frac{EI}{mR^4}}\right) = \left(1 \times \sqrt{\frac{100000}{5 \times 10^4}}\right) = 1.41421\ \text{rad/s}$$

In which,

$$A_i = \sum_{j=i}^{N} m_j\left(x_{j+1}^2 - x_j^2\right)$$

Let N = 1, for 1 element, left side of the beam no element, so $x_j = 0$,

$$A_i = m_j\left(l^2 - 0\right) = 5\left(10^2\right) = 500\,\text{kgm}$$

By substituting the above values in stiffness matrix we get,

$$K_{ij} = \begin{bmatrix} 1242.86 & 6014.29 & -1242.86 & 6064.29 \\ 6014.29 & 40571.4 & -6014.29 & 19904.8 \\ -1242.86 & -6014.29 & 1242.86 & -6064.29 \\ 6064.29 & 19904.8 & -6064.29 & 40238.1 \end{bmatrix}$$

By applying cantilever boundary condition for mass and stiffness matrix first two rows and columns will be eliminated, we get 2×2 matrix,

$$m_{ij} = \begin{bmatrix} 18.5714 & -26.1905 \\ -26.1905 & 47.619 \end{bmatrix}$$

$$K_{ij} = \begin{bmatrix} 1242.86 & -6064.29 \\ -6064.29 & 40238.1 \end{bmatrix}$$

Eigenvalues and Eigenvectors of the matrix can be determined as,
The characteristic equation, $\left|K_{ij} - Dm_{ij}\right| = 0$.

$$\begin{vmatrix} 1242.86 & -6064.29 \\ -6064.29 & 40238.1 \end{vmatrix} - D \begin{vmatrix} 18.5714 & -26.1905 \\ -26.1905 & 47.619 \end{vmatrix} = 0$$

By solving, we get

Eigenvalue, $D_1 = 27.4$ and corresponding Eigenvectors, –0.2854 and –0.0392.
Eigenvalue, $D_2 = 2436.3$ and corresponding Eigenvectors, 0.3982 and 0.3034.

Natural frequency can be determined by,

$$f_1 = \sqrt{(D_1)} = 5.23255\,\text{Hz}$$
$$f_2 = \sqrt{(D_2)} = 49.3581\,\text{Hz}$$

Also, Natural frequency in terms of non dimensional form can be determined by,

$$\omega_1 = \frac{f_1}{\sqrt{\frac{EI}{mR^4}}} = 3.69997$$

$$\omega_2 = \frac{f_2}{\sqrt{\frac{EI}{mR^4}}} = 34.9014$$

For comparison, the fem code in Appendix B gives $\omega_1 = 3.68165$ and $\omega_2 = 22.181$ with $N = 50$ elements.

Case 2: $\lambda = 12$ (Non dimensional rotating speed)
The stiffness matrix is given by,

$$K_{ij} = EI\begin{bmatrix} \frac{12}{l^3} & \frac{6}{l^2} & \frac{-12}{l^3} & \frac{6}{l^2} \\ \frac{6}{l^2} & \frac{4}{l} & \frac{-6}{l^2} & \frac{2}{l} \\ \frac{-12}{l^3} & \frac{-6}{l^2} & \frac{12}{l^3} & \frac{-6}{l^2} \\ \frac{6}{l^2} & \frac{2}{l} & \frac{-6}{l^2} & \frac{4}{l} \end{bmatrix} + \frac{\Omega^2 A_i}{2}\begin{bmatrix} \frac{6}{5l} & \frac{1}{10} & \frac{-6}{5l} & \frac{1}{10} \\ \frac{1}{10} & \frac{2l}{15} & \frac{-1}{10} & \frac{-l}{30} \\ \frac{-6}{5l} & \frac{-1}{10} & \frac{6}{5l} & \frac{-1}{10} \\ \frac{1}{10} & \frac{-l}{30} & \frac{-1}{10} & \frac{2l}{15} \end{bmatrix}$$

$$-m_i\Omega^2\begin{bmatrix} \frac{3}{5}x_i + \frac{6l}{35} & \frac{l}{10}x_i + \frac{l^2}{28} & \frac{-3}{5}x_i - \frac{6l}{35} & \frac{-l^2}{70} \\ \frac{l}{10}x_i + \frac{l^2}{28} & \frac{l^2}{30}x_i + \frac{l^3}{105} & \frac{-l}{10}x_i - \frac{l^2}{28} & \frac{-l^2}{60}x_i - \frac{l^3}{140} \\ \frac{-3}{5}x_i - \frac{6l}{35} & \frac{-l}{10}x_i - \frac{l^2}{28} & \frac{3}{5}x_i + \frac{6l}{35} & \frac{l^2}{70} \\ \frac{-l^2}{70} & \frac{-l^2}{60}x_i - \frac{l^3}{140} & \frac{l^2}{70} & \frac{l^2}{30}x_i + \frac{l^3}{105} \end{bmatrix}$$

Rotating speed (Ω) can be calculated by,

$$\Omega = \left(\lambda\sqrt{\frac{EI}{mR^4}}\right) = \left(12 \times \sqrt{\frac{100000}{5 \times 10^4}}\right) = 16.9705\,\text{rad/s}$$

In which,

$$A_i = \sum_{j=i}^{N} m_j\left(x_{j+1}^2 - x_j^2\right)$$

Let N = 1, for 1 element, left side of the beam no element, so $x_j = 0$,

$$A_i = m_j\left(l^2 - 0\right) = 5\left(10^2\right) = 500\,\text{kgm}$$

By substituting the above values in stiffness matrix we get,

$$K_{ij} = \begin{bmatrix} 7371.43 & 8057.14 & -7371.43 & 15257.1 \\ 8057.14 & 122286.0 & -8057.14 & 6285.71 \\ -7371.43 & -8057.14 & 7371.43 & -15257.1 \\ 15257.1 & 6285.71 & -15257.1 & 74285.7 \end{bmatrix}$$

By applying cantilever boundary condition for mass and stiffness matrix first two rows and columns will be eliminated, we get 2×2 matrix,

$$m_{ij} = \begin{bmatrix} 18.5714 & -26.1905 \\ -26.1905 & 47.619 \end{bmatrix}$$

$$K_{ij} = \begin{bmatrix} 7371.43 & -15257.1 \\ -15257.1 & 74285.7 \end{bmatrix}$$

Eigenvalues and Eigenvectors of the matrix can be determined as
The characteristic equation, $\left|K_{ij} - Dm_{ij}\right| = 0$.

$$\begin{vmatrix} 7371.43 & -15257.1 \\ -15257.1 & 74285.7 \end{vmatrix} - D \begin{vmatrix} 18.5714 & -26.1905 \\ -26.1905 & 47.619 \end{vmatrix} = 0$$

By solving, we get

Eigenvalue, $D_1 = 366.66$ and corresponding Eigenvectors, –0.2689 and –0.0268.
Eigenvalue, $D_2 = 4327.8$ and corresponding Eigenvectors, 0.4095 and 0.3048.

Natural frequency can be determined by,

$$f_1 = \sqrt{(D_1)} = 19.1473\,\text{Hz}$$
$$f_2 = \sqrt{(D_2)} = 65.7859\,\text{Hz}$$

Also, Natural frequency in terms of non dimensional form can be determined by,

$$\omega_1 = \frac{f_1}{\sqrt{\frac{EI}{mR^4}}} = 13.5392$$

$$\omega_2 = \frac{f_2}{\sqrt{\frac{EI}{mR^4}}} = 46.5176$$

For comparison, the fem code in Appendix B gives $\omega_1 = 13.1702$ and $\omega_2 = 37.6031$ with $N = 50$ elements.

Example 1.14 To find the rotating and non-rotating natural frequencies of the cantilever uniform beam for 2 element with 2 degree of freedom. Assume EI(flexural rigidity) = 100000 Nm^2, m (mass per unit length) = 5 Kg/m and R (radius of beam) = 10 m.

Solution: Given Data,
$N = 2$
$EI = 100000\,\text{Nm}^2$
$DOF = 2$
$m = 5\,\text{Kg/m}$
$R = 10\,\text{m}$
Let, length of each element is given by,

$$l = \frac{R}{N} = \frac{10}{2} = 5\,\text{m}$$

The mass matrix is given by,

$$m_{ij} = m\begin{bmatrix} \frac{13}{35}l & \frac{11}{210}l^2 & \frac{9}{70}l & \frac{-13}{420}l^2 \\ \frac{11}{210}l^2 & \frac{1}{105}l^3 & \frac{13}{420}l^2 & \frac{-1}{140}l^3 \\ \frac{9}{70}l & \frac{13}{420}l^2 & \frac{13}{35}l & \frac{-11}{210}l^2 \\ \frac{-13}{420}l^2 & \frac{-1}{140}l^3 & \frac{-11}{210}l^2 & \frac{l^3}{105} \end{bmatrix}$$

By substituting the values of l and m in above matrix we get,

$$m_{ij} = \begin{bmatrix} 9.28571 & 6.54762 & 3.21429 & -3.86905 \\ 6.54762 & 5.95238 & 3.86905 & -4.46429 \\ 3.21429 & 3.86905 & 9.28571 & -6.54762 \\ -3.86905 & -4.46429 & -6.54762 & 5.95238 \end{bmatrix}$$

Case 1: $\lambda = 1$ (Non dimensional rotating speed)
The stiffness matrix is given by,

$$K_{ij} = EI\begin{bmatrix} \frac{12}{l^3} & \frac{6}{l^2} & \frac{-12}{l^3} & \frac{6}{l^2} \\ \frac{6}{l^2} & \frac{4}{l} & \frac{-6}{l^2} & \frac{2}{l} \\ \frac{-12}{l^3} & \frac{-6}{l^2} & \frac{12}{l^3} & \frac{-6}{l^2} \\ \frac{6}{l^2} & \frac{2}{l} & \frac{-6}{l^2} & \frac{4}{l} \end{bmatrix} + \frac{\Omega^2 A_i}{2}\begin{bmatrix} \frac{6}{5l} & \frac{1}{10} & \frac{-6}{5l} & \frac{1}{10} \\ \frac{1}{10} & \frac{2l}{15} & \frac{-1}{10} & \frac{-l}{30} \\ \frac{-6}{5l} & \frac{-1}{10} & \frac{6}{5l} & \frac{-1}{10} \\ \frac{1}{10} & \frac{-l}{30} & \frac{-1}{10} & \frac{2l}{15} \end{bmatrix}$$

$$-m_i\Omega^2\begin{bmatrix} \frac{3}{5}x_i + \frac{6l}{35} & \frac{l}{10}x_i + \frac{l^2}{28} & \frac{-3}{5}x_i - \frac{6l}{35} & \frac{-l^2}{70} \\ \frac{l}{10}x_i + \frac{l^2}{28} & \frac{l^2}{30}x_i + \frac{l^3}{105} & \frac{-l}{10}x_i - \frac{l^2}{28} & \frac{-l^2}{60}x_i - \frac{l^3}{140} \\ \frac{-3}{5}x_i - \frac{6l}{35} & \frac{-l}{10}x_i - \frac{l^2}{28} & \frac{3}{5}x_i + \frac{6l}{35} & \frac{l^2}{70} \\ \frac{-l^2}{70} & \frac{-l^2}{60}x_i - \frac{l^3}{140} & \frac{l^2}{70} & \frac{l^2}{30}x_i + \frac{l^3}{105} \end{bmatrix}$$

Rotating speed (Ω) can be calculated by,

$$\Omega = \left(\lambda\sqrt{\frac{EI}{mR^4}}\right) = \left(1 \times \sqrt{\frac{100000}{5 \times 10^4}}\right) = 1.41421\,\text{rad/s}$$

In which,

$$A_i = \sum_{j=i}^{N} m_j\left(x_{j+1}^2 - x_j^2\right)$$

Let N = 2, for 1st element $x_j = 0$,

$$A_1 = m_j\left(l^2 - 0\right) = 5\left(5^2\right) = 500\,\text{kgm}$$

By substituting the above values in stiffness matrix we get,

$$K_1 = \begin{bmatrix} 9711.43 & 24041.1 & -9711.43 & 24053.6 \\ 24041.1 & 80321.4 & -24041.1 & 39925.6 \\ -9711.43 & -24041.1 & 9711.43 & -24103.57 \\ 24053.6 & 39925.6 & -24103.57 & 80321.4 \end{bmatrix}$$

K_1 is first element stiffness matrix.
Let N = 2, For 2nd element $x_j = 5$,

$$A_2 = m_j\left(l^2 - 0\right) = 5\left(10^2 - 5^2\right) = 375\,\text{kgm}$$

By substituting the above values in second element stiffness matrix we get,

$$K_2 = \begin{bmatrix} 9651.43 & 24153.5706 & -9651.43 & 24041.1 \\ 24153.5706 & 80071.4 & -24003.6 & 39967.3 \\ -9651.43 & -24003.6 & 9651.43 & -24041.1 \\ 24041.1 & 39967.3 & -24041.1 & 80071.4 \end{bmatrix}$$

K_2 is second element stiffness matrix.
Total stiffness matrix,

$$K_{ij} = K_1 + K_2$$

$$K_{ij} = \begin{bmatrix} 9711.43 & 24041.1 & -9711.43 & 24053.6 & 0 & 0 \\ 24041.1 & 80321.4 & -24041.1 & 39925.6 & 0 & 0 \\ -9711.43 & -24041.1 & 19362.9 & -50.0 & -9651.43 & 24041.1 \\ 24053.6 & 39925.6 & -50.0 & 160476.0 & -24003.6 & 39967.3 \\ 0 & 0 & -9651.43 & -24003.6 & 9651.43 & -24041.1 \\ 0 & 0 & 24041.1 & 39967.3 & -24041.1 & 80071.4 \end{bmatrix}$$

By applying cantilever boundary condition for mass and stiffness matrix first two rows and columns are eliminated, we get 4×4 matrix,

$$m_{ij} = \begin{bmatrix} 9.28571 & 6.54762 & 3.21429 & -3.86905 \\ 6.54762 & 5.95238 & 3.86905 & -4.46429 \\ 3.21429 & 3.86905 & 9.28571 & -6.54762 \\ -3.86905 & -4.46429 & -6.54762 & 5.95238 \end{bmatrix}$$

$$K_{ij} = \begin{bmatrix} 19362.9 & -50.0 & -9651.43 & 24041.1 \\ -50.0 & 160476.0 & -24003.6 & 39967.3 \\ -9651.43 & -24003.6 & 9651.43 & -24041.1 \\ 24041.1 & 39967.3 & -24041.1 & 80071.4 \end{bmatrix}$$

Eigenvalues and Eigenvectors of the matrix can be determined as

The characteristic equation, $|K_{ij} - Dm_{ij}| = 0$.

$$\left| \begin{matrix} 19362.9 & -50.0 & -9651.43 & 24041.1 \\ -50.0 & 160476.0 & -24003.6 & 39967.3 \\ -9651.43 & -24003.6 & 9651.43 & -24041.1 \\ 24041.1 & 39967.3 & -24041.1 & 80071.4 \end{matrix} \right| - D$$

$$\left| \begin{matrix} 9.28571 & 6.54762 & 3.21429 & -3.86905 \\ 6.54762 & 5.95238 & 3.86905 & -4.46429 \\ 3.21429 & 3.86905 & 9.28571 & -6.54762 \\ -3.86905 & -4.46429 & -6.54762 & 5.95238 \end{matrix} \right| = 0$$

By solving, we get

Eigenvalue, $D_1 = 27$ and corresponding Eigenvectors, 0.0964, 0.0329, 0.2828 and 0.0388.

Eigenvalue, $D_2 = 1001$ and corresponding Eigenvectors, 0.2060, –0.0124, –0.2861 and –0.1376.

Eigenvalue, $D_3 = 11335$ and corresponding Eigenvectors, 0.0323, –0.2429, 0.3177 and 0.3065.

Eigenvalue, $D_4 = 95234$ and corresponding Eigenvectors, 0.1351, 0.2777, 0.5334 and 1.0311.

Natural frequency can be determined by,

$f_1 = \sqrt{(}D_1) = 5.20899\,\text{Hz}$

$$f_2 = \sqrt{(D_2)} = 31.6347\,\text{Hz}$$

$$f_3 = \sqrt{(D_3)} = 106.467\,\text{Hz}$$

$$f_4 = \sqrt{(D_4)} = 308.600\,\text{Hz}$$

Also, Natural frequency in terms of non dimensional form can be determined by,

$$\omega_1 = \frac{f_1}{\sqrt{\frac{EI}{mR^4}}} = 3.68331$$

$$\omega_2 = \frac{f_2}{\sqrt{\frac{EI}{mR^4}}} = 22.3691$$

$$\omega_3 = \frac{f_3}{\sqrt{\frac{EI}{mR^4}}} = 75.2838$$

$$\omega_4 = \frac{f_4}{\sqrt{\frac{EI}{mR^4}}} = 218.213$$

For comparison, the fem code in Appendix B gives $\omega_1 = 3.68165$, $\omega_2 = 22.181$, $\omega_3 = 61.8418$ and $\omega_4 = 121.051$ with $N = 50$ elements.

Case 2: $\lambda = 12$ (Non dimensional rotating speed)
The stiffness matrix is given by,

$$K_{ij} = EI\begin{bmatrix} \frac{12}{l^3} & \frac{6}{l^2} & \frac{-12}{l^3} & \frac{6}{l^2} \\ \frac{6}{l^2} & \frac{4}{l} & \frac{-6}{l^2} & \frac{2}{l} \\ \frac{-12}{l^3} & \frac{-6}{l^2} & \frac{12}{l^3} & \frac{-6}{l^2} \\ \frac{6}{l^2} & \frac{2}{l} & \frac{-6}{l^2} & \frac{4}{l} \end{bmatrix} + \frac{\Omega^2 A_i}{2}\begin{bmatrix} \frac{6}{5l} & \frac{1}{10} & \frac{-6}{5l} & \frac{1}{10} \\ \frac{1}{10} & \frac{2l}{15} & \frac{-1}{10} & \frac{-l}{30} \\ \frac{-6}{5l} & \frac{-1}{10} & \frac{6}{5l} & \frac{-1}{10} \\ \frac{1}{10} & \frac{-l}{30} & \frac{-1}{10} & \frac{2l}{15} \end{bmatrix}$$

$$-m_i\Omega^2\begin{bmatrix} \frac{3}{5}x_i + \frac{6l}{35} & \frac{l}{10}x_i + \frac{l^2}{28} & \frac{-3}{5}x_i - \frac{6l}{35} & \frac{-l^2}{70} \\ \frac{l}{10}x_i + \frac{l^2}{28} & \frac{l^2}{30}x_i + \frac{l^3}{105} & \frac{-l}{10}x_i - \frac{l^2}{28} & \frac{-l^2}{60}x_i - \frac{l^3}{140} \\ \frac{-3}{5}x_i - \frac{6l}{35} & \frac{-l}{10}x_i - \frac{l^2}{28} & \frac{3}{5}x_i + \frac{6l}{35} & \frac{l^2}{70} \\ \frac{-l^2}{70} & \frac{-l^2}{60}x_i - \frac{l^3}{140} & \frac{l^2}{70} & \frac{l^2}{30}x_i + \frac{l^3}{105} \end{bmatrix}$$

Rotating speed (Ω) can be calculated by,

$$\Omega = \left(\lambda\sqrt{\frac{EI}{mR^4}}\right) = \left(12 \times \sqrt{\frac{100000}{5 \times 10^4}}\right) = 16.97052\,\text{rad/s}$$

In which,

$$A_i = \sum_{j=i}^{N} m_j \left(x_{j+1}^2 - x_j^2\right)$$

Let N = 2, for 1st element $x_j = 0$,

$$A_1 = m_j\left(l^2 - 0\right) = 5\left(5^2\right) = 500\,\text{kgm}$$

By substituting the above values in stiffness matrix we get,

$$K_1 = \begin{bmatrix} 25645.7 & 29914.3 & -25645.7 & 31714.3 \\ 29914.3 & 126286.0 & -29914.3 & 29285.7 \\ -25645.7 & -29914.3 & 25645.7 & -31714.246 \\ 31714.3 & 29285.7 & -31714.246 & 126286.0 \end{bmatrix}$$

K_1 is first element stiffness matrix.
Let N = 2, For 2nd element $x_j = 5$,

$$A_2 = m_j\left(l^2 - 0\right) = 5\left(10^2 - 5^2\right) = 375\,\text{kgm}$$

By substituting the above values in second element stiffness matrix we get,

$$K_2 = \begin{bmatrix} 17005.7 & 24514.2838 & -17005.7 & 29914.3 \\ 24514.2838 & 90285.7 & -24514.3 & 35285.7 \\ -17005.7 & -24514.3 & 17005.7 & -29914.3 \\ 29914.3 & 35285.7 & -29914.3 & 90285.7 \end{bmatrix}$$

K_2 is second element stiffness matrix.
Total stiffness matrix,
$K_{ij} = K_1 + K_2$

$$K_{ij} = \begin{bmatrix} 25645.7 & 29914.3 & -25645.7 & 31714.3 & 0 & 0 \\ 29914.3 & 126286.0 & -29914.3 & 29285.7 & 0 & 0 \\ -25645.7 & -29914.3 & 42651.4 & -7200.0 & -17005.7 & 29914.3 \\ 31714.3 & 29285.7 & -7200.0 & 228571.0 & -24514.3 & 35285.7 \\ 0 & 0 & -17005.7 & -24514.3 & 17005.7 & -29914.3 \\ 0 & 0 & 29914.3 & 35285.7 & -29914.3 & 90285.7 \end{bmatrix}$$

By applying cantilever boundary condition for mass and stiffness matrix first two rows and columns are eliminated, we get 4×4 matrix,

$$m_{ij} = \begin{bmatrix} 9.28571 & 6.54762 & 3.21429 & -3.86905 \\ 6.54762 & 5.95238 & 3.86905 & -4.46429 \\ 3.21429 & 3.86905 & 9.28571 & -6.54762 \\ -3.86905 & -4.46429 & -6.54762 & 5.95238 \end{bmatrix}$$

$$K_{ij} = \begin{bmatrix} 42651.4 & -7200.0 & -17005.7 & 29914.3 \\ -7200.0 & 228571.0 & -24514.3 & 35285.7 \\ -17005.7 & -24514.3 & 17005.7 & -29914.3 \\ 29914.3 & 35285.7 & -29914.3 & 90285.7 \end{bmatrix}$$

Eigenvalues and Eigenvectors of the matrix can be determined as,

The characteristic equation, $\left|K_{ij} - Dm_{ij}\right| = 0.$

$$\begin{vmatrix} 42651.4 & -7200.0 & -17005.7 & 29914.3 \\ -7200.0 & 228571.0 & -24514.3 & 35285.7 \\ -17005.7 & -24514.3 & 17005.7 & -29914.3 \\ 29914.3 & 35285.7 & -29914.3 & 90285.7 \end{vmatrix} - D$$

$$\begin{vmatrix} 9.28571 & 6.54762 & 3.21429 & -3.86905 \\ 6.54762 & 5.95238 & 3.86905 & -4.46429 \\ 3.21429 & 3.86905 & 9.28571 & -6.54762 \\ -3.86905 & -4.46429 & -6.54762 & 5.95238 \end{vmatrix} = 0$$

By solving, we get

Eigenvalue, $D_1 = 350$ and corresponding Eigenvectors, 0.1115, 0.0289, 0.2640 and 0.0310.

Eigenvalue, $D_2 = 2890$ and corresponding Eigenvectors, 0.1973, –0.0131, –0.3128 and –0.1493.

Eigenvalue, $D_3 = 16750$ and corresponding Eigenvectors, 0.0366, –0.2372, 0.3225 and 0.3270.

Eigenvalue, $D_4 = 104650$ and corresponding Eigenvectors, 0.1354, 0.2829, 0.5251 and 1.0234.

Natural frequency can be determined by,

$$f_1 = \sqrt{(}D_1) = 18.7585\,\text{Hz}$$

$$f_2 = \sqrt{(}D_2) = 53.7832\,\text{Hz}$$

$$f_3 = \sqrt{(}D_3) = 129.423\,\text{Hz}$$

$$f_4 = \sqrt{(}D_4) = 323.502\,\text{Hz}$$

Also, Natural frequency in terms of non dimensional form can be determined by,

$$\omega_1 = \frac{f_1}{\sqrt{\frac{EI}{mR^4}}} = 13.2642$$

$$\omega_2 = \frac{f_2}{\sqrt{\frac{EI}{mR^4}}} = 38.0305$$

$$\omega_3 = \frac{f_3}{\sqrt{\frac{EI}{mR^4}}} = 91.5159$$

$$\omega_4 = \frac{f_4}{\sqrt{\frac{EI}{mR^4}}} = 228.751$$

For comparison, the fem code in Appendix B gives $\omega_1 = 13.1702$, $\omega_2 = 37.6031$, $\omega_3 = 79.6145$ and $\omega_4 = 140.535$ with $N = 50$ elements.

Chapter 2
Stiff String Basis Functions

The prediction of natural frequencies of a rotating beam is an important practical problem and is often done using the finite element method [1, 2]. The previous chapter presented the basics of the finite element method as applied to rotating beams. An accurate approach to develop a finite element is to select shape functions which satisfy the static part of the homogenous governing differential equation for the problem [3]. In this chapter, we develop new shape functions using the exact solution of the governing static differential equation of a stiff-string [4]. In order to simplify the analysis required to derive the shape functions, the centrifugal force is assumed as a constant for an element, which leads to the rotating beam equation becoming the stiff-string equation within the element due to the constant applied tension. Fortunately, the stiff-string equation captures the effect of the centrifugal force and also has an analytical solution.

2.1 Stiff String Equation

The partial differential equation for free vibration of a rotating beam was derived in the previous chapter as

$$(EI(x)\, w'')'' + m(x)\, \ddot{w} - (T(x)\, w')' = 0 \tag{2.1}$$

where, $T(x) = \int_x^L m(x)\, \Omega^2\, (R + x)\, dx + F$ is the centrifugal tensile load at a distance x from the axis of rotation, $EI(x)$ is the flexural stiffness, $m(x)$ is the mass per unit length, w is the bending displacement, Ω is the rotation speed, R is the hub radius, L is the beam length and F is the axial force at the end of the beam. Ignoring the inertia

R. Ganguli, *Finite Element Analysis of Rotating Beams*,
Foundations of Engineering Mechanics, DOI 10.1007/978-981-10-1902-9_2

term in Eq. (2.1) yields the static homogeneous equation, which for a uniform beam reduces to,

$$(EI\, w'')'' - (T(x)\, w')' = 0 \tag{2.2}$$

where $T(x) = m\Omega^2(R(L-x) + \frac{L^2-x^2}{2}) + F$. The complicated expression for the $T(x)$ term makes it possible to only obtain series solutions of Eq. (2.2). Most works on rotating beams use cubic shape functions which result from the solution $EIw'''' = 0$ which means that the second term in Eq. (2.2) is completely ignored. Instead, let us consider an approximation which computes the centrifugal stiffening terms in an approximate sense: $T(x) = T = constant$. This approximation effectively reduces the rotating beam Eq. (2.1) to the stiff-string equation given by [4]

$$(EI\, w'')'' + m\ddot{w} - Tw'' = 0 \tag{2.3}$$

The static homogenous form of Eq. (2.3) is

$$(EI\, w'')'' - Tw'' = 0 \tag{2.4}$$

Until this point, no finite element discretization has been introduced. The process of assuming a constant tension as an approximation to the centrifugal stiffening effect may appear to be rather crude. However, if we consider the beam to be divided into N finite elements, T could be assumed to be constant within the element. The constant tension approximation would then become increasingly realistic as the number of elements increase. For an ith element along the beam, the relation between local co-ordinate ($\bar{x}$) and global co-ordinate (x) from Fig. 2.1 is given by $x = x_i + \bar{x}$ where $x_i = \sum_{j=1}^{i-1} l_j$. For a uniform mesh used in this paper, $x_i = (i-1)l$. Using $x = x_i + \bar{x}$ and assuming $EI = EI_i =$ constant for an element, and the tension within the element is a constant(T_i), Eq. (2.4) can be expressed as

$$\frac{d^4 w}{d\bar{x}^4} - C_i{}^2 \frac{d^2 w}{d\bar{x}^2} = 0 \tag{2.5}$$

where $C_i = \sqrt{\frac{T_i}{EI_i}}$. Equation (2.5), is the governing static homogenous differential equation of a stiff-string in terms of the element co-ordinate for element i. The constant T_i in the expression of C_i for an element is approximated by taking the maximum centrifugal tension which an element experiences. The maximum centrifugal tension T_i for an ith element can be expressed as

$$T_i = \int_{x_i}^{L} m_i(x)\Omega^2(R+x)dx + F = \sum_{j=i}^{j=N} \int_{x_j}^{x_{j+1}} m_j(x)\Omega^2(R+x)dx + F \tag{2.6}$$

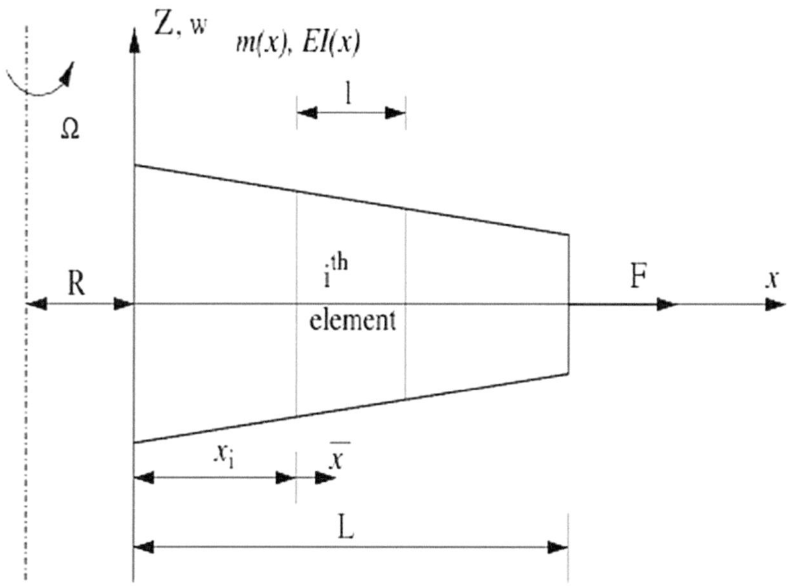

Fig. 2.1 Rotating tapered beam element geometry

Here x_i is the location of the left edge of the element i and $x_{N+1} = L$. The solution of Eq. (2.5) is used as the displacement field.

$$w(x) = a_0 + a_1\bar{x} + a_2 e^{-C_i\bar{x}} + a_3 e^{C_i\bar{x}} \tag{2.7}$$

2.2 Stiff String Basis Functions

Consider the two noded, 4 degree of freedom beam finite element shown in Fig. 2.2. The boundary conditions for the element of length l are given by $w(0) = w_1$, $\frac{dw(0)}{d\bar{x}} = \theta_1 = w_2$, $w(l) = w_3$, $\frac{dw(l)}{d\bar{x}} = \theta_2 = w_4$. Putting Eq. (2.7) into the element boundary conditions yields: $w_1 = a_0 + a_2 + a_3$, $w_2 = a_1 - Ca_2 + Ca_3$, $w_3 = a_0 + a_1 l + a_2 e^{-Cl} + a_3 e^{Cl}$ and $w_4 = a_1 - a_2 C e^{-Cl} + a_3 C e^{Cl}$. Here we have dropped the subscript i in C as the entire discussion here is relevant within the element. Solving for a_0, a_1, a_2 and a_3 in terms of the nodal displacements and slopes using the above expressions, w can be approximated by

$$w = w_1\, N_1 + w_2\, N_2 + w_3\, N_3 + w_4\, N_4 \tag{2.8}$$

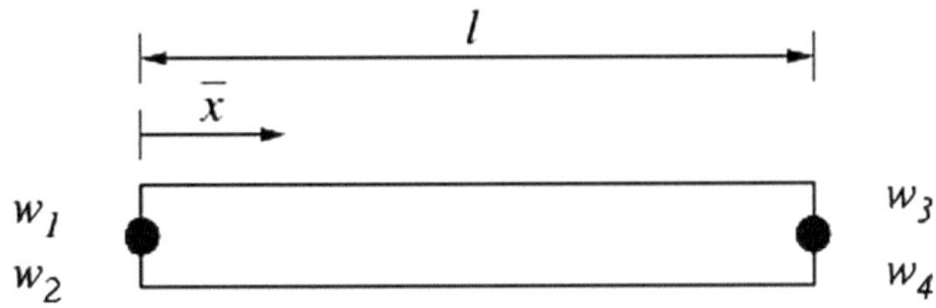

Fig. 2.2 Beam element

where N_1, N_2, N_3 and N_4 are the shape functions and are given as

$$N_1 = \frac{R_1(\bar{x})}{D}, N_2 = \frac{R_2(\bar{x})}{CD}, N_3 = \frac{R_3(\bar{x})}{D}, N_4 = \frac{R_4(\bar{x})}{CD} \tag{2.9}$$

where

$$D = -4 + 2e^{Cl} + 2e^{-Cl} + Ce^{-Cl}l - Cle^{Cl} \tag{2.10}$$

$$\begin{aligned} R_1(\bar{x}) = &-(-e^{Cl} - e^{-Cl} - Ce^{-Cl}l + 2 + Cle^{Cl} + C\bar{x}e^{-Cl} - C\bar{x}e^{Cl} \\ &-e^{-C\bar{x}+Cl} + e^{-C\bar{x}} + e^{C\bar{x}} - e^{C\bar{x}-Cl}) \end{aligned} \tag{2.11}$$

$$\begin{aligned} R_2(\bar{x}) = &\, e^{Cl} - Cle^{Cl} - e^{-Cl} - Ce^{-Cl}l + C\bar{x}e^{Cl} + C\bar{x}e^{-Cl} - 2C\bar{x} \\ &- e^{-C\bar{x}+Cl} + e^{-C\bar{x}+Cl}Cl + e^{-C\bar{x}} + e^{C\bar{x}-Cl} \\ &+ e^{C\bar{x}-Cl}Cl - e^{C\bar{x}} \end{aligned} \tag{2.12}$$

$$\begin{aligned} R_3(\bar{x}) = &\,(e^{Cl} + e^{-Cl} - 2 + C\bar{x}e^{-Cl} - C\bar{x}e^{Cl} - e^{-C\bar{x}+Cl} + e^{-C\bar{x}} \\ &+ e^{C\bar{x}} - e^{C\bar{x}-Cl}) \end{aligned} \tag{2.13}$$

$$\begin{aligned} R_4(\bar{x}) = &\,(2Cl - e^{Cl} + e^{-Cl} - 2C\bar{x} + C\bar{x}e^{Cl} + C\bar{x}e^{-Cl} - e^{-C\bar{x}}Cl \\ &- e^{-C\bar{x}} + e^{-C\bar{x}+Cl} + e^{C\bar{x}} - e^{C\bar{x}}Cl - e^{C\bar{x}-Cl}) \end{aligned} \tag{2.14}$$

We call these the stiff string basis functions. For vibration analysis of a stiff string, these basis functions satisfy the static homogenous part of the governing differential equation and therefore yield all the favorable properties discussed in [3]. For a rotating beam, their use is an approximation which will be justified by numerical studies later in this chapter. Note that the stiff string basis functions are now also a function of the non-dimensional rotational speed, element mass and stiffness, mass of outboard elements, beam length and location of the element due to their dependence on $C = \sqrt{\frac{T_i}{EI_i}}$. These shape functions therefore capture the effect of the rotation speed on the element displacements as well as the fact that different locations contribute differently to the centrifugal stiffening effect.

The analytical limits of the stiff string basis functions as the rotation speed tends to zero is shown in Eqs. (2.15) and (2.16) to become the Hermite cubics.

$$\lim_{C\to 0} N_1 = \frac{2x^3 - 3x^2 l + l^3}{l^3}, \quad \lim_{C\to 0} N_2 = \frac{x^3 - 2x^2 l + xl^2}{l^2} \tag{2.15}$$

$$\lim_{C\to 0} N_3 = \frac{-2x^3 + 3x^2 l}{l^3}, \quad \lim_{C\to 0} N_4 = \frac{-x^2 l + x^3}{l^2} \tag{2.16}$$

As the rotation speed tends to infinity, the basis functions N_1 and N_3 become linear and N_2 and N_4 approach zero, as given in Eq. (2.1).

$$\lim_{C\to\infty} N_1 = 1 - \frac{\bar{x}}{l}, \quad \lim_{C\to\infty} N_2 = 0, \quad \lim_{C\to\infty} N_3 = \frac{\bar{x}}{l}, \quad \lim_{C\to\infty} N_4 = 0 \tag{2.17}$$

Variation of the shape functions along the elements ($N = 1$) is shown in Figs. 2.3 and 2.4 with the conventional Hermite cubic and the stiff string basis functions at a

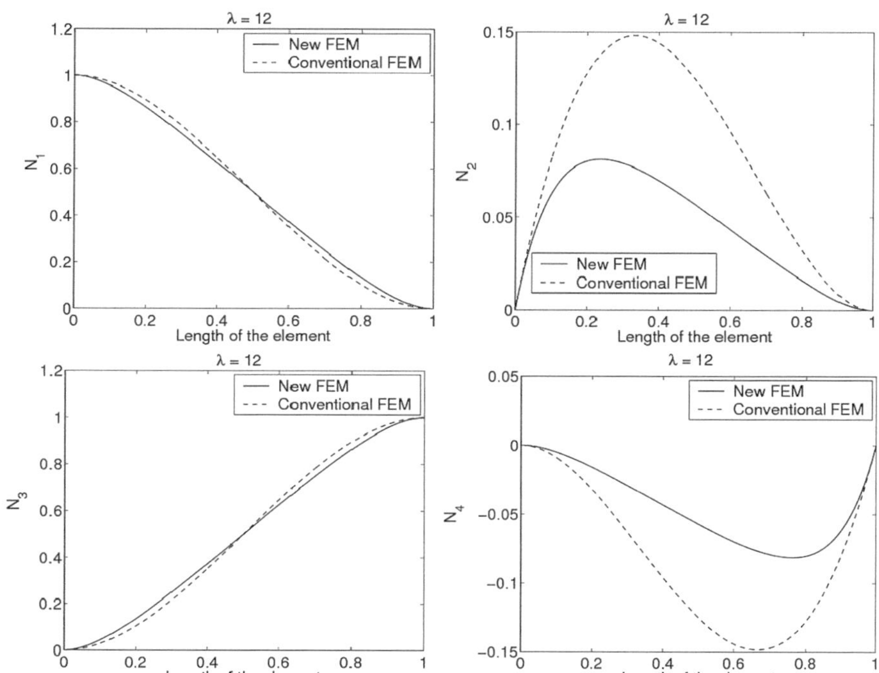

Fig. 2.3 Variation of shape functions along the elements ($N = 1$, $\lambda = 12$) with the new and conventional finite element at high rotation speed

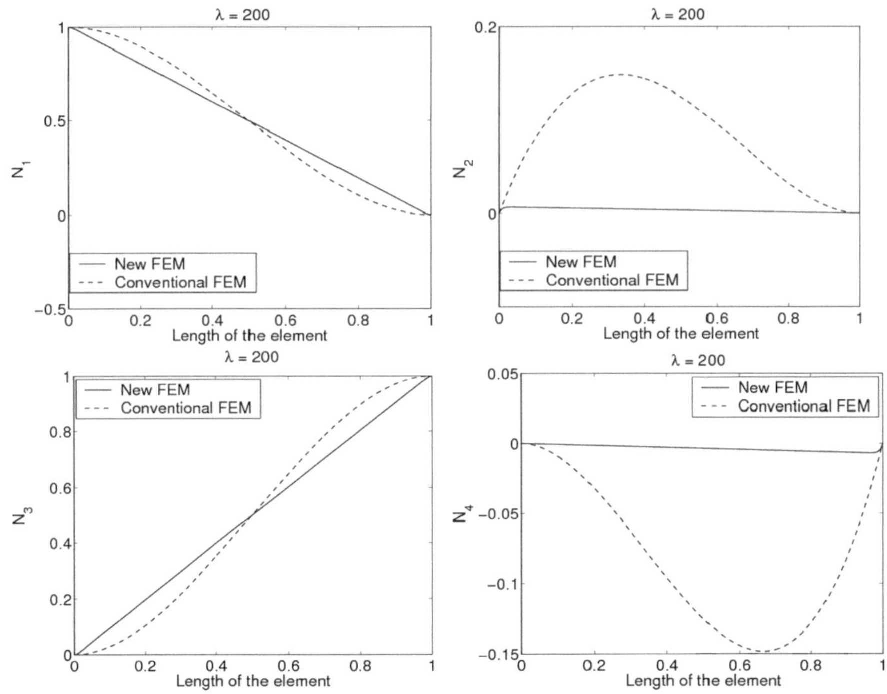

Fig. 2.4 Variation of shape functions along the elements ($N = 1$, $\lambda = 200$) with the new and conventional finite element at very high rotation speed

high ($\lambda = 12$) and very high ($\lambda = 200$) rotation speed, respectively ($\lambda^2 = \frac{m\omega^2 L^4}{EI}$). The difference between the two shape functions is clear. These shape functions therefore capture the effect of the rotation speed on the element displacements. As the rotation speed rises, the stiff string basis functions become almost linear and the rotating beam approaches a rotating string, as is clear from Eq. (2.17). One can also observe from Eq. (2.7) that when $T_i = 0$, w can be represented as a cubic polynomial and when $T_i \to \infty$, the second term dominates and w is a linear function.

The stiff string basis functions are used to develop the finite element equations for free vibration of the rotating beam and numerical results are obtained for a uniform beam and a tapered beam, in the next two sections.

2.3 Uniform Rotating Beam

Tables 2.1 and 2.2 show a comparison of non-dimensional natural frequencies of a rotating uniform cantilever and hinged beam, respectively, with results from [1, 5, 6]. Convergence for the first five modes was achieved using 75 uniform finite elements and the results compare well.

Table 2.1 Comparison of non-dimensional natural frequencies of cantilever uniform beam

Mode	Present-FEM	Wang and Wereley [6]	Wright et al. [5]	Hodges et al. [1]
$\lambda = 12$				
1	13.1702	13.1702	13.1702	13.1702
2	37.6031	37.6031	37.6031	37.6031
3	79.6145	79.6145	79.6145	79.6145
4	140.534	140.534	140.534	N/A
5	220.537	220.536	220.536	N/A

Table 2.2 Comparison of non-dimensional natural frequencies of hinged uniform beam

Mode	Present-FEM	Wang and Wereley [6]	Wright et al. [5]
$\lambda = 12$			
1	12.0000	12.0000	12.0000
2	33.7603	33.7603	33.7603
3	70.8373	70.8373	70.8373
4	126.431	126.431	126.431
5	201.123	201.122	201.122

The new element is now compared with the conventional element with cubic basis functions discussed in Chap. 1. A convergence study is done at two different rotation speeds ($\lambda = 12, \lambda = 200$) on the first three modes, since they are critical for dynamic modeling and modes higher than three show little effect of rotation [2]. For $\lambda = 12$ results in Fig. 2.5, the convergence of the first mode is extremely good, though the second and third mode show slower convergence. For $\lambda = 200$ results in Fig. 2.6, convergence of the second mode also shows an improvement.

2.4 Tapered Rotating Beam

Consider the tapered beam used in [1] with $m(\xi) = m_0\,(1 - 0.5\,\xi)$, and $EI(\xi) = EI_0\,(1 - 0.5\,\xi)^3$. Here m_0 and EI_0 correspond to the value of mass per unit length and flexural rigidity at the thick end of the beam ($\xi = 0$), respectively. Table 2.3 shows the present results and those in [1, 6]. The comparison is very good.

Fig. 2.5 Convergence of the natural frequencies with $\lambda = 12$

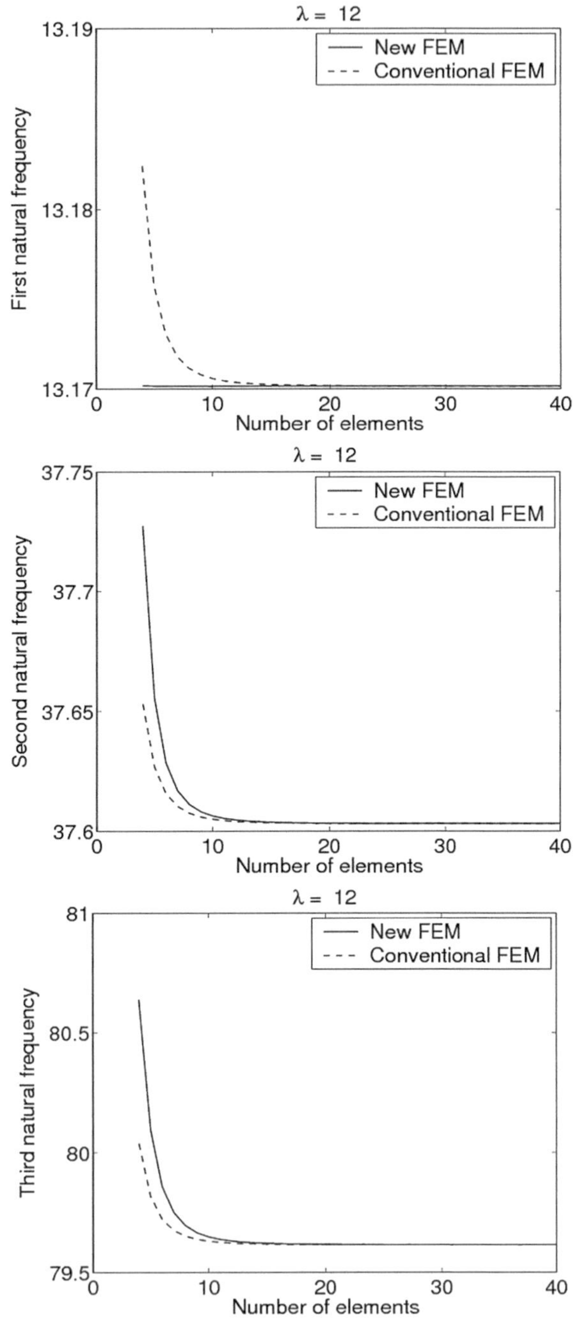

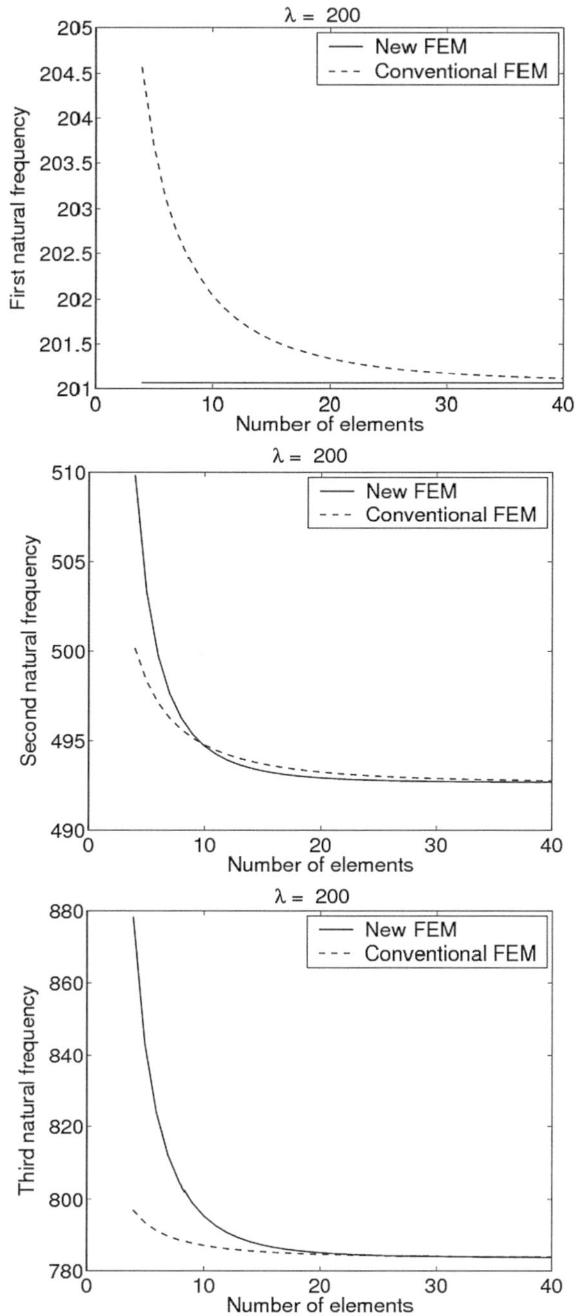

Fig. 2.6 Convergence of the natural frequencies with $\lambda = 200$

Table 2.3 Comparison of non-dimensional natural frequencies of tapered cantilever beam under different rotation speeds

λ	First mode			Second mode			Third mode		
	Present	Ref. [6]	Ref. [1]	Present	Ref. [6]	Ref. [1]	Present	Ref. [6]	Ref. [1]
1	3.9866	3.9866	3.9866	18.4740	18.4740	18.4740	47.4173	47.4173	47.4173
2	4.4368	4.4368	4.4368	18.9366	18.9366	18.9366	47.8717	47.8716	47.8716
3	5.0927	5.0927	5.0927	19.6839	19.6839	19.6839	48.6191	48.6190	48.6190
4	5.8788	5.8788	5.8788	20.6852	20.6852	20.6852	49.6457	49.6456	49.6456
5	6.7434	6.7434	6.7345	21.9053	21.9053	21.9053	50.9339	50.9338	50.9338
6	7.6551	7.6551	7.6551	23.3093	23.3093	23.3093	52.4633	52.4633	52.4633
7	8.5956	8.5956	8.5956	24.8647	24.8647	24.8647	54.2125	54.2124	54.2124
8	9.5540	9.5540	9.5540	26.5437	26.5437	26.5437	56.1596	56.1595	56.1595
9	10.5239	10.5239	10.5239	28.3227	28.3227	28.3227	58.2834	58.2833	58.2833
10	11.5015	11.5015	11.5015	30.1827	30.1827	30.1827	60.5640	60.5639	60.5639
11	12.4845	12.4845	12.4845	32.1085	32.1085	32.1085	62.9830	62.9829	62.9829
12	13.4711	13.4711	13.4711	34.0877	34.0877	34.0877	65.5238	65.5237	65.5237

2.5 Hybrid Basis Functions

Using the stiff string basis function improves FEM performance at the first mode but does not improve performance for higher modes. It is possible to get better results by combining the stiff string basis function with the cubic function. The function assumed for the hybrid basis FE formulation is a linear combination of terms is given as

$$w(\bar{x}) = a_0 + a_1\bar{x} + a_2\bar{x}^2 + a_3\bar{x}^3 + a_4e^{-C_i\bar{x}} + a_5e^{C_i\bar{x}} \tag{2.18}$$

Differentiating Eq. (2.18) with respect to $\bar{x}$ and is given by

$$\frac{dw}{d\bar{x}} = a_1 + 2a_2\bar{x} + 3a_3\bar{x}^2 - a_4C_ie^{-C_i\bar{x}} + a_5C_ie^{C_i\bar{x}} \tag{2.19}$$

It is clearly evident from Eqs. (2.18) and (2.19) that the present basis function is capable of representing a rigid body mode [8] in translation and rotation when only a_0 and a_1 are non-zero respectively. In the present problem the generalized strain and stresses can be attributed to curvature and bending moment and the present interpolation function is capable of representing a constant strain [8] state i.e., capable of representing a constant curvature when only a_2 is non-zero. Hence, Eq. (2.18) represents a complete interpolation function [8].

Consider the three noded, 6 degree of freedom finite element shown in Fig. 2.7. The boundary conditions for the element of length l are given by $w(0) = w_1$, $\frac{dw(0)}{d\bar{x}} = \theta_1 = w_2$, $w(l/2) = w_3$, $\frac{dw(l/2)}{d\bar{x}} = \theta_2 = w_4$, $w(l) = w_5$, $\frac{dw(l)}{d\bar{x}} = \theta_3 = w_6$ Putting Eq. (2.18)

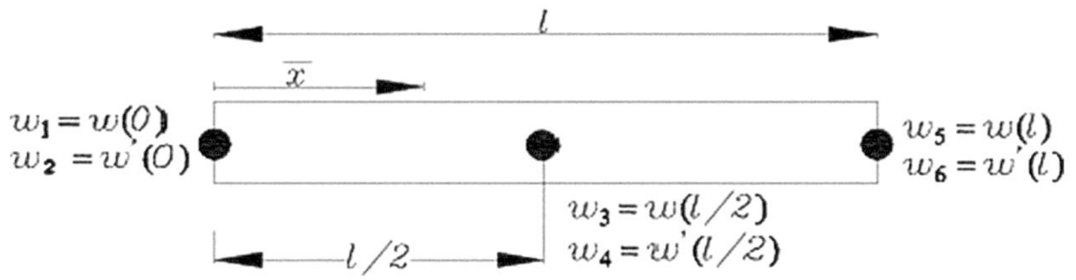

Fig. 2.7 Beam element

into the element boundary conditions yields the following six equations. For further study we have dropped the subscript *i* in C as the entire discussion here is relevant within the element.

$$w_1 = a_0 + a_4 + a_5 \tag{2.20}$$

$$w_2 = a_1 - Ca_4 + Ca_5 \tag{2.21}$$

$$w_3 = a_0 + a_1(l/2) + a_2(l/2)^2 + a_3(l/2)^3 + a_4 e^{-C(l/2)} + a_5 e^{C(l/2)} \tag{2.22}$$

$$w_4 = a_1 + 2a_2(l/2) + 3a_3(l/2)^2 - a_4 Ce^{-Cl/2} + a_5 Ce^{Cl/2} \tag{2.23}$$

$$w_5 = a_0 + a_1 l + a_2 l^2 + a_3 l^3 + a_4 e^{-Cl} + a_5 e^{Cl} \tag{2.24}$$

$$w_6 = a_1 + 2a_2 l + 3a_3 l^2 - a_4 Ce^{-Cl} + a_5 Ce^{Cl} \tag{2.25}$$

Solving for a_0, a_1, a_2, a_3, a_4 and a_5 in terms of the degrees of freedom considered using Eqs. (2.20)–(2.25), w can be approximated by

$$w = w_1\,N_1 + w_2\,N_2 + w_3\,N_3 + w_4\,N_4 + w_5\,N_5 + w_6\,N_6 \tag{2.26}$$

where $N_1, N_2, N_3, N_4, N_5, N_6$ are the interpolating functions derived using the present hybrid basis functions, and are given below.

$$N_1 = \frac{\bar{R}_1(\bar{x})}{-l^2\,D(\bar{x})},\ N_2 = \frac{\bar{R}_2\,(\bar{x})}{-l^2 D(\bar{x})},\ N_3 = \frac{\bar{R}_3\,(\bar{x})}{l\,D(\bar{x})},\ N_4 = \frac{\bar{R}_4\,(\bar{x})}{l^2 D(\bar{x})},$$

$$N_5 = \frac{\bar{R}_5\,(\bar{x})}{-l^2 D(\bar{x})},\ N_6 = \frac{\bar{R}_6\,(\bar{x})}{l^2 D(\bar{x})}$$

where

$$
\begin{aligned}
\bar{R}_1(\bar{x}) = {} & 24\,C\bar{x}l^2e^{Cl} - 5\,e^{C(-\bar{x}+l)}l^3C + 5\,C^2\bar{x}l^3e^{-Cl} + e^{C\bar{x}}l^3C - 64\,C\bar{x}^3e^{1/2\,Cl} \\
& - 64\,C\bar{x}^3e^{-1/2\,Cl} + e^{-C\bar{x}}l^3C - 5\,C^2\bar{x}l^3e^{Cl} + 24\,e^{-1/2\,C(-2\,\bar{x}+l)}l^2 \\
& + 24\,e^{C(-\bar{x}+l)}l^2 + 8\,C^2\bar{x}l^3e^{-1/2\,Cl} - 24\,C\bar{x}l^2e^{1/2\,Cl} - 24\,C\bar{x}l^2e^{-1/2\,Cl} \\
& + 4\,C^2\bar{x}^2l^2e^{-1/2\,Cl} - 4\,C^2\bar{x}^2l^2e^{1/2\,Cl} + 96\,C\bar{x}^2le^{1/2\,Cl} + 96\,C\bar{x}^2le^{-1/2\,Cl} \\
& - 8\,C^2\bar{x}^3le^{-1/2\,Cl} + 4\,l^4C^2e^{1/2\,Cl} + l^4C^2e^{Cl} - l^4C^2e^{-Cl} - 4\,l^4C^2e^{-1/2\,Cl} \\
& + 34\,l^3C - 5\,e^{-C(-\bar{x}+l)}l^3C - 5\,l^3Ce^{-Cl} - 5\,l^3Ce^{Cl} - 8\,e^{1/2\,C(-2\,\bar{x}+l)}l^3C \\
& - 8\,C^2\bar{x}^2l^2e^{-Cl} + 8\,C^2\bar{x}^2l^2e^{Cl} - 36\,C\bar{x}^2le^{Cl} - 4\,C^2\bar{x}^3le^{Cl} - 120\,C\bar{x}^2l \\
& + 96\,C\bar{x}^3 - 8\,e^{-1/2\,C(-2\,\bar{x}+l)}l^3C - 24\,l^2e^{1/2\,Cl} + 24\,l^2e^{-1/2\,Cl} - 8\,C^2\bar{x}l^3e^{1/2\,Cl} \\
& + 8\,C^2\bar{x}^3le^{1/2\,Cl} - 24\,e^{1/2\,C(-2\,\bar{x}+l)}l^2 - 36\,C\bar{x}^2le^{-Cl} - 24\,e^{-C(-\bar{x}+l)}l^2 \\
& + 16\,C\bar{x}^3e^{-Cl} + 16\,C\bar{x}^3e^{Cl} + 4\,C^2\bar{x}^3le^{-Cl} \\
& + 24\,C\bar{x}l^2e^{-Cl}
\end{aligned}
\tag{2.27}
$$

$$
\begin{aligned}
\bar{R}_2(\bar{x}) = {} & -l^4e^{-C(-\bar{x}+l)}C + 24\,\bar{x}l^2e^{Cl} + 40\,\bar{x}^3Cl + l^4Ce^{-Cl} + 72\,\bar{x}^2le^{1/2\,Cl} \\
& + 32\,\bar{x}^3e^{-1/2\,Cl} + l^4Ce^{Cl} - 24\,\bar{x}l^2e^{-Cl} - l^4e^{C(-\bar{x}+l)}C - 5\,l^3e^{-C(-\bar{x}+l)} \\
& - 2\,l^4e^{1/2\,C(-2\,\bar{x}+l)}C + 5\,l^3e^{C(-\bar{x}+l)} - 4\,\bar{x}^3lCe^{Cl} - 4\,l^3e^{1/2\,C(-2\,\bar{x}+l)} \\
& + 4\,l^3e^{-1/2\,C(-2\,\bar{x}+l)} - 2\,l^4e^{-1/2\,C(-2\,\bar{x}+l)}C + 5\,l^3e^{-Cl} + 34\,\bar{x}l^3C \\
& - 72\,\bar{x}^2l^2C - 12\,\bar{x}l^3Ce^{-1/2\,Cl} - 2\,\bar{x}l^4C^2e^{-1/2\,Cl} - 12\,\bar{x}l^3Ce^{1/2\,Cl} \\
& + 2\,\bar{x}l^4C^2e^{1/2\,Cl} + 28\,\bar{x}^2l^2Ce^{-1/2\,Cl} + 6\,\bar{x}^2l^3C^2e^{-1/2\,Cl} + 28\,\bar{x}^2l^2Ce^{1/2\,Cl} \\
& - 6\,\bar{x}^2l^3C^2e^{1/2\,Cl} - 16\,\bar{x}^3lCe^{-1/2\,Cl} - 4\,\bar{x}^3C^2l^2e^{-1/2\,Cl} - 16\,\bar{x}^3lCe^{1/2\,Cl} \\
& + 4\,\bar{x}^3l^2C^2e^{1/2\,Cl} - 48\,\bar{x}l^2e^{1/2\,Cl} + 2\,l^4Ce^{1/2\,Cl} + 2\,l^4Ce^{-1/2\,Cl} + 48\,\bar{x}l^2e^{-1/2\,Cl} \\
& - 72\,\bar{x}^2le^{-1/2\,Cl} + l^3e^{C\bar{x}} - l^3e^{-C\bar{x}} + 4\,l^3e^{1/2\,Cl} - 4\,l^3e^{-1/2\,Cl} - 32\,\bar{x}^3e^{1/2\,Cl} \\
& - 5\,\bar{x}l^3Ce^{Cl} - 5\,\bar{x}l^3Ce^{-Cl} + 8\,\bar{x}^2l^2Ce^{Cl} + 8\,\bar{x}^2l^2Ce^{-Cl} - 16\,\bar{x}^3e^{-Cl} - 5\,l^3e^{Cl} \\
& - 4\,\bar{x}^3lCe^{-Cl} - 36\,\bar{x}^2le^{Cl} + 36\,\bar{x}^2le^{-Cl} + 16\,\bar{x}^3e^{Cl}
\end{aligned}
\tag{2.28}
$$

$$
\begin{aligned}
\bar{R}_3(\bar{x}) = {} & -24\,C\bar{x}^2e^{Cl} - 4\,e^{C(-\bar{x}+l)}l^2C - 4\,C^2\bar{x}l^2e^{Cl} - 16\,C^2\bar{x}l^2e^{1/2\,Cl} + 16\,C^2\bar{x}l^2e^{-1/2\,Cl} \\
& + 16\,C^2\bar{x}^2le^{1/2\,Cl} - 16\,C^2\bar{x}^2le^{-1/2\,Cl} + 24\,e^{C(-\bar{x}+l)}l - 24\,e^{-C\bar{x}}l + 24\,e^{C\bar{x}}l \\
& - 24\,e^{-C(-\bar{x}+l)}l - 24\,e^{Cl}l + 24\,e^{-Cl}l + 4\,l^2Ce^{-Cl} + 4\,l^2Ce^{Cl} + 16\,l^2Ce^{1/2\,Cl} \\
& + 16\,l^2Ce^{-1/2\,Cl} + 8\,Cl^2 - 24\,C\bar{x}^2e^{-Cl} - 48\,C\bar{x}l \\
& + 24\,C\bar{x}le^{-Cl} + 4\,C^2\bar{x}l^2e^{-Cl} + 4\,C^2\bar{x}^2le^{Cl} - 4\,C^2\bar{x}^2le^{-Cl} + 24\,C\bar{x}le^{Cl} \\
& - 16\,e^{1/2\,C(-2\,\bar{x}+l)}l^2C - 4\,e^{-C\bar{x}}l^2C - 4\,e^{C\bar{x}}l^2C + 48\,C\bar{x}^2 - 16\,e^{-1/2\,C(-2\,\bar{x}+l)}l^2C \\
& - 4\,e^{-C(-\bar{x}+l)}l^2C
\end{aligned}
\tag{2.29}
$$

$$\begin{aligned}
\bar{R}_4(\bar{x}) = {} & 2\,l^4 e^{-C(-\bar{x}+l)} C - 16\,\bar{x}^3 Cl - 2\,l^4 Ce^{-Cl} + 96\,\bar{x}^2 le^{1/2\,Cl} + 64\,\bar{x}^3 e^{-1/2\,Cl} \\
& - 2\,l^4 Ce^{Cl} + 2\,l^4 e^{C(-\bar{x}+l)} C + 8\,l^3 e^{-C(-\bar{x}+l)} - 8\,l^3 e^{C(-\bar{x}+l)} - 24\,\bar{x}^3 lCe^{Cl} \\
& + 16\,l^3 e^{1/2\,C(-2\,\bar{x}+l)} - 16\,l^3 e^{-1/2\,C(-2\,\bar{x}+l)} + 2\,\bar{x}l^4 C^2 e^{Cl} - 2\,l^4 e^{-C\bar{x}} C \\
& - 2\,l^4 e^{C\bar{x}} C - 8\,l^3 e^{-Cl} + 6\,\bar{x}^2 l^3 C^2 e^{-Cl} - 4\,\bar{x}^3 C^2 l^2 e^{-Cl} + 4\,l^4 C \\
& - 16\,\bar{x}l^3 C + 24\,\bar{x}^2 l^2 C + 16\,\bar{x}l^3 Ce^{-1/2\,Cl} + 16\,\bar{x}l^3 Ce^{1/2\,Cl} - 48\,\bar{x}^2 l^2 Ce^{-1/2\,Cl} \\
& - 48\,\bar{x}^2 l^2 Ce^{1/2\,Cl} + 32\,\bar{x}^3 lCe^{-1/2\,Cl} + 32\,\bar{x}^3 lCe^{1/2\,Cl} - 96\,\bar{x}^2 le^{-1/2\,Cl} + 8\,l^3 e^{C\bar{x}} \\
& - 8\,l^3 e^{-C\bar{x}} - 16\,l^3 e^{1/2\,Cl} + 16\,l^3 e^{-1/2\,Cl} - 64\,\bar{x}^3 e^{1/2\,Cl} - 8\,\bar{x}l^3 Ce^{Cl} \\
& - 8\,\bar{x}l^3 Ce^{-Cl} + 36\,\bar{x}^2 l^2 Ce^{Cl} + 36\,\bar{x}^2 l^2 Ce^{-Cl} - 32\,\bar{x}^3 e^{-Cl} + 8\,l^3 e^{Cl} \\
& - 24\,\bar{x}^3 lCe^{-Cl} - 48\,\bar{x}^2 le^{Cl} + 48\,\bar{x}^2 le^{-Cl} + 4\,\bar{x}^3 C^2 l^2 e^{Cl} + 32\,\bar{x}^3 e^{Cl} \\
& - 2\,\bar{x}l^4 C^2 e^{-Cl} - 6\,\bar{x}^2 l^3 C^2 e^{Cl}
\end{aligned} \tag{2.30}$$

$$\begin{aligned}
\bar{R}_5(\bar{x}) = {} & 24\,e^{1/2\,C(-2\,\bar{x}+l)} l^2 - 24\,e^{-1/2\,C(-2\,\bar{x}+l)} l^2 - 8\,e^{1/2\,C(-2\,\bar{x}+l)} l^3 C + 8\,\bar{x}^3 C^2 le^{-1/2\,Cl} \\
& - 96\,C\bar{x}^2 le^{-1/2\,Cl} + 12\,C\bar{x}^2 le^{Cl} - 8\,\bar{x}^3 C^2 le^{1/2\,Cl} + 12\,C\bar{x}^2 le^{-Cl} - 4\,C^2 \bar{x}^2 l^2 e^{Cl} \\
& - 16\,\bar{x}^3 Ce^{Cl} - C^2 \bar{x}l^3 e^{-Cl} - 16\,\bar{x}^3 Ce^{-Cl} + e^{C(-\bar{x}+l)} l^3 C + 4\,C^2 \bar{x}^2 l^2 e^{-Cl} \\
& - 48\,C\bar{x}l^2 + 168\,C\bar{x}^2 l + 8\,C^2 \bar{x}l^3 e^{-1/2\,Cl} + 8\,l^3 Ce^{-1/2\,Cl} - 8\,C^2 \bar{x}l^3 e^{1/2\,Cl} \\
& - 4\,\bar{x}^3 C^2 le^{-Cl} - l^3 Ce^{Cl} - 96\,\bar{x}^3 C - 96\,C\bar{x}^2 le^{1/2\,Cl} + e^{-C(-\bar{x}+l)} l^3 C \\
& + 24\,C\bar{x}l^2 e^{1/2\,Cl} + 24\,C\bar{x}l^2 e^{-1/2\,Cl} + 20\,C^2 \bar{x}^2 l^2 e^{1/2\,Cl} - 20\,C^2 \bar{x}^2 l^2 e^{-1/2\,Cl} \\
& + C^2 \bar{x}l^3 e^{Cl} + 4\,\bar{x}^3 C^2 le^{Cl} + 8\,l^3 Ce^{1/2\,Cl} - 8\,e^{-1/2\,C(-2\,\bar{x}+l)} l^3 C + 10\,l^3 C \\
& - l^3 Ce^{-Cl} - 24\,e^{-C\bar{x}} l^2 + 24\,e^{C\bar{x}} l^2 + 24\,l^2 e^{-1/2\,Cl} - 24\,l^2 e^{1/2\,Cl} \\
& + 64\,\bar{x}^3 Ce^{1/2\,Cl} + 64\,\bar{x}^3 Ce^{-1/2\,Cl} - 5\,e^{-C\bar{x}} l^3 C - 5\,e^{C\bar{x}} l^3 C
\end{aligned} \tag{2.31}$$

$$\begin{aligned}
\bar{R}_6(\bar{x}) = {} & -40\,\bar{x}^3 Cl - 24\,\bar{x}^2 le^{1/2\,Cl} - 32\,\bar{x}^3 e^{-1/2\,Cl} - l^3 e^{-C(-\bar{x}+l)} - 2\,l^4 e^{1/2\,C(-2\,\bar{x}+l)} C \\
& + l^3 e^{C(-\bar{x}+l)} + 4\,\bar{x}^3 lCe^{Cl} + 4\,l^3 e^{1/2\,C(-2\,\bar{x}+l)} - 4\,l^3 e^{-1/2\,C(-2\,\bar{x}+l)} - l^4 e^{-C\bar{x}} C \\
& - l^4 e^{C\bar{x}} C - 2\,l^4 e^{-1/2\,C(-2\,\bar{x}+l)} C + l^3 e^{-Cl} + 2\,l^4 C - 10\,\bar{x}l^3 C + 48\,\bar{x}^2 l^2 C \\
& + 4\,\bar{x}l^3 Ce^{-1/2\,Cl} + 2\,\bar{x}l^4 C^2 e^{-1/2\,Cl} + 4\,\bar{x}l^3 Ce^{1/2\,Cl} - 2\,\bar{x}l^4 C^2 e^{1/2\,Cl} \\
& - 20\,\bar{x}^2 l^2 Ce^{-1/2\,Cl} - 6\,\bar{x}^2 l^3 C^2 e^{-1/2\,Cl} - 20\,\bar{x}^2 l^2 Ce^{1/2\,Cl} + 6\,\bar{x}^2 l^3 C^2 e^{1/2\,Cl} \\
& + 16\,\bar{x}^3 lCe^{-1/2\,Cl} + 4\,\bar{x}^3 C^2 l^2 e^{-1/2\,Cl} + 16\,\bar{x}^3 lCe^{1/2\,Cl} - 4\,\bar{x}^3 l^2 C^2 e^{1/2\,Cl} \\
& + 2\,l^4 Ce^{1/2\,Cl} + 2\,l^4 Ce^{-1/2\,Cl} + 24\,\bar{x}^2 le^{-1/2\,Cl} + 5\,l^3 e^{C\bar{x}} - 5\,l^3 e^{-C\bar{x}} \\
& - 4\,l^3 e^{1/2\,Cl} + 4\,l^3 e^{-1/2\,Cl} + 32\,\bar{x}^3 e^{1/2\,Cl} + \bar{x}l^3 Ce^{Cl} + \bar{x}l^3 Ce^{-Cl} - 4\,\bar{x}^2 l^2 Ce^{Cl} \\
& - 4\,\bar{x}^2 l^2 Ce^{-Cl} + 16\,\bar{x}^3 e^{-Cl} - l^3 e^{Cl} + 4\,\bar{x}^3 lCe^{-Cl} + 12\,\bar{x}^2 le^{Cl} \\
& - 12\,\bar{x}^2 le^{-Cl} - 16\,\bar{x}^3 e^{Cl}
\end{aligned} \tag{2.32}$$

$$D(\bar{x}) = -36\,Cl - 48\,e^{-1/2\,Cl} + 24\,e^{-Cl} + 48\,e^{1/2\,Cl} - 24\,e^{Cl} + 8\,lCe^{-1/2\,Cl} + 8\,lCe^{1/2\,Cl} + 10\,lCe^{-Cl} - 4\,l^2C^2e^{1/2\,Cl} + 10\,lCe^{Cl} + 4\,C^2l^2e^{-1/2\,Cl} + C^2l^2e^{-Cl} - C^2l^2e^{Cl} \tag{2.33}$$

The analytical limits of the hybrid basis functions as the rotation speed tends to zero is shown in Eqs. (2.34)–(2.39) which are identical to the shape functions obtained using quintic polynomial.

$$\lim_{C\to 0} N_1 = \frac{-23\,\bar{x}^2l^3 + 24\,\bar{x}^5 - 68\,\bar{x}^4l + 66\,\bar{x}^3l^2 + l^5}{l^5} \tag{2.34}$$

$$\lim_{C\to 0} N_2 = \frac{\bar{x}\left(-12\,\bar{x}^3l - 6\,\bar{x}l^3 + l^4 + 13\,\bar{x}^2l^2 + 4\,\bar{x}^4\right)}{l^4} \tag{2.35}$$

$$\lim_{C\to 0} N_3 = \frac{16\bar{x}^2\left(l^2 + \bar{x}^2 - 2\,l\bar{x}\right)}{l^4} \tag{2.36}$$

$$\lim_{C\to 0} N_4 = \frac{-8\bar{x}^2\left(5\,\bar{x}^2l + l^3 - 4\,\bar{x}l^2 - 2\,\bar{x}^3\right)}{l^4} \tag{2.37}$$

$$\lim_{C\to 0} N_5 = \frac{\bar{x}^2\left(7\,l^3 - 24\,\bar{x}^3 + 52\,\bar{x}^2l - 34\,\bar{x}l^2\right)}{l^5} \tag{2.38}$$

$$\lim_{C\to 0} N_6 = \frac{-8\,\bar{x}^4l - \bar{x}^2l^3 + 5\,\bar{x}^3l^2 + 4\,\bar{x}^5}{l^4} \tag{2.39}$$

Variation of the shape functions corresponding to end degrees of freedom along the element ($N = 1$) is shown in Figs. 2.8 and 2.9 with the conventional Hermite cubic, fifth order and the stiff string basis functions at low and high speeds of $\lambda = 12$ and $\lambda = 100$, respectively. The fifth order polynomial here is based on the 6 degree of freedom element in Fig. 2.1 with three nodes for each element. It is observed that at low rotation speed the present basis functions are similar to the fifth order basis functions as it is observed from analytical limits given in Eqs. (2.34)–(2.39). Also, at low rotation speeds, the stiff-string basis functions are similar to that of the Hermite cubic basis functions, as shown in Ref. [7]. As the rotation speed increases the effect of rotation on shape functions and their deviation from the other basis functions can be clearly observed in Figs. 2.10 and 2.11. At high rotation speeds the stiff string basis function approach linear variation for displacement and become zero for slopes. The new hybrid element captures the slope behavior of the stiff string functions at high rotation speed while retaining the polynomial like displacement behavior for the displacements.

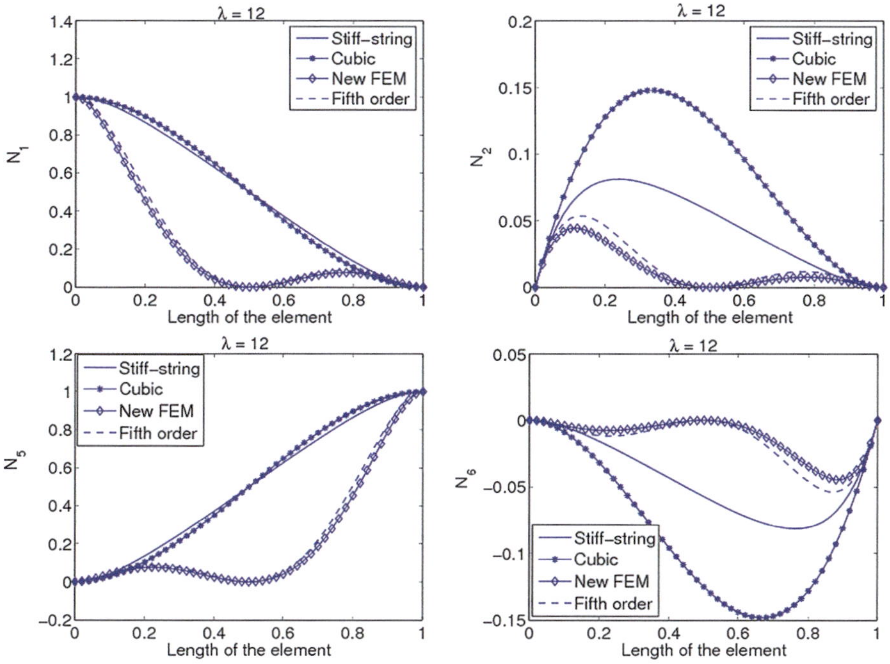

Fig. 2.8 Variation of shape functions ($N = 1, \lambda = 12$) with the new, stiff-string, conventional cubic and fifth order finite elements

2.6 Finite Element

The present improved basis functions are now used to develop the finite element equations for free vibration of the rotating beam. The mass and stiffness matrices can be obtained using the energy expressions. The kinetic energy for a rotating beam is given by

$$T = \frac{1}{2}\int_0^L m(x)\,[\dot{w}(x,\ t)]^2 dx \tag{2.40}$$

where, $\dot{w}(x,\ t)$ is the derivative of $w(x,\ t)$ with respect to time t. The potential/strain energy is given by

$$U = \frac{1}{2}\int_0^L EI(x)\,[w''(x,\ t)]^2 dx + \frac{1}{2}\int_0^L T(x)\,[w'(x,\ t)]^2 dx \tag{2.41}$$

The mass and stiffness matrices ($\mathbf{M}_i$ and $\mathbf{K}_i$) for a beam element can be obtained from the above energy expressions. The calculations for these matrices involve calculating the following integrals:

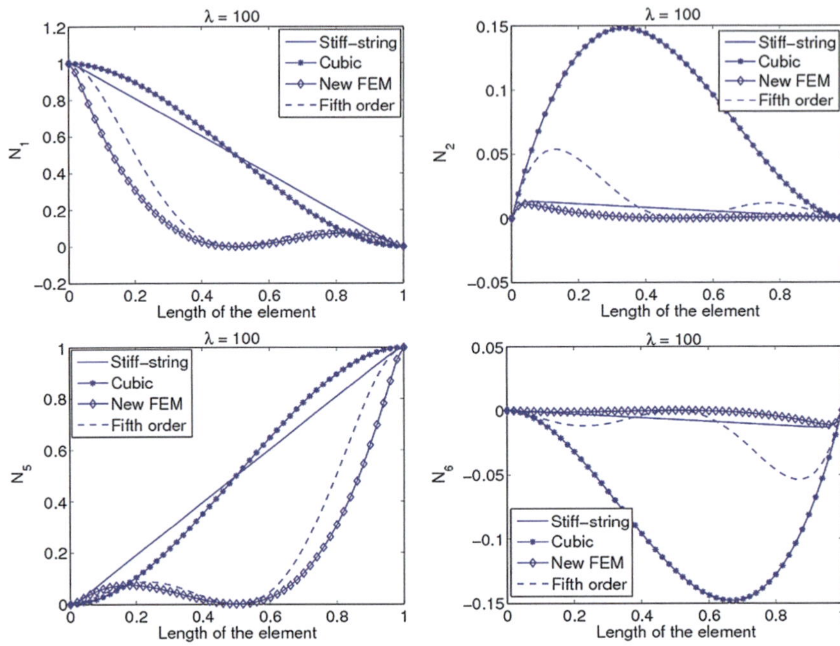

Fig. 2.9 Variation of shape functions ($N = 1$, $\lambda = 100$) with the new, stiff-string, conventional cubic and fifth order finite elements

$$\mathbf{M}_i = \int_0^l m_i(\bar{x})\,\mathbf{N}^T\,\mathbf{N}d\bar{x} \tag{2.42}$$

$$\mathbf{K}_i = \int_0^l EI_i(\bar{x})(\mathbf{N}'')^T\,\mathbf{N}''d\bar{x} + \int_0^l T_i(\bar{x})(\mathbf{N}')^T\,\mathbf{N}'d\bar{x} \tag{2.43}$$

where

$$T_i(\bar{x}) = \sum_{j=i}^{N} \int_{x_j}^{x_{j+1}} m_j(\bar{x})\Omega^2\bar{x}d\bar{x} - \int_{x_i}^{x_i+\bar{x}} m_i(\bar{x})\Omega^2\bar{x}d\bar{x} \tag{2.44}$$

Here $x_i = (i-1)l$ as a uniform mesh is used. Note that no approximations are made regarding the spatial variation in $T(x)$ when deriving the element energy expressions. These assumptions are only made to derive the present hybrid basis functions. The element matrices are then assembled and the boundary conditions applied to get the global stiffness matrix **K** and global mass matrix **M**. The natural frequencies (ω) are then obtained by solving the eigenvalue problem given as

$$\mathbf{K}\Phi = \omega^2\,\mathbf{M}\Phi \tag{2.45}$$

Several numerical examples are now considered and the predictions of the new element are compared with the published literature.

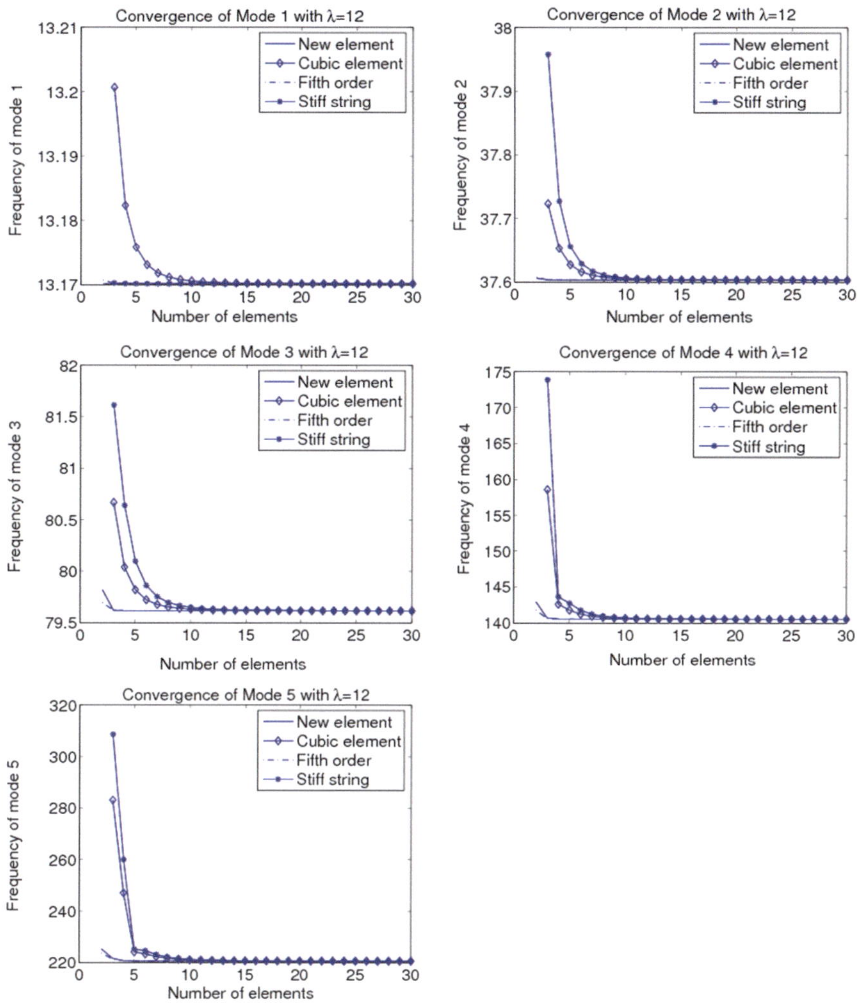

Fig. 2.10 Convergence of the natural frequencies for uniform rotating cantilever beam with $\lambda = 12$

2.6.1 *Uniform Rotating Beam*

For uniform beams, $m(x)$ and $EI(x)$ are constant throughout the span. Table 2.4 shows a comparison of non-dimensional natural frequencies of a rotating uniform cantilever beam with results from Hodges and Rutkowsky [1], Wright et al. [5] and Wang and Wereley [6]. Convergence for the first five modes was achieved using 10 uniform finite elements in contrast to the stiff-string basis functions [7] which requires 75 uniform finite elements, primarily because of slow convergence of the higher modes.

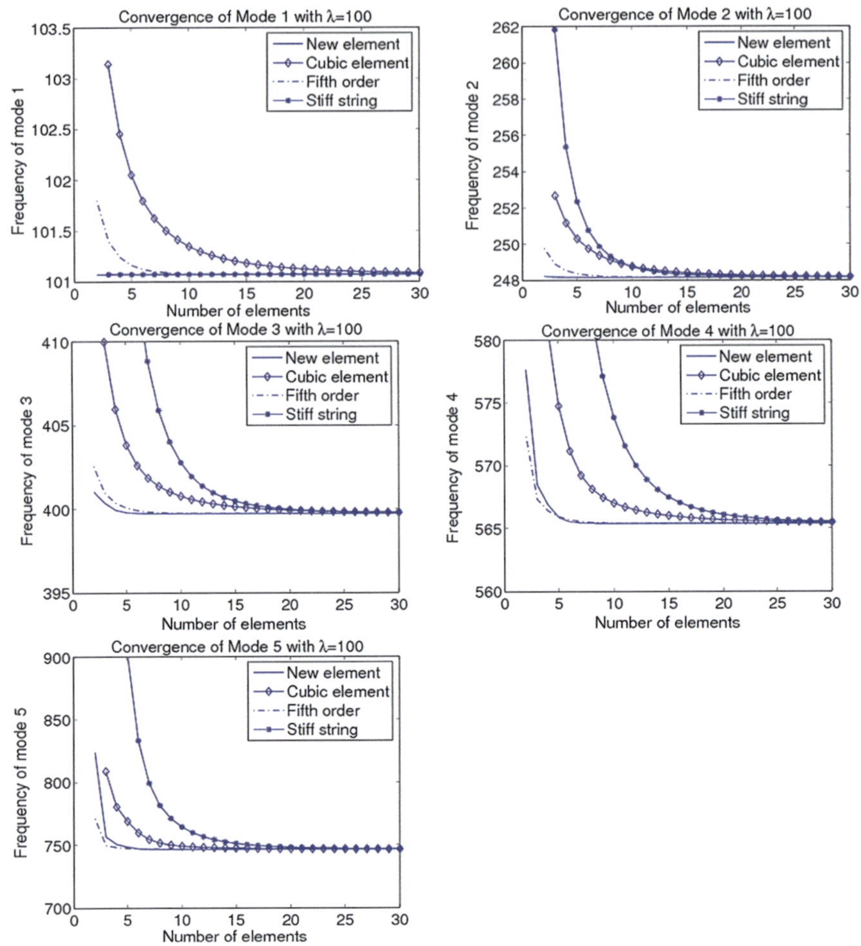

Fig. 2.11 Convergence of the natural frequencies for uniform rotating cantilever beam with $\lambda = 100$

Similar results are obtained for rotating hinged uniform beam and compared with the results from published literature in Table 2.5.

A convergence study is done at two different rotation speeds ($\lambda = 12$ and 100) on the first five modes, since they are commonly used in structural applications. The convergence behaviour of the present basis functions at low and high rotation speeds can be seen in Figs. 2.10 and 2.11 along with the other basis functions. We see that the stiff string basis functions work very well for the fundamental mode. However, their convergence is slow for all the other modes. In contrast, the polynomial work well at the higher modes. It is observed that the convergence of the present basis functions is similar to that of the fifth order basis functions at $\lambda = 12$ as both the shape functions are identical to each other at $\lambda = 12$. As the rotation speed increases, the present basis

Table 2.4 Comparison of non-dimensional natural frequencies of rotating uniform cantilever beam

Mode	Present-FEM	Wang and Wereley [6]	Wright et al. [5]	Hodges et al. [1]
$\lambda = 12$				
1	13.1702	13.1702	13.1702	13.1702
2	37.6031	37.6031	37.6031	37.6031
3	79.6145	79.6145	79.6145	79.6145
4	140.534	140.534	140.534	N/A
5	220.537	220.536	220.536	N/A

Table 2.5 Comparison of non-dimensional natural frequencies of rotating uniform hinged beam

Mode	Present-FEM	Wang and Wereley [6]	Wright et al. [5]
$\lambda = 12$			
1	12.0000	12.0000	12.0000
2	33.7603	33.7603	33.7603
3	70.8373	70.8373	70.8373
4	126.431	126.431	126.431
5	201.123	201.122	201.122

bunctions are better able to capture the rotation effects and significant improvement in convergence for the higher modes can be clearly seen from the Fig. 2.11. Especially, the convergence of the first three modes is significantly improved at high rotation speeds.

An important point to note is that the present FEM needs only one element for the first mode to converge to an accuracy of less than 0.0001 at any high rotation speed. The present FEM also requires only 10 uniform finite elements for the convergence of first five modes to the accuracy of 0.0001 at $\lambda = 12$. Thus the element is effective at both low and high speeds.

Table 2.6 shows the number of elements required to converge the first five frequencies to an accuracy of less than four decimal places with various basis functions considered at a rotation speed of $\lambda = 100$. It is evident that the hybrid basis functions requires fewer number of elements to converge to the desired accuracy as compared to the other basis functions which can require a very high number of elements. It should however be mentioned that the stiff string and cubic elements have one node lesser than the new element and fifth order element. The total number of degrees of freedom for these cases is therefore also given in Table 2.7. It is to be noted that the degrees of freedom required by the new element are much less compared to that for cubic and stiff string cases, and also quite less compared to the fifth order polynomial.

The mode shapes corresponding to $\lambda = 12$ and 100 are shown in Fig. 2.12. At high rotation speeds, the fundamental mode becomes linear as the centrifugal stiffening effect overwhelms the flexural stiffness effect for the mode.

Table 2.6 Comparison of number of elements required with various basis functions to get the converged frequency to the accuracy of <0.0001 for rotating uniform cantilevered beam ($\lambda = 100$)

Mode no.	New element	Fifth order	Stiff string	Cubic
1	1	20	4	129
2	8	23	128	187
3	11	26	163	183
4	14	28	203	198
5	17	31	333	286

Table 2.7 Comparison of degrees of freedom with various basis functions to get the converged frequency to the accuracy of <0.0001 for rotating uniform cantilevered beam ($\lambda = 100$)

Mode no.	New element	Fifth order	Stiff string	Cubic
1	4	80	8	258
2	32	92	256	374
3	44	104	326	366
4	56	112	406	396
5	68	124	666	572

2.7 Tapered Rotating Beam

In order to validate the robustness of the new finite element with the hybrid basis functions, two different types of linearly tapered beams are chosen from the published literature. These beams and the results are given by Hodges and Rutkowsky [1] and Wright et al. [5].

In general, we assume that variation of mass along the beam length is defined as

$$m(\xi) = m_0\,(1 - \alpha\,\xi) \tag{2.46}$$

where m_0 correspond to value of mass per unit length at the thick end of the beam ($\xi = 0$), α is the taper parameter such that $0 < \alpha < 1$. Note that $\alpha \neq 1$, which results in singularity at $\xi = 1$. Flexural stiffness variation along the length of beam element is defined as

$$EI(\xi) = EI_0\,(1 - \beta_1\,\xi - \beta_2\,\xi^2 - \beta_3\,\xi^3 - \beta_4\,\xi^4) \tag{2.47}$$

where EI_0 corresponds to the value of flexural rigidity at the thick end of the beam ($\xi = 0$). Here β_i, $i = 1, 4$, are taper parameters for the stiffness distribution. These parameters can be determined by using α for beams with a rectangular cross-sectional

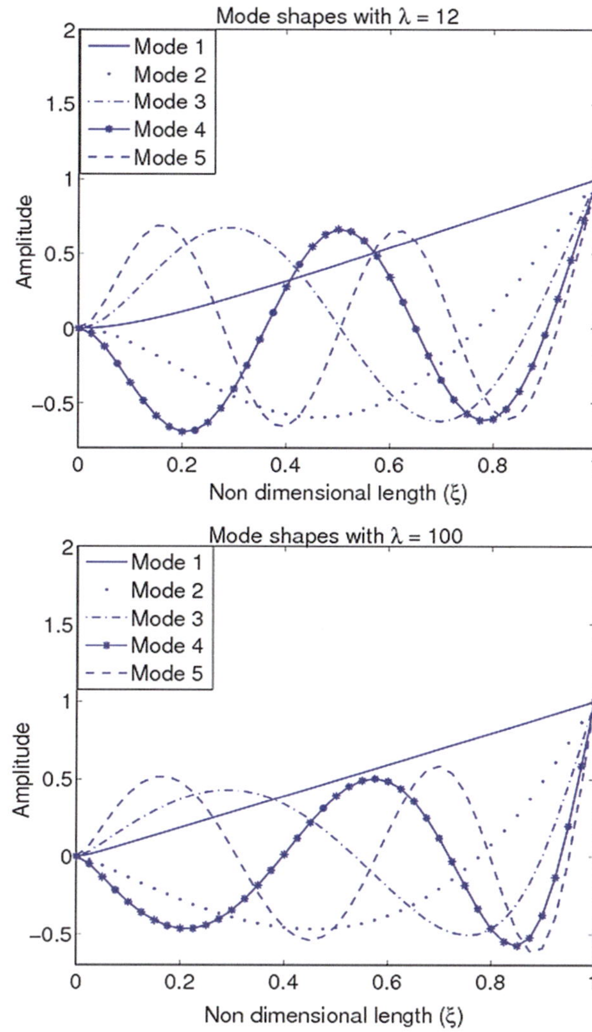

Fig. 2.12 Mode shapes of uniform rotating cantilever beam with $\lambda = 12$ and 100

area and thickness varying along the beam length. However, as with the example studied by Wright et al. [5], the taper parameters for mass and flexural stiffness are not necessarily related. They are independent variables. However, these parameters should not result in a singularity for flexural stiffness at $\xi = 1$. Two numerical examples based on special cases of Eqs. (3.40) and (3.41) are considered next (Table 2.8).

Table 2.8 Comparison of non-dimensional natural frequencies of hinged tapered beam linear (Example 1)

Mode	Present-FEM	Wang and Wereley [6]	Wright et al. [5]
$\lambda = 12$			
1	12.0000	12.0000	12.0000
2	30.7745	30.7741	30.7745
3	63.1722	63.1758	63.1722
4	112.090	112.040	112.090
5	178.016	178.978	178.105

Table 2.9 Comparison of non-dimensional natural frequencies of cantilever tapered beam (Example 2)

Mode	Present-FEM	Wang and Wereley [6]	Wright et al. [5]
$\lambda = 12$			
1	14.0313	14.0313	14.0313
2	35.9064	35.9060	35.9064
3	72.8565	72.8604	72.8565
4	126.401	126.336	126.401
5	198.880	198.243	198.880

Example 1 (Linear Mass, Linear Stiffness, Cantilevered Beam)

In the first example, the tapered beam used by Wright et al. [5] is considered. For this particular problem, the taper is such that the mass per unit length is $m(\xi) = m_0\,(1 - 0.8\,\xi)$ and the bending flexural rigidity is $EI(\xi) = EI_0\,(1 - 0.95\,\xi)$.

The results obtained for this case are compared with those of Wright et al. [5] and Wang and Wereley [6]. Table 2.9 shows the comparison of our results with their works for the first five modes at $\lambda = 12$. Our results compare very well with the published results using only 10 tapered finite elements. Wang and Wereley [6] used a single spectral finite element which requires 350 terms in the Frobenius power series solution. Wright et al. [5] method is also based on a similar principle of using a power series as that of Wang and Wereley [6].

Convergence Study for Tapered Rotating Beam (Example 1)

The convergence behaviour for the tapered rotating beam considered in this example can be clearly seen from the Figs. 2.13 and 2.14. Again the significant improvement in convergence using the new shape functions is clearly evident. The hybrid shape functions work well both at high and low rotation speeds. For the fundamental mode, the new shape functions act like stiff string polynomials, and are also able to act like polynomials for the higher modes.

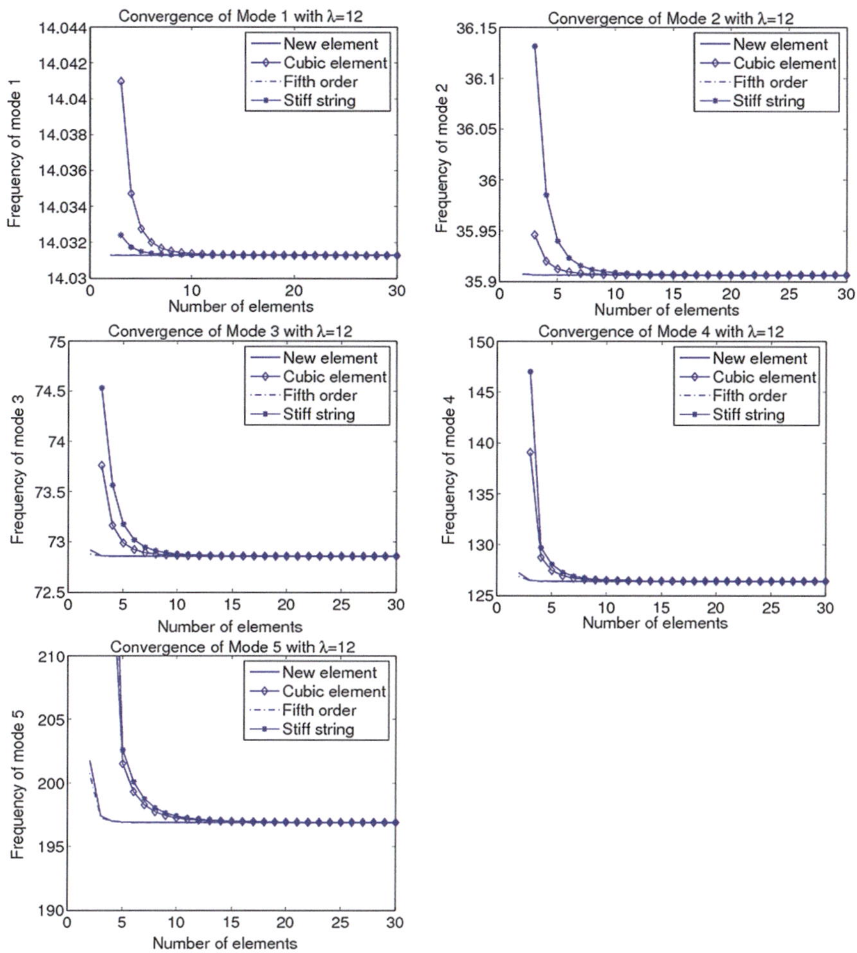

Fig. 2.13 Convergence of the natural frequencies for tapered rotating cantilever beam with $\lambda = 12$

Example 2 (Linear Mass, Linear Stiffness, Hinged Beam)

In the third example, the tapered beam used by Wright et al. [6] with hinged boundary conditions is considered. For this particular problem, the taper is such that both the mass per unit length is $m(\xi) = m_0\,(1 - 0.8\,\xi)$, and the bending flexural rigidity is $EI(\xi) = EI_0\,(1 - 0.95\,\xi)$. The results obtained using the present FEM for first five modes are compared with those of published literature in Table 2.10. It can therefore be concluded that the present FEM works well for beams with various mass and stiffness tapers.

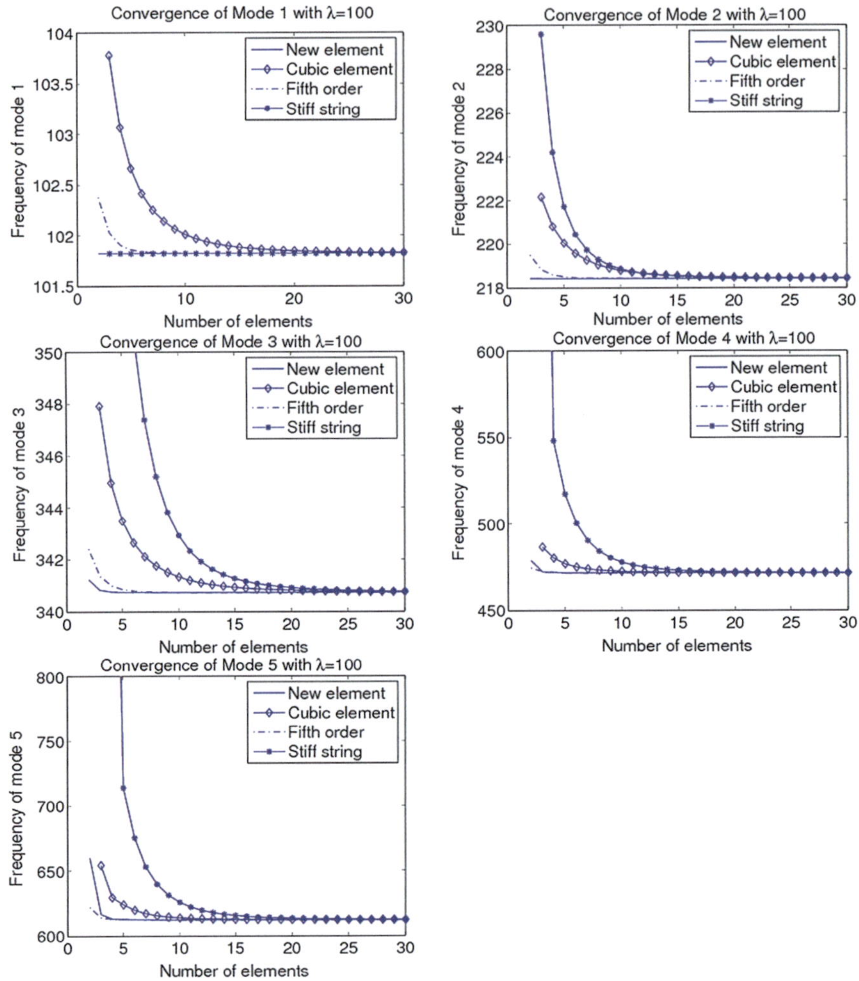

Fig. 2.14 Convergence of the natural frequencies for tapered rotating cantilever beam with $\lambda = 100$

Example 3 (Linear Mass, Cubic Stiffness, Cantilevered Beam)

In this example, the taper is such that the variation of the mass per unit length is $m(\xi) = m_0\,(1 - 0.5\,\xi)$, and the bending flexural rigidity is $EI(\xi) = EI_0\,(1 - 0.5\,\xi)^3$. This type of tapered beam is used by Hodges and Rutkowsky [1] for analysis. These equations cover all the beams of a solid rectangular cross section with a constant width and linearly varying depth. The results obtained for this case are compared with those obtained by Wang and Wereley [6], and Hodges and Rutkowsky [1] in Table 2.11. Results for higher modes are not available in the published literature. Therefore, the comparison of only three modes is shown in this table. It can be observed that the comparison is very good.

Table 2.10 Comparison of non-dimensional natural frequencies of hinged tapered beam (Example 2)

Mode	Present-FEM	Wang and Wereley [6]	Wright et al. [5]
$\lambda = 12$			
1	12.0000	12.0000	12.0000
2	30.7745	30.7741	30.7745
3	63.1722	63.1758	63.1722
4	112.090	112.040	112.090
5	178.016	178.978	178.105

Table 2.11 Comparison of non-dimensional natural frequencies of cantilever tapered beam (Example 3)

Mode	Present-FEM	Ref. [6]	Ref. [1]
$\lambda = 12$			
1	13.4711	13.4711	13.4711
2	34.0877	34.0877	34.0877
3	65.5237	65.5237	65.5237

2.8 Summary

In the present chapter, new shape functions are derived for rotating beams, by using the exact solution of the homogenous part of the governing static differential equation of a stiff string. In this case, the shape functions are not only functions of the element length, but also they are functions of rotation speed, element location across the beam, element mass and stiffness, mass of outboard elements, and length of the beam. The element shows superior convergence of the first two modes at high rotation speed over the conventional Hermite cubics and also presents a new shape function for rotating beams which capture the effect of centrifugal force and element location. The poor convergence of the fundamental mode at high rotation speeds using the cubic polynomials is solved by using the stiff string basis functions. This new element is also applied to determine the natural frequencies of uniform and tapered rotating beams and the results compare very well with the published results.

In the present chapter, hybrid basis functions are also derived for rotating beams by using the linear combination of terms from the exact solution of static homogenous differential equation of a stiff string and that of a non-rotating beam. The shape functions are not only functions of the element length, but also they are functions of rotation speed, element location across the beam, element mass and stiffness, mass of outboard elements, and length of the beam. The new element shows superior convergence for the first five modes considered compared to the stiff-string, conventional fifth order and Hermite cubic based finite elements and also presents a new shape function for rotating beams which capture the effect of centrifugal force and element location. The new element alleviates the convergence difficulties posed by the stiff-string functions in predicting the frequencies corresponding to higher modes at high rotation speeds. It also overcomes the difficulty which the polynomials have in predicting the fundamental mode at high rotation speeds. The robustness of the proposed element is demonstrated by determining the natural frequencies of the uniform and tapered rotating beams and the corresponding results shows an excellent corelation with the published results.

References

1. Hodges DJ, Rutkowsky MJ (1981) Free vibration analysis of rotating beams by a variable order finite element method. AIAA J 19(11):1459–1466
2. Pesheck E, Pierre C, Shaw SW (2002) Modal reduction of a nonlinear rotating beam through nonlinear normal modes. ASME J Vib Acoust 124(2):229–236
3. Chakraborty A, Gopalakrishnan S, Reddy JN (2003) A new beam finite element for the analysis of functionally graded materials. Int J Mech Sci 45(3):519–539
4. Fletcher H (1964) Normal vibration frequencies of a stiff string. J Acoust Soc Am 36(1):203–210
5. Wright AD, Smith CE, Thresher RW, Wang JLC (1982) Vibration modes of centrifugally stiffened beams. J Appl Mech 49(2):197–202
6. Wang G, Wereley NM (2004) Free vibration analysis of rotating blades with uniform tapers. AIAA J 42(12):2429–2437
7. Gunda JB, Ganguli R (2008) Stiff-string basis functions for vibration analysis of high speed rotating beams. J Appl Mech 75(2):0245021–0245025
8. Bathe KJ (1996) Finite element procedures. Prentice Hall, Upper Saddle River

Chapter 3
Rational Interpolation Functions

In this chapter, we seek to develop new interpolating functions which satisfy the static part of the homogenous governing differential equation for rotating beams. Since the differential equation does not have an exact solution, it is imposed as a constraint equation during the derivation of the shape functions.

3.1 Governing Differential Equation

The governing partial differential equation with variable coefficients for out-of-plane (transverse) bending vibration of an Euler-Bernoulli rotating beam is given by,

$$(EI(x)\,w'')'' + m(x)\,\ddot{w} - (T(x)\,w')' = 0 \tag{3.1}$$

where,

$$T(x) = \int_x^L m(x)\,\Omega^2\,(R + x)\,dx + F \tag{3.2}$$

is the centrifugal tensile load at a distance x from the axis of rotation. Ignoring the inertia term yields the static case.

The homogenous form of the above partial differential Eq. (3.1) is given by

$$(EI(x)\,w'')'' - (T(x)\,w')' = 0 \tag{3.3}$$

For a uniform beam$(EI(x) = EI, m(x) = m)$ Eq. (3.3) can be expressed

$$EI\,w'''' - (T(x)\,w')' = 0 \tag{3.4}$$

R. Ganguli, *Finite Element Analysis of Rotating Beams*,
Foundations of Engineering Mechanics, DOI 10.1007/978-981-10-1902-9_3

Further simplifying the above expression, it can be expressed as

$$EI\, w'''' - T(x)\, w'' - w'\, (T(x))' = 0 \tag{3.5}$$

Assuming root offset(R) and axial force(F) applied at the end of the beam to be zero, the Eq. (3.2) for a uniform beam can be reduced

$$T(x) = m\, \Omega^2 \int_x^L x\, dx = \frac{m\, \Omega^2}{2}\, (L^2 - x^2) \tag{3.6}$$

Differentiating Eq. (3.6) with respect to x we get

$$(T(x))' = -m\, \Omega^2\, x \tag{3.7}$$

Substituting Eqs. (3.6) and (3.7) in Eq. (3.5) and further simplifying we get

$$EI\, w'''' + \frac{m\, \Omega^2\, x^2}{2}\, w'' - \frac{m\, \Omega^2\, L^2}{2}\, w'' + m\, \Omega^2\, x\, w' = 0 \tag{3.8}$$

Dividing by EI, Eq. (3.8) can be expressed as

$$w'''' + \frac{m\, \Omega^2\, x^2}{2\, EI}\, w'' - \frac{m\, \Omega^2\, L^2}{2\, EI}\, w'' + \frac{m\, \Omega^2\, x}{EI}\, w' = 0 \tag{3.9}$$

Defining $a = \frac{m\Omega^2}{EI}$, and $b = \frac{m\, \Omega^2\, L^2}{2\, EI}$, the Eq. (3.9) can be expressed as

$$w'''' + \frac{a\, x^2}{2}\, w'' - b\, w'' + a\, x\, w' = 0 \tag{3.10}$$

For an ith element across the beam, the relation between local co-ordinate($\bar{x}$) and global co-ordinate(x) from Fig. 9.1 is given by

$$x = x_i + \bar{x} \tag{3.11}$$

where $x_i = \sum_{j=1}^{i-1} l_j$. For a uniform mesh used in this chapter, $x_i = (i-1)l$.
Using Eq. (3.11), Eq. (3.10) can be written within element i as

$$\frac{d^4 w}{d\bar{x}^4} + \frac{a(x_i + \bar{x})^2}{2} \frac{d^2 w}{d\bar{x}^2} - b \frac{d^2 w}{d\bar{x}^2} + a(x_i + \bar{x}) \frac{dw}{d\bar{x}} = 0 \tag{3.12}$$

The two Eqs. (3.10) and (3.12) represent the static part of the homogenous governing differential equation for the uniform rotating beam in the global and element co-ordinates, respectively.

In the local coordinate system for non-rotating beams ($\Omega = 0$, $a = b = 0$), the Eq. (3.12) becomes

$$\frac{d^4 w}{d\bar{x}^4} = 0 \tag{3.13}$$

3.2 Hermite Shape Functions

The solution of the simple Euler-Bernouli governing differential Eq. (3.13) is the cubic polynomial.

Consider the cubic polynomial

$$w(\bar{x}) = \bar{a}_0 + \bar{a}_1\bar{x} + \bar{a}_2\bar{x}^2 + \bar{a}_3\bar{x}^3 \tag{3.14}$$

Putting the cubic polynomial (3.14) into the element boundary conditions (shown in the Fig. 3.1) yields the Hermite cubic interpolation functions. Using these shape functions we can write the displacement within the element as [1]

$$w = N_{1c}\, w_1 + N_{2c}\, w_2 + N_{3c}\, w_3 + w_4\, N_{4c} \tag{3.15}$$

where $\mathbf{N_c}$ is given by

$$\left[\frac{l^3 - 3\bar{x}^2 l + 2\bar{x}^3}{l^3} \quad \frac{\bar{x}(l^2 - 2\bar{x}l + \bar{x}^2)}{l^2} \quad \frac{\bar{x}^2(3l - 2\bar{x})}{l^3} \quad -\frac{\bar{x}^2(l - \bar{x})}{l^2}\right] \tag{3.16}$$

Note that if Eq. (3.15) is put in Eq. (3.13), it is exactly satisfied. However, substituting Eq. (3.15) in Eq. (3.12), we get

$$w_1\, R_1 + w_2\, R_2 + w_3\, R_3 + w_4\, R_4 = 0 \tag{3.17}$$

where

$$R_1 = \frac{3(4a\bar{x}^3 + (6x_i - 3l)a\bar{x}^2 + (2a{x_i}^2 - 4alx_i - 4b)\bar{x} - al{x_i}^2 + 2bl)}{l^3} \tag{3.18}$$

$$R_2 = \frac{6a\bar{x}^3 + (9x_i - 6l)a\bar{x}^2 + (3a{x_i}^2 - 8alx_i - 6b + al^2)\bar{x} - 2al{x_i}^2 + 4bl + ax_i l^2}{l^2} \tag{3.19}$$

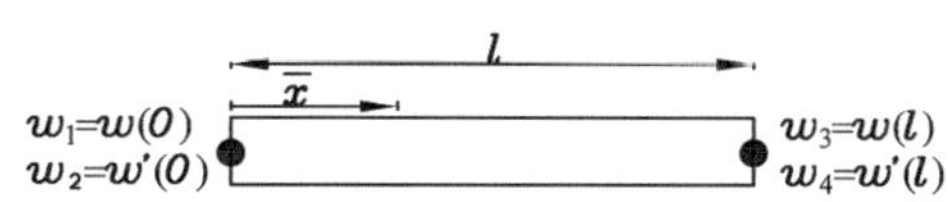

Fig. 3.1 Beam element

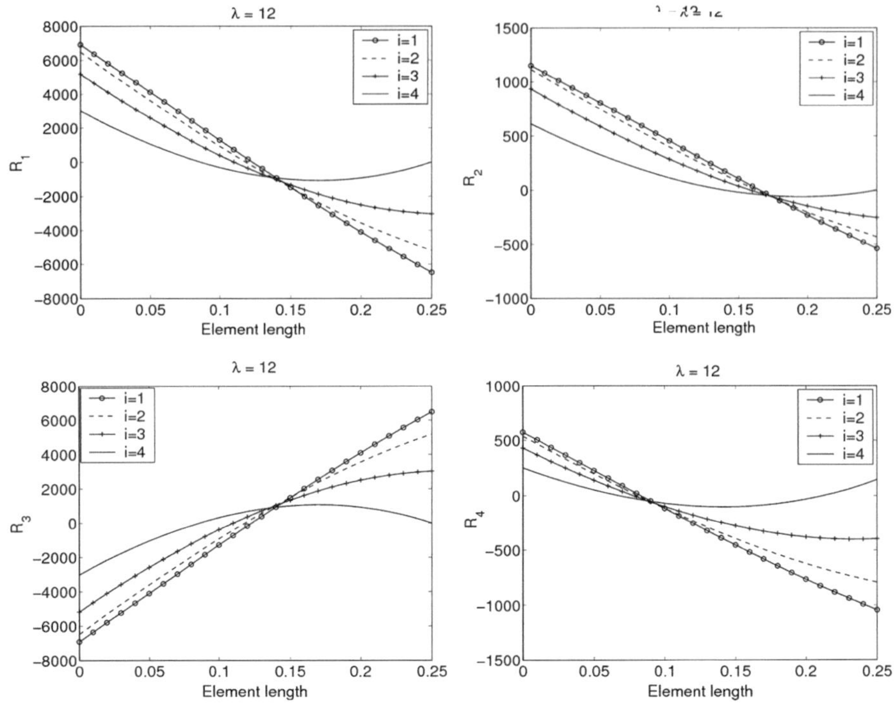

Fig. 3.2 Variation of residues resulting from the use of cubic polynomials across the element length (m) ($N = 4, \lambda = 12$)

$$R_3 = \frac{3(4a\bar{x}^3 + (6x_i - 3l)a\bar{x}^2 + (2ax_i{}^2 - 4alx_i - 4b)\bar{x} - alx_i{}^2 + 2bl}{l^3} \tag{3.20}$$

$$R_4 = \frac{6a\bar{x}^3 + (9ax_i - 3al)\bar{x}^2 + (3ax_i{}^2 - 4alx_i - 6b)\bar{x} - alx_i{}^2 + 2bl}{l^2} \tag{3.21}$$

Since w_1, w_2, w_3, w_4 are arbitrary, in order to satisfy the Eq. (3.17), R_1, R_2, R_3, and R_4 must be zero across the element, which is possible only if a = b = 0 which means $\Omega = 0$. A non-zero rotational speed results in non-zero residues (R_i) across the element. The variation of the residues for $\lambda = 12$ is shown in the Fig. 3.2 with $N = 4$. Different elements have different values of R_i because of the x_i term in Eqs. (3.18)–(3.21). Ideally, the residues R_i should be zero across the element. However, since the Hermite shape functions do not satisfy Eq. (3.12), the residues have non-zero values. In the following section, we seek to create a new element where Eq. (3.12) is satisfied by the interpolating functions.

3.3 New Shape Functions

Unfortunately, Eq. (3.12) does not have a closed form exact solution. A power series solution can be obtained but that is not directly useful for the conventional finite element method. We therefore use the unconventional approach of satisfying the equation at the element level using the interpolating function. To derive the shape functions which satisfy the governing static differential equation of the rotating beams as an additional constraint, the field variable is assumed in the form of

$$w = a_0 + a_1\,\bar{x} + a_2\,\bar{x}^2 + a_3\,\bar{x}^3 + a_4\,\bar{x}^4 \tag{3.22}$$

Here a_0, a_1, a_2, a_3, a_4 are arbitrary constants. These five constants can be determined using the static homogenous differential equation and the four element boundary conditions. Putting Eq. (3.22) into Eq. (3.12) and using the element boundary conditions yields the following five Eqs. (3.23)–(3.27).

$$\begin{aligned} &a_1(ax_i + a\bar{x}) + a_2(a{x_i}^2 + 4ax_i\bar{x} + 3a\bar{x}^2 - 2b) \\ &+ a_3(3a{x_i}^2\bar{x} + 9ax_i\bar{x}^2 + 6a\bar{x}^3 - 6b\bar{x}) \\ &+ a_4(24 + 6a{x_i}^2\bar{x}^2 + 16ax_i\bar{x}^3 + 10a\bar{x}^4 - 12b\bar{x}^2) = 0 \end{aligned} \tag{3.23}$$

$$a_0 = w_1 \tag{3.24}$$

$$a_1 = w_2 \tag{3.25}$$

$$a_0 + a_1 l + a_2 l^2 + a_3 l^3 + a_4 l^4 = w_3 \tag{3.26}$$

$$a_1 + 2\,a_2 l + 3\,a_3 l^2 + 4\,a_4 l^3 = w_4 \tag{3.27}$$

Solving for a_0, a_1, a_2, a_3, a_4 in terms of the nodal displacements and slopes using Eqs. (3.23)–(3.27), w can be approximated by

$$w = w_1\,N_1 + w_2\,N_2 + w_3\,N_3 + w_4\,N_4 \tag{3.28}$$

where N_1, N_2, N_3, N_4 are the interpolating functions derived using the governing static differential equation of the beam, and are given below.

$$N_1 = \frac{\bar{R}_1(\bar{x})}{l\,D(\bar{x})},\ N_2 = \frac{\bar{R}_2\,(\bar{x})}{D(\bar{x})},\ N_3 = \frac{\bar{R}_3\,(\bar{x})}{l\,D(\bar{x})},\ N_4 = \frac{\bar{R}_4\,(\bar{x})}{D(\bar{x})}$$

where

$$\begin{aligned} D(\bar{x}) = {} & l^2(10a\bar{x}^4 + (16x_i - 12l)a\bar{x}^3 + ((3l^2 + 6{x_i}^2 - 18lx_i)a - 12b)\bar{x}^2 \\ & + ((4x_i l^2 - 6l{x_i}^2)a + 12bl)\bar{x} + 24 + a{x_i}^2 l^2 - 2bl^2) \end{aligned}$$

$$\begin{aligned}\bar{R}_1(\bar{x}) = {} & 8a\bar{x}^7 + (14x_i - 21l)a\bar{x}^6 + ((12l^2 + 6x_i{}^2 - 36lx_i)a - 12b)\bar{x}^5 \\ & +((10l^3 + 20l^2x_i - 15lx_i{}^2)a + 30bl)\bar{x}^4 \\ & +((16l^3x_i + 8l^2x_i{}^2 - 12l^4)a - 16bl^2 + 48)\bar{x}^3 \\ & +((3l^5 + 6l^3x_i{}^2 - 18x_il^4)a - 12bl^3 - 72l)\bar{x}^2 \\ & +((4l^5x_i - 6x_i{}^2l^4)a + 12bl^4)\bar{x} \\ & +24l^3 - 2bl^5 + al^5x_i{}^2\end{aligned}$$

$$\begin{aligned}\bar{R}_2(\bar{x}) = {} & 4a\bar{x}^6 + (7x_i - 14l)a\bar{x}^5 + ((3x_i{}^2 - 24lx_i + 18l^2)a - 6b)\bar{x}^4 \\ & +((30l^2x_i - 10lx_i{}^2 - 10l^3)a + 20bl)\bar{x}^3 \\ & +((12l^2x_i{}^2 - 16l^3x_i + 2l^4)a - 24bl^2 + 24)\bar{x}^2 \\ & +((3l^4x_i - 6l^3x_i{}^2)a + 12bl^3 - 48l)\bar{x} \\ & +24l^2 - 2bl^4 + l^4ax_i{}^2)\end{aligned}$$

$$\begin{aligned}\bar{R}_3(\bar{x}) = {} & -\bar{x}^2(8a\bar{x}^5 + (14x_i - 21l)a\bar{x}^4 + ((6x_i{}^2 + 12l^2 - 36lx_i)a - 12b)\bar{x}^3 \\ & +((20l^2x_i - 15lx_i{}^2)a + 30bl)\bar{x}^2 + (8al^2x_i{}^2 - 16bl^2 + 48)\bar{x} - 72l)\end{aligned}$$

$$\begin{aligned}\bar{R}_4(\bar{x}) = {} & \bar{x}^2(4a\bar{x}^5 + (x_i - l)7a\bar{x}^4 - ((3x_i{}^2 - 12lx_i + 3l^2)a - 6b)\bar{x}^3 \\ & +((l^2x_i - lx_i{}^2)5a + 10bl)\bar{x}^2 + (24 + 2al^2x_i{}^2 - 4bl^2)\bar{x} - 24l)\end{aligned}$$

The new shape functions obtained above yield the Hermite polynomials when $\Omega = 0$ or $a = b = 0$. For $a = b = 0$, we get $D(\bar{x}) = 24l^2$, $\bar{R}_1(\bar{x}) = 48\bar{x}^3 - 72l\bar{x}^2 + 24l^3$, $\bar{R}_2(\bar{x}) = \bar{x}(24\bar{x}^2 - 48l\bar{x} + 24l^2)$, $\bar{R}_3(\bar{x}) = -\bar{x}^2(48\bar{x} - 72l)$, $\bar{R}_4(\bar{x}) = \bar{x}^2(24\bar{x} - 24l)$ which yield the Hermite cubics as the shape functions.

Note that these new shape functions are rational functions which are the ratio of two polynomials. Rational functions are often used as interpolating functions in numerical analysis [2] but their use in finite element analysis appears to be limited [3, 4]. Note that the shape functions are now also a function of the non-dimensional rotational speed, element mass and stiffness, beam length and location of the element. These shape functions therefore capture the effect of the rotation speed on the element displacements as well as the fact that different locations contribute differently to the centrifugal stiffening effect. Variation of the shape functions along the elements ($N = 4$) is shown in Figs. 3.3 and 3.4 with conventional and the new FEM at a low and high rotation speed, respectively. The shape functions are identical to the Hermite polynomial at low rotation speed, as seen in Fig. 3.3. The difference between the shape functions is clear for the higher rotation speed, as seen in Fig. 3.4. These shape functions therefore capture the effect of the rotation speed and element location on the element displacements. It is satisfying to see that the rational interpolation functions approach the Hermite cubics at low rotation speeds and show gradual change from the cubics at higher rotation speeds.

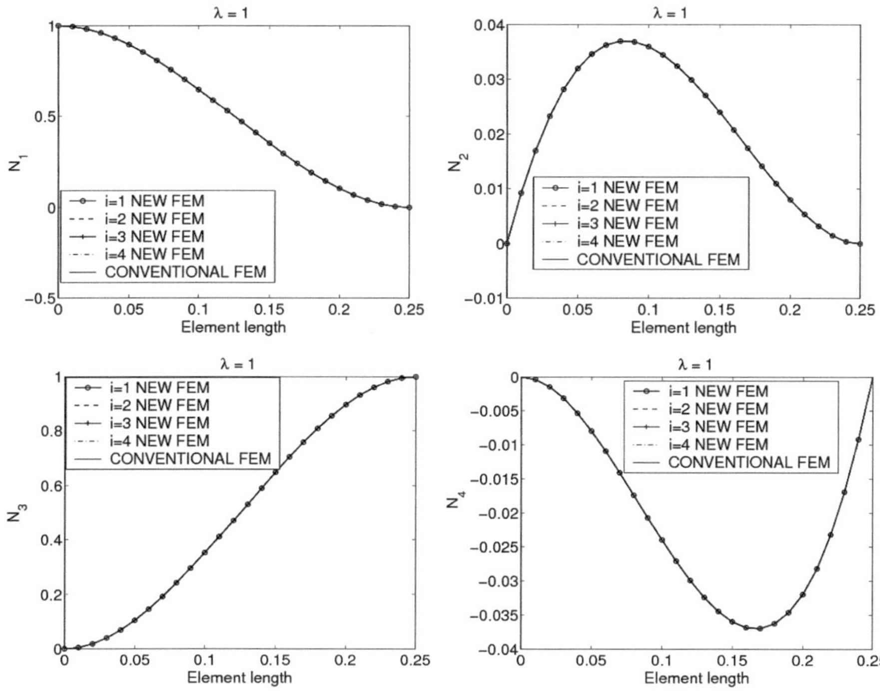

Fig. 3.3 Variation of shape functions along the element length (m) ($N = 4, \lambda = 1$) with the new and conventional finite element at low rotation speed

3.4 Static Finite Element Analysis

The governing static differential equation of the rotating beams (Eq. 3.10) can be imagined as a problem of a non-rotating beam having a spatially varying axial force applied along the length of the beam. The axial force depends on the rotation speed of the beam. Consider a uniform cantilever beam having an axial force applied on it which depends on the value of λ and is subjected to tip loading. The governing equation for the static case in finite element form is given by

$$\mathbf{K}\,\mathbf{w} = \mathbf{Q} \tag{3.29}$$

The conventional finite element given by Hermite cubic shape functions and the present finite element is applied to determine the tip deflection of the axially loaded beam.

The exact solution (Eq. (3.30)) of the governing static homogenous differential Eq. (3.10) is obtained using Maple [5] in terms of the *Whittaker* functions (Eqs. (3.31) and (3.32)) [6, 7] and further expressed as a power series, in which the terms up to the order of $O(x^{100})$ were retained.

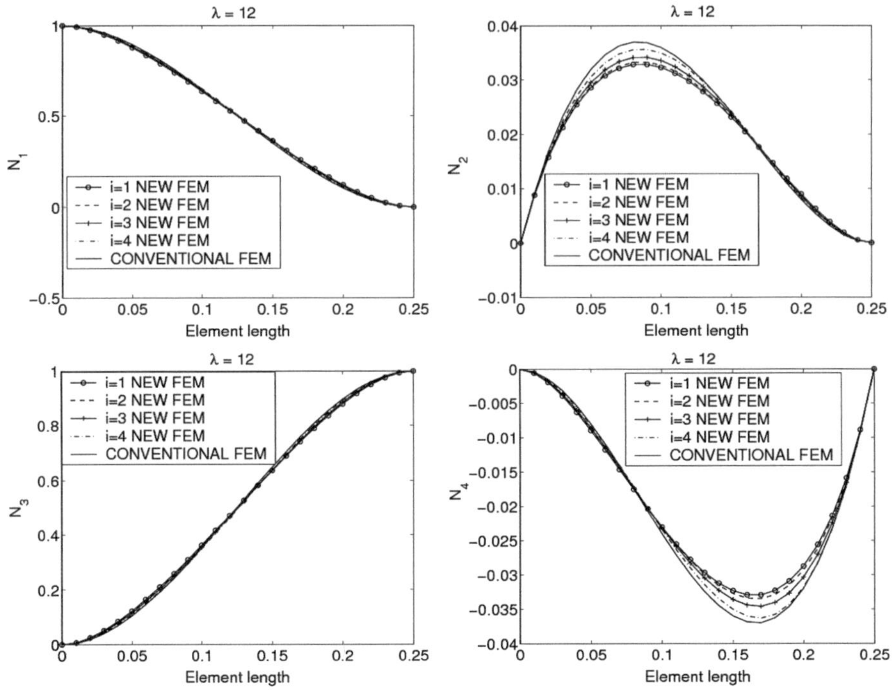

Fig. 3.4 Variation of shape functions along the element length (m) ($N = 4$, $\lambda = 12$) with the new and conventional finite element at high rotation speed

$$w(x) = C_1 + C_2 \int \frac{S_1(x)}{\sqrt{x}} dx + C_3 \int \frac{S_2(x)}{\sqrt{x}} dx + C_4 \int \frac{S_2(x) \int \frac{S_1(x)}{\sqrt{x}} dx + S_1(x) \int \frac{S_2(x)}{\sqrt{x}} dx}{\sqrt{x}} dx \tag{3.30}$$

where C_1, C_2, C_3, C_4 are arbitrary constants to be determined using the boundary conditions of the problem and

$$S_1(x) = WhittakerM\left(\frac{ib\sqrt{2}}{4\sqrt{a}}, \frac{1}{4}, \frac{i\sqrt{2}\sqrt{a}x^2}{2}\right) \tag{3.31}$$

$$S_2(x) = WhittakerW\left(\frac{ib\sqrt{2}}{4\sqrt{a}}, \frac{1}{4}, \frac{i\sqrt{2}\sqrt{a}x^2}{2}\right) \tag{3.32}$$

The Whittaker functions are special functions which are the solutions of the Whittaker differential equation [6, 7]. There is a singularity in the exact solution for $\lambda = 0$, however, a numerical solution can be obtained by choosing λ to be a very small value. The absolute error value is obtained from the conventional and new FEM with reference to the exact solution by taking two different number of elements (N = 4, N = 8). Parametric values for the beam chosen are $L = 1$ m, $m = 6.4$ Kg/m,

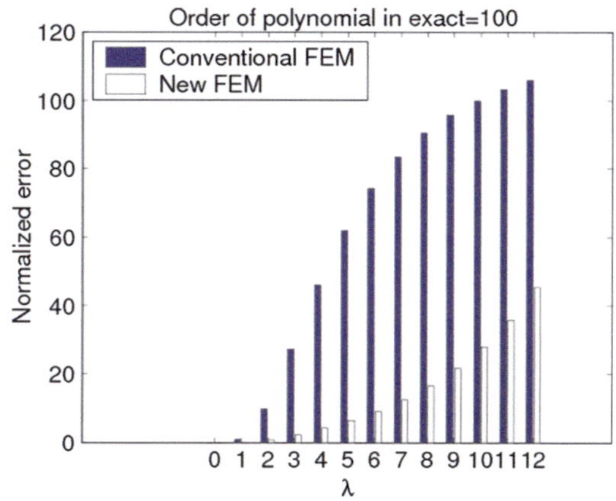

Fig. 3.5 Normalized error in tip deflection between analytical series solution and FEM (Number of Elements = 4)

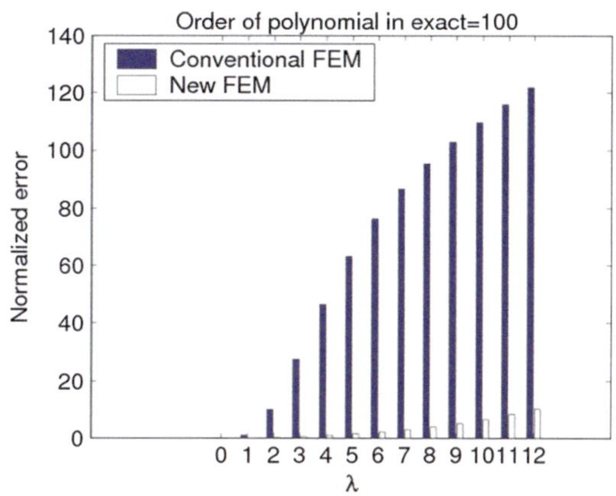

Fig. 3.6 Normalized error in tip deflection between analytical series solution and FEM (Number of Elements = 8)

$EI = 1.2 * 10^5\,\text{Nm}^2$, P(applied transverse tip load) $= 1000\,\text{N}$. It can be observed from the Figs. 3.5 and 3.6 that the error obtained in the tip deflection is lesser with the new FEM. Here the error is normalized with the error value at $\lambda = 1$ using the conventional finite element with $N = 4$. Hence the advantage of the element for which the interpolation function satisfies the governing static differential equation can be clearly seen for the static problem.

The error in tip deflection (w) is given by the formula

$$Error_{fem} = abs(w_{fem} - w_{analytical}) \tag{3.33}$$

Note that here fem in Eq. (3.33) refers to both the conventional and the present fem. Normalised error at various amplitudes of λ is calculated by dividing the Eq. (3.33) with that of the value using the conventional fem at $\lambda = 1$ using $N = 4$ as a reference value for normalization.

3.5 Dynamic Finite Element Analysis

The mass and stiffness matrices can be obtained using the energy expressions. The kinetic energy for a rotating beam is given by

$$T = \frac{1}{2}\int_0^R m(x)\,[\dot{w}(x,\ t)]^2 dx \tag{3.34}$$

where $\dot{w}(x,\ t)$ is derivative of $w(x,\ t)$ with respect to time t. The potential/strain energy is given by

$$U = \frac{1}{2}\int_0^R EI(x)\,[w''(x,\ t)]^2 dx + \frac{1}{2}\int_0^R T(x)\,[w'(x,\ t)]^2 dx \tag{3.35}$$

where, $T(x)$ is defined in Eq. (3.2).

The mass and stiffness matrices ($\mathbf{M}_i$ and $\mathbf{K}_i$) for a beam element can be obtained from the above energy expressions. The calculations for these matrices involve solving the following integrals:

$$\mathbf{M}_i = \int_0^l m_i(\bar{x})\,\mathbf{N}^T\,\mathbf{N} d\bar{x} \tag{3.36}$$

$$\mathbf{K}_i = \int_0^l EI_i(\bar{x})(\mathbf{N}'')^T\,\mathbf{N}'' d\bar{x} + \int_0^l T_i(\bar{x})(\mathbf{N}')^T\,\mathbf{N}' d\bar{x} \tag{3.37}$$

Where

$$T_i(\bar{x}) = \sum_{j=i}^{N}\int_{x_j}^{x_{j+1}} m_j(\bar{x})\Omega^2\bar{x}d\bar{x} - \int_{x_i}^{x_i+\bar{x}} m_i(\bar{x})\Omega^2\bar{x}d\bar{x} \tag{3.38}$$

Here $x_i = (i-1)l$. The element matrices are then assembled and the boundary conditions applied to get the global stiffness matrix $\mathbf{K}$ and global mass matrix $\mathbf{M}$. The natural frequencies are then obtained by solving the eigenvalue problem.

$$\mathbf{K}\Phi = \omega^2\,\mathbf{M}\Phi \tag{3.39}$$

Several numerical examples are now considered.

3.5.1 Uniform Beam

For uniform beams, $m(x)$ and $EI(x)$ are constant throughout the span. Table 3.1 shows a comparison of non-dimensional natural frequencies of a rotating uniform cantilever beam with results from Hodges and Rutkowsky [8], Wright et al. [9] and Wang and Wereley [10]. Two values of non-dimensional rotation speed, λ, are chosen

Table 3.1 Comparison of non-dimensional natural frequencies of cantilever uniform beam

Mode	Present-FEM	Wang and Wereley [10]	Wright et al. [9]	Hodges et al. [8]
$\lambda = 0$				
1	3.5160	3.5160	3.5160	3.5160
2	22.0345	22.0345	22.0345	22.0345
3	61.6972	61.6972	61.6972	61.6972
4	120.902	120.902	120.902	N/A
5	199.862	199.860	199.860	N/A
$\lambda = 12$				
1	13.1702	13.1702	13.1702	13.1702
2	37.6032	37.6031	37.6031	37.6031
3	79.6146	79.6145	79.6145	79.6145
4	140.535	140.534	140.534	N/A
5	220.539	220.536	220.536	N/A

Table 3.2 Comparison of non-dimensional natural frequencies of hinged uniform beam

Mode	Present-FEM	Wang and Wereley [10]	Wright et al. [9]
$\lambda = 0$			
1	0.0000	0.0000	0.0000
2	15.4182	15.4182	15.4182
3	49.9649	49.9649	49.9649
4	104.248	104.248	104.248
5	178.271	178.270	178.270
$\lambda = 12$			
1	12.0000	12.0000	12.0000
2	33.7603	33.7603	33.7603
3	70.8374	70.8373	70.8373
4	126.432	126.431	126.431
5	201.125	201.122	201.122

for comparison. Convergence for the first five modes was achieved using 40 uniform finite elements to get the desired accuracy (tolerance of $<=$ 0.001 for fifth mode). Similar results are obtained for hinged uniform beam and compared with the results from published literature in Table 3.2.

3.5.2 Tapered Rotating Beam

Two different types of linearly tapered beams have been chosen from the published literature by Hodges and Rutkowsky [8] and Wright et al. [9]. Convergence for the five modes was achieved using 40 elements.

In general, we assume that variation of mass along the beam length is defined as

$$m(\xi) = m_0\,(1 - \alpha\,\xi) \tag{3.40}$$

where m_0 correspond to value of mass per unit length at the thick end of the beam ($\xi = 0$), α is the taper parameter such that $0 < \alpha < 1$. $\alpha \neq 1$, which results in singularity at $\xi = 1$. Flexural stiffness variation along the length of beam element is defined as

$$EI(\xi) = EI_0\,(1 - \beta_1\,\xi - \beta_2\,\xi^2 - \beta_3\,\xi^3 - \beta_4\,\xi^4) \tag{3.41}$$

where EI_0 correspond to value of flexural rigidity at the thick end of the beam ($\xi = 0$). Here $\beta_i, i = 1, 4$, are taper parameters for the stiffness distribution. These parameters can be determined by α for beams with a rectangular cross-sectional area and thickness varying along the beam length. However, as with the example studied by Wright et al. [9], the taper parameters for mass and flexural stiffness are not necessarily related. They are independent variables. However, these parameters should not result in a singularity for flexural stiffness at $\xi = 1$. Two numerical examples based on special cases of Eqs. (3.40) and (3.41) are considered below.

Example 1 (Linear Mass, Cubic Stiffness, Cantilevered Beam)

In this example, the taper is such that the variations of the mass per unit length $m(\xi)$, and the bending flexural rigidity $EI(\xi)$ at a non-dimensional distance ξ from thick end are governed by the following expressions:

$$m(\xi) = m_0\,(1 - 0.5\,\xi) \tag{3.42}$$

and

$$EI(\xi) = EI_0\,(1 - 0.5\,\xi)^3 \tag{3.43}$$

Taper parameters are

$$\alpha = 0.5, \qquad \beta_1 = 3\alpha = 1.5,$$

$$\beta_2 = -3\alpha^2 = -0.75, \qquad \beta_3 = \alpha^3 = 0.125, \qquad \beta_4 = 0$$

This type of tapered beam is used by Hodges and Rutkowsky [8] for analysis. These equations cover all the beams of a solid rectangular cross section with a constant width and linearly varying depth. The results obtained for this case are compared

Table 3.3 Comparison of non-dimensional natural frequencies of tapered hinged beam under different rotation speeds (Example 3)

λ	Fourth mode			Fifth mode		
	Present	Ref. [10]	Ref. [9]	Present	Ref. [10]	Ref. [9]
0	97.1707	97.1599	97.1704	163.003	163.277	163.002
1	97.2826	97.2717	97.2823	163.112	163.388	163.111
2	97.6175	97.6062	97.6172	163.440	163.723	163.438
3	98.1730	98.1611	98.1727	163.984	164.281	163.983
4	98.9453	98.9324	98.9450	164.743	165.059	164.741
5	99.9290	99.9147	99.9287	165.713	166.055	165.771
6	101.118	101.102	101.117	166.890	167.268	166.889
7	102.504	102.485	102.504	168.271	168.695	168.269
8	104.080	104.058	104.079	169.849	170.333	169.847
9	105.835	105.809	105.835	171.619	172.180	171.618
10	107.762	107.730	107.762	173.575	174.236	173.573
11	109.850	109.810	109.850	175.710	176.500	175.708
12	112.091	112.040	112.090	178.017	178.978	178.105

with those obtained by Wang and Wereley [10], and Hodges and Rutkowsky [8] in Table 3.3. Results for higher modes are not available in the published literature, therefore comparison of only three modes is shown in this table. It can be observed that the comparison is very good.

Example 2 (Linear Mass, Linear Stiffness, Cantilevered Beam)

In the second example, the tapered beam used by Wright et al. [9] is considered. For this particular problem, the taper is such that both the mass per unit length $m(\xi)$, and the bending flexural rigidity $EI(\xi)$ vary linearly along the length of the beam so that,

$$m(\xi) = m_0\,(1 - 0.8\,\xi) \tag{3.44}$$

and

$$EI(\xi) = EI_0\,(1 - 0.95\,\xi) \tag{3.45}$$

Taper parameters are

$$\alpha = 0.8, \qquad \beta_1 = 0.95, \qquad \beta_2 = \beta_3 = \beta_4 = 0$$

As mentioned by Wright et al. [9], this beam design was used in wind turbine blades. The results obtained for this case are compared with those of Wright et al. [9]

Table 3.4 Comparison of non-dimensional natural frequencies of tapered hinged beam under different rotation speeds (Example 3)

λ	First mode			Second mode			Third mode		
	Present	Ref. [10]	Ref. [9]	Present	Ref. [10]	Ref. [9]	Present	Ref. [10]	Ref. [9]
0	0.0	0.0	0.0	16.7328	16.7328	16.7328	48.4692	48.4696	48.4691
1	1.0	1.0	1.0	16.8711	16.8711	16.8711	48.5872	48.5876	48.5872
2	2.0	2.0	2.0	17.2794	17.2793	17.2794	48.9396	48.9399	48.9395
3	3.0	3.0	3.0	17.9390	17.9389	17.9388	49.5210	49.5214	49.5210
4	4.0	4.0	4.0	18.8233	18.8233	18.8233	50.3234	50.3239	50.3234
5	5.0	5.0	5.0	19.9022	19.9022	19.9022	51.3360	51.3366	51.3360
6	6.0	6.0	6.0	21.1456	21.1456	21.1456	52.5464	52.5471	52.5463
7	7.0	7.0	7.0	22.5260	22.5259	22.5260	53.9406	53.9415	53.9405
8	8.0	8.0	8.0	24.0195	24.0194	24.0195	55.5043	55.5054	55.5042
9	9.0	9.0	9.0	25.6060	25.6059	25.6060	57.2231	57.2245	57.2230
10	10.0	10.0	10.0	27.2692	27.2690	27.2692	59.0828	59.0847	59.0828
11	11.0	11.0	11.0	28.9956	28.9953	28.9956	61.0701	61.0727	61.0700
12	12.0	12.0	12.0	30.7745	30.7741	30.7745	63.1723	63.1758	63.1722

Table 3.5 Comparison of non-dimensional natural frequencies of tapered cantilever beam under different rotation speeds (Example 1)

λ	First mode			Second mode			Third mode		
	Present	Ref. [10]	Ref. [8]	Present	Ref. [10]	Ref. [8]	Present	Ref. [10]	Ref. [8]
0	3.8239	3.8238	3.8238	18.3173	18.3173	18.3173	47.2649	47.2648	47.2648
1	3.9866	3.9866	3.9866	18.4740	18.4740	18.4740	47.4173	47.4173	47.4173
2	4.4368	4.4368	4.4368	18.9366	18.9366	18.9366	47.8717	47.8716	47.8716
3	5.0927	5.0927	5.0927	19.6839	19.6839	19.6839	48.6190	48.6190	48.6190
4	5.8788	5.8788	5.8788	20.6852	20.6852	20.6852	49.6457	49.6456	49.6456
5	6.7434	6.7434	6.7345	21.9053	21.9053	21.9053	50.9339	50.9338	50.9338
6	7.6551	7.6551	7.6551	23.3093	23.3093	23.3093	52.4633	52.4633	52.4633
7	8.5956	8.5956	8.5956	24.8647	24.8647	24.8647	54.2125	54.2124	54.2124
8	9.5540	9.5540	9.5540	26.5437	26.5437	26.5437	56.1596	56.1595	56.1595
9	10.5239	10.5239	10.5239	28.3227	28.3227	28.3227	58.2834	58.2833	58.2833
10	11.5016	11.5015	11.5015	30.1828	30.1827	30.1827	60.5640	60.5639	60.5639
11	12.4845	12.4845	12.4845	32.1086	32.1085	32.1085	62.9830	62.9829	62.9829
12	13.4711	13.4711	13.4711	34.0877	34.0877	34.0877	65.5238	65.5237	65.5237

and Wang and Wereley [10]. Table 3.4 shows the comparison of our results with their works for the first three modes and Table 3.5 shows the comparison for fourth and fifth modes.

Example 3 (Linear Mass, Linear Stiffness, Hinged Beam)

In the third example, the tapered beam used by Wright et al. [9] with hinged boundary conditions is considered. For this particular problem, the taper is such that both the mass per unit length $m(\xi)$, and the bending flexural rigidity $EI(\xi)$ vary linearly along the length of the beam so that,

$$m(\xi) = m_0\,(1 - 0.8\,\xi) \tag{3.46}$$

and

$$EI(\xi) = EI_0\,(1 - 0.95\,\xi) \tag{3.47}$$

Taper parameters are

$$\alpha = 0.8, \qquad \beta_1 = 0.95, \qquad \beta_2 = \beta_3 = \beta_4 = 0$$

The results obtained using present FEM for first three modes are compared with those of published literature in Table 3.6. Table 3.7 presents comparison of fourth and fifth modes of same beam. Again the results compare well.

We also obtained numerical results using the cubic elements for these cases. In general, we found that the same number of elements were needed to obtain convergence for the cubic as for the rational elements. This is likely to be due to the fact

Table 3.6 Comparison of non-dimensional natural frequencies of tapered cantilever beam under different rotation speeds (Example 2)

λ	First mode			Second mode			Third mode		
	Present	Ref. [10]	Ref. [9]	Present	Ref. [10]	Ref. [9]	Present	Ref. [10]	Ref. [9]
0	5.2738	5.2738	5.2738	24.0041	24.0041	24.0041	59.9702	59.9708	59.9701
1	5.3903	5.3903	5.3903	24.1069	24.1069	24.1069	60.0696	60.0703	60.0696
2	5.7249	5.7249	5.7249	24.4130	24.4129	24.4130	60.3670	60.3676	60.3669
3	6.2402	6.2402	6.2402	24.9149	24.9148	24.9149	60.8591	60.8598	60.8590
4	6.8928	6.8928	6.8928	25.6013	25.6013	25.6013	61.5412	61.5420	61.5412
5	7.6443	7.6443	7.6443	26.4581	26.4580	26.4581	62.4069	62.4078	62.4069
6	8.4653	8.4653	8.4653	27.4693	27.4692	27.4693	63.4483	63.4494	63.4483
7	9.3347	9.3347	9.3347	28.6185	28.6184	28.6185	64.6566	64.6579	64.6566
8	10.2379	10.2379	10.2379	29.8894	29.8893	29.8894	66.0223	66.0238	66.0222
9	11.1650	11.1651	11.1650	31.2669	31.2667	31.2669	67.5352	67.5370	67.5351
10	12.1092	12.1092	12.1092	32.7370	32.7367	32.7369	69.1853	69.1875	69.1851
11	13.0657	13.0657	13.0657	34.2870	34.2868	34.2871	70.9623	70.9653	70.9622
12	14.0313	14.0313	14.0313	35.9064	35.9060	35.9064	72.8566	72.8604	72.8565

Table 3.7 Comparison of non-dimensional natural frequencies of tapered cantilever beam under different rotation speeds (Example 2)

λ	Fourth mode			Fifth mode		
	Present	Ref. [10]	Ref. [9]	Present	Ref. [10]	Ref. [9]
0	112.910	112.892	112.909	183.026	183.473	183.024
1	113.009	112.992	113.009	183.126	183.576	183.124
2	113.308	113.290	113.307	183.426	183.887	183.424
3	113.803	113.784	113.803	183.925	184.404	183.923
4	114.493	114.472	114.492	184.621	185.127	184.619
5	115.373	115.351	115.372	185.511	186.055	185.509
6	116.439	116.414	116.439	186.593	187.186	186.591
7	117.686	117.658	117.685	187.864	188.520	187.862
8	119.107	119.075	119.107	189.319	190.055	189.316
9	120.697	120.659	120.696	190.953	191.793	190.950
10	122.447	122.403	122.446	192.762	193.734	192.759
11	124.351	124.298	124.350	194.740	195.882	194.737
12	126.401	126.336	126.401	196.882	198.243	196.880

that the interpolating functions are close to the cubic as shown in Figs. 3.3 and 3.4. In addition, the higher modes have less effect of centrifugal stiffening and so many elements are needed to get the higher mode frequencies to converge. However, the element does show superior behavior in the static case and also presents a new interpolating functions for rotating beams which capture the effect of centrifugal force and element location.

3.6 Summary

In the present chapter, new shape functions are derived for rotating beams, by using the homogenous part of the governing static differential equation of rotating beams as an additional constraint. In this case, the shape functions are not only functions of the element length, but also they are functions of non-dimensional rotational speed, element location across the beam, element mass and stiffness, and length of the beam. The new shape functions are rational functions and become Hermite cubics as the rotation speed becomes zero. These rational interpolation functions are able to capture the effect of centrifugal stiffening and element location on the displacement. This new element is first applied to the static problem by considering a cantilever beam with an applied tip load and having a varying axial force along the length of the beam, which is a function of the rotational speed. The analytical series solution of the governing static homogenous differential equation of the rotating beam, is obtained in terms of Whittaker functions. The converged series solution value of tip deflection is compared with that of present finite element method and the classical finite element method which uses cubic polynomials. The present finite element gives a very good approximation of the tip deflection when compared with the classical finite element method when less elements are used. In the dynamic analysis, this new element is applied to determine the natural frequencies of the uniform, tapered rotating beams and the results compare very well with the published results. The shape functions derived in this chapter provide a theoretically elegant alternative for formulation of the rotating beam finite element. While most current works in this area use the dynamic stiffness or spectral method, the current approach provides good results using the conventional finite element method. It is therefore practical for implementation in existing conventional finite element codes by simple modification of the shape functions.

References

1. Reddy JN (1993) An Introduction to the Finite Element Method, 2nd edn. McGraw-Hill, New York
2. Werner H, Bunger HJ (eds) (1983) Pade' approximation and its applications, Lecture notes in mathematics, no. 1071, Germany, Bad Honnef, Springer, New York
3. Malsch EA, Dasgupta B (2001) Shape functions for concave quadrilaterals. In: Bathe KJ (ed) Computational Fluid and Solid Mechanics, vol 2. Elsevier, Amsterdam, pp 1617–1619
4. Kwon Y, Kim SJ, Kim S (2003) Finite element modeling of high Deborah number planar contraction flows with rational function interpolation of the Leonov model. J Korea-Aust Rheol 15(3):131–150
5. Heck A (2003) Introduction to Maple, 3rd edn. Springer, New York
6. Abramowitz M, Stegun IA (1965) Handbook of Mathematical Functions, chapter 13. Dover publications, New York
7. Luke YL (1969) The Special Functions and Their Approximations, volume 1, chapter 4. Academic Press, New York
8. Hodges DJ, Rutkowski MJ (1981) Free vibration analysis of rotating beams by a variable order finite element method. AIAA J 19(11):1459–1466
9. Wright AD, Smith CE, Thresher RW, Wang JLC (1982) Vibration modes of centrifugally stiffened beams. J Appl Mech 49(1):197–202
10. Wang G, Wereley NM (2004) Free vibration analysis of rotating blades with uniform tapers. AIAA J 42(12):2429–2437

Chapter 4
Fourier-p Superelement

Typically, the conventional finite element method (CFEM) for rotating beams uses cubic polynomials as interpolating functions and convergence is achieved by increasing the number of elements. Since dynamic analysis requires at least the first five modes, capturing these modes accurately requires many elements which leads to a large size eigenvalue problem. The resulting large-degree-of-freedom FEM model is impractical in a real-time dynamic simulation or control problems for which finite element models are often used [1, 2]. As a result, the model order must be reduced via static or dynamic procedures to a practical number of degrees of freedom, which is constrained by simulation time, or the control interval in a digital control system [3]. Furthermore, the use of many elements requires careful development of the finite element mesh for non-uniform rotor blades and increase in computation time due to assembly [4].

To address the shortcomings of CFEM relating to large number of elements and the consequently large size eigenvalue problem, another approach of FEM called the spectral finite element method (SFEM, also called dynamic stiffness method) has evolved for obtaining the same accuracy using fewer number of elements [3, 5, 6]. Banerjee [5] considered uniform beams and Wang and Wereley [3] extended the concept to tapered beams. Banerjee used many elements to model a tapered beam whereas Wang and Wereley developed a tapered element and used only one spectral element. However, a very large number of terms in the power series solution were needed. For example, Wang and Wereley used as many as 350 terms in the Frobenius power series method to find the frequencies of non-uniform rotating beams [3]. In the SFEM, the shape functions are duplicated from exact wave propagation solutions using the governing equation [7, 8]. However, since SFEM uses the solution obtained in frequency domain, the natural frequencies are obtained by solving transcendental equations instead of solving the eigenvalue problems as in CFEM, which can be quite complicated.

R. Ganguli, *Finite Element Analysis of Rotating Beams*,
Foundations of Engineering Mechanics, DOI 10.1007/978-981-10-1902-9_4

Since practical beams are non-uniform, it is advantageous to develop accurate finite element models for them using a minimum number of finite elements. This could be done by constructing a superelement with accurate interpolating functions. Using a single element leads to easy handling of the polynomial variations in mass and stiffness variation across the beam. The superelements have been widely applied for problems to reduce analysis and assembly time drastically for beam, plate, and box-beam problems [9, 10] and also permit the assembly of these substructures into a master structure. They often use a combinations of trigonometric and polynomial functions as interpolating functions [9, 11]. This is different from p-version FEM where only polynomial shape functions are used. Typically, upto five modes are required for dynamic analysis of rotating beams and accurate predictions upto the fifth frequency is therefore important and can require very high values of polynomial order. In p-version FEM, it is not practical to increase p (order of interpolating polynomial functions) to very high values. This is because high order polynomial functions are well known to be ill-conditioned [12], e.g., the computer can hardly find any difference between x^{10} and x^{11} in the interval $0 < x < 1$. This problem of p-version FEM can be addressed by using trigonometric functions along with polynomials which leads to the Fourier p-version of FEM [13, 14]. It is found in these works that higher modes converge much faster using combinations of trigonometric functions and polynomials than when using polynomials alone. The Fourier-p approach can therefore be used to develop a superelement for a rotating beam.

In this chapter, a superelement is developed using a combination of polynomials and Fourier series as shape functions, which results in an efficient formulation which can be used for practical vibration analysis and control problems for rotating structures such as helicopter rotor blades. A linear mass and quartic stiffness distribution which can represent rotor blades is included as a part of the FEM formulation. Superelements for the blades also permit easy integration with finite element models of the rotor hub and fuselage for helicopters and the rotor hub and tower for wind turbines.

4.1 Governing Equation of Rotating Beams

A schematic of a rotating tapered beam is shown in Fig. 4.1. Here $m(x)$ and $EI(x)$ are the mass and flexural stiffness per unit length at a distance x from the axis of rotation, Ω is the rotational speed, $w(x,t)$ and $f(x,t)$ are the displacement in the Z direction and force per unit length, respectively. $T(x)$ is the centrifugal tensile load at a distance x from the axis of rotation, F is an axial force applied at the end of beam, L is the length of the beam and R is the distance of the beam root from the axis of rotation. In the present analysis, R is assumed to be zero. Such beams are good models for long slender structures such as helicopter rotor blades and wind turbine rotor blades whose cross-section dimensions are much smaller than the length [15, 16].

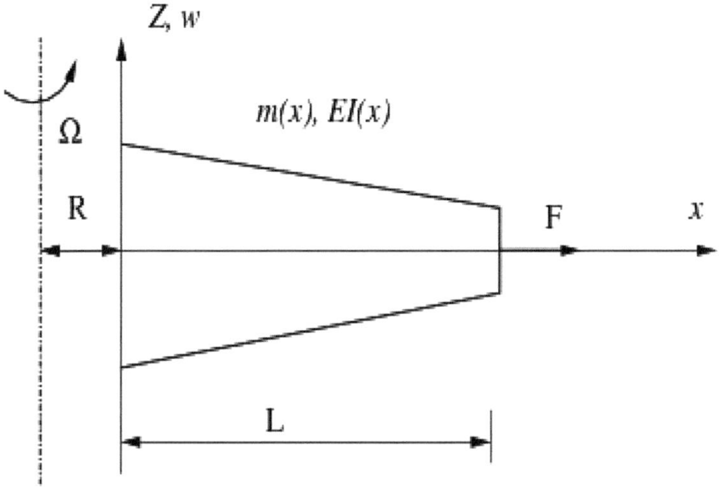

Fig. 4.1 Rotating tapered beam element geometry

The governing partial differential equation with variable coefficients for out-of-plane (transverse) bending vibration of an Euler–Bernoulli rotating beam is given by

$$(EI(x)\,w'')'' + m(x)\,\ddot{w} - (T(x)\,w')' = f(x,t) \tag{4.1}$$

where

$$T(x) = \int_x^L m(x)\,\Omega^2\,(R + x)\,dx + F \tag{4.2}$$

where w' and w'' are first and second derivatives of w with respect to x, respectively.

4.2 Shape Functions

Consider the transverse bending (flapping) vibrations of rotating beams. For bending elements, the shape functions are required to give displacement and slope continuity at the element interfaces. When considering the bending of a beam of unit length, the appropriate shape functions which are obtained by considering the quintic polynomial are given by:

$$N(1,1) = 1 - \frac{331}{9}\,\xi^2 + \frac{326}{3}\,\xi^3 - 112\,\xi^4 + \frac{352}{9}\,\xi^5 \tag{4.3}$$

$$N(2,1) = \xi\,L(1 - \frac{22}{3}\,\xi + 17\,\xi^2 - 16\,\xi^3 + \frac{16}{3}\,\xi^4) \tag{4.4}$$

$$N(3,1) = \frac{128}{3}\xi^2 - \frac{1280}{9}\xi^3 + \frac{1408}{9}\xi^4 - \frac{512}{9}\xi^5 \quad (4.5)$$

$$N(4,1) = \frac{-128}{9}\xi^2 + \frac{256}{3}\xi^3 - 128\,\xi^4 + \frac{512}{9}\xi^5 \quad (4.6)$$

$$N(5,1) = \frac{25}{3}\xi^2 - \frac{466}{9}\xi^3 + \frac{752}{9}\xi^4 - \frac{352}{9}\xi^5 \quad (4.7)$$

$$N(6,1) = \xi\, L(-\xi + \frac{19}{3}\xi^2 - \frac{32}{3}\xi^3 + \frac{16}{3}\xi^4) \quad (4.8)$$

Here $\xi = x/L$ is the non-dimensional length of beam element.

For C^1 continuity requirement, the Fourier version is either $1 - \cos j\pi\xi$ or $(\xi - \xi^2)\sin j\pi\xi$ [17]. Both functions and their first derivatives vanish at $\xi = 0$ and $\xi = 1$. For convenience, the first function is called as the cosine version and the second function as the sine version. Though the cosine version is simpler, it produces zero shear forces at the nodes and is too flexible for shear connections. The cosine version is not recommended when the structure has just one element. Therefore, the sine version is used in this analysis. The enhanced shape functions are given by:

$$N(1,1) = 1 - \frac{331}{9}\xi^2 + \frac{326}{3}\xi^3 - 112\,\xi^4 + \frac{352}{9}\xi^5 \quad (4.9)$$

$$N(2,1) = \xi\, L(1 - \frac{22}{3}\xi + 17\,\xi^2 - 16\,\xi^3 + \frac{16}{3}\xi^4) \quad (4.10)$$

$$N(3,1) = \frac{128}{3}\xi^2 - \frac{1280}{9}\xi^3 + \frac{1408}{9}\xi^4 - \frac{512}{9}\xi^5 \quad (4.11)$$

$$N(4,1) = \frac{-128}{9}\xi^2 + \frac{256}{3}\xi^3 - 128\,\xi^4 + \frac{512}{9}\xi^5 \quad (4.12)$$

$$N(j+4,1) = (\xi - \xi^2)\sin j\pi\xi \quad (4.13)$$

$$N(k+5,1) = \frac{25}{3}\xi^2 - \frac{466}{9}\xi^3 + \frac{752}{9}\xi^4 - \frac{352}{9}\xi^5 \quad (4.14)$$

$$N(k+6,1) = \xi\, L(-\xi + \frac{19}{3}\xi^2 - \frac{32}{3}\xi^3 + \frac{16}{3}\xi^4) \quad (4.15)$$

where k indicates the number of sine terms used in the approximating polynomial function and the value of j varies from 1 to k.

The total number of degrees of freedom in this method is equal to that of the CFEM model plus the number of sine terms (k). The CFEM has two degrees of freedom at each node, namely, the deflection and rotation (slope). The sine terms correspond to the internal degrees of freedom of the element while the quintic polynomials correspond to the nodal degrees of freedom and two internal degrees of freedom at $L/4$ and $3L/4$. The number of sine terms can be increased for getting the solution to converge without increasing the number of elements. Thus, the mixed trigonometric and polynomial shape functions can be used to obtain convergence with one element only.

4.3 Superelement Matrices

The kinetic energy for a rotating beam is given by:

$$\mathscr{T} = \frac{1}{2} \int_0^L m(x)\, [\dot{w}(x,\, t)]^2 dx \tag{4.16}$$

where $\dot{w}(x,\, t)$ is derivative of $w(x,\, t)$ with respect to time t. The potential/strain energy is given by:

$$\mathscr{U} = \frac{1}{2} \int_0^L EI(x)\, [w''(x,\, t)]^2 dx + \frac{1}{2} \int_0^L T(x)\, [w'(x,\, t)]^2 dx \tag{4.17}$$

where $T(x)$ is defined in Eq. (4.2).

The mass and stiffness matrices (**M** and **K**) for such a beam element can be obtained from the above energy expressions. The calculations for these matrices involve solving the following integrals:

$$\mathbf{M} = \int_0^L m(x)\, \mathbf{N}\, \mathbf{N}^T dx \tag{4.18}$$

$$\mathbf{K} = \int_0^L E\, I(x)\, \mathbf{N}''\, (\mathbf{N}'')^T dx + \int_0^L T(x)\, \mathbf{N}'\, (\mathbf{N}')^T dx \tag{4.19}$$

Here superscript T in Eqs. (4.18) and (4.19) denotes the transpose of the matrix and **N** are the shape functions. Since $m(x)$, $EI(x)$ and $T(x)$ are inside the integrals in Eqs. (4.18) and (4.19), the superelement mass and stiffness matrices can be evaluated easily for a given mass and stiffness distribution. Also, the element length is equal to the beam length. Note that for conventional FEM, the element stiffness matrix needs to be calculated at each element because of the centrifugal term, leads to considerable analysis time. The natural frequencies are obtained by solving the eigen value problem.

$$\mathbf{K}\Phi = \omega^2\, \mathbf{M}\Phi$$

4.4 Numerical Results

The natural frequencies are calculated using this model for different rotational speeds using a single element with 10 Fourier(sine) terms.

Table 4.1 Comparison of non-dimensional natural frequencies of cantilever uniform beam

Mode	Present	Wang and Wereley [3]	Wright et al. [6]	Hodges et al. [18]
$\lambda = 0$				
1	3.5160	3.5160	3.5160	3.5160
2	22.0345	22.0345	22.0345	22.0345
3	61.6972	61.6972	61.6972	61.6972
4	120.902	120.902	120.902	N/A
5	199.860	199.860	199.860	N/A
$\lambda = 12$				
1	13.1702	13.1702	13.1702	13.1702
2	37.6031	37.6031	37.6031	37.6031
3	79.6145	79.6145	79.6145	79.6145
4	140.534	140.534	140.534	N/A
5	220.537	220.536	220.536	N/A

4.4.1 Uniform Rotating Beam

Table 4.1 shows an interesting comparison of non-dimensional natural frequencies of a uniform cantilever beam with results from Hodges and Rutkowsky [18], Wright et al. [6] and Wang and Wereley [3]. Two values of non-dimensional rotation speed, λ, are chosen for comparison. Here $\lambda^2 = \frac{\Omega^2}{EI_o/m_oL^4}$ where EI_o and m_o are the reference flexural stiffness and mass per unit length, respectively. All the frequencies predicted by superelement agree very well with the published results.

In this chapter, the whole beam is considered as a single element and 10 Fourier (sine) terms are used with quintic polynomials as shape functions. Thus, the total number of degrees of freedom is 14 after application of cantilever boundary conditions and it gives fairly good accuracy for frequencies of up to the fifth mode. It is found that the same accuracy level for up to fifth mode frequency using h-FEM requires at least 50 finite elements, i.e., 102 degrees of freedom (2 degrees of freedom for every node) or 100 degree of freedom after applying cantilever boundary conditions.

Similar results are obtained for a hinged uniform beam and compared with the results from published literature in Table 4.2. These results also show excellent agreement.

4.4.2 Tapered Rotating Beam

For a better approximation to the practical rotor blade, we analyze it as a tapered beam. Although any type of tapered beam can be analyzed using the present approach,

Table 4.2 Comparison of non-dimensional natural frequencies of hinged uniform beam

Mode	Present	Wang and Wereley [3]	Wright et al. [6]
$\lambda = 0$			
1	0.0000	0.0000	0.0000
2	15.4182	15.4182	15.4182
3	49.9649	49.9649	49.9649
4	104.248	104.248	104.248
5	178.270	178.270	178.270
$\lambda = 12$			
1	12.0000	12.0000	12.0000
2	33.7603	33.7603	33.7603
3	70.8373	70.8373	70.8373
4	126.431	126.431	126.431
5	201.123	201.122	201.122

for illustrative purposes two different types of linearly tapered cantilever beams are selected from the published literature by Hodges and Rutkowsky [18] and Wright et al. [6].

The beam element used in this analysis is a tapered element. The element stiffness and mass matrices are calculated exactly for a tapered element. Thus the tapered beam is not idealized using several piece-wise uniform beam elements as often done in conventional formulation. Therefore, this approach gives more accurate results with one element.

In general, we assume that variation of mass along the beam length is defined as

$$m(x) = m_0\,(1 - \alpha\,\xi) \tag{4.20}$$

where m_0 corresponds to the value of mass per unit length at the thick end of the beam ($\xi = 0$), α is the taper parameter such that $0 < \alpha < 1$. $\alpha \neq 1$, which results in a singularity at $\xi = 1$. Flexural stiffness variation along the length of beam element is defined as

$$EI(x) = EI_0\,(1 - \beta_1\,\xi - \beta_2\,\xi^2 - \beta_3\,\xi^3 - \beta_4\,\xi^4) \tag{4.21}$$

where EI_0 correspond to the value of flexural rigidity at the thick end of the beam ($\xi = 0$). Here β_i, $i = 1$ to 4 are taper parameters for stiffness distribution. These parameters can be determined by α for beams with a rectangular cross-sectional area and thickness varying along the beam length. However, as with the example studied by Wright et al. [6], the taper parameters for mass and flexural stiffness are not necessarily related. They are independent variables. However, these parameters should not result in a singularity for flexural stiffness at $\xi = 1$.

Example 1 (Linear Mass, Cubic Stiffness, Cantilevered Beam)

In this example, the taper is such that the variations of the mass per unit length $m(x)$, and the bending flexural rigidity $EI(x)$ at a distance x from the thick end are governed by the following expressions:

$$m(x) = m_0\,(1 - 0.5\,\xi) \tag{4.22}$$

and

$$EI(x) = EI_0\,(1 - 0.5\,\xi)^3 \tag{4.23}$$

Taper parameters are

$$\alpha = 0.5, \qquad \beta_1 = 3\alpha = 1.5,$$

$$\beta_2 = -3\alpha^2 = -0.75, \qquad \beta_3 = \alpha^3 = 0.125, \qquad \beta_4 = 0$$

These equations cover all beams having a solid rectangular cross section with constant width and linearly varying depth. The results obtained for this case are compared with those obtained by Wang and Wereley [3] and Hodges and Rutkowsky [18] in Table 4.3. Since results for higher modes are not available in the published literature, a comparison of only three modes is shown in this table. The results obtained using superelement show excellent agreement with the published results.

Table 4.3 Comparison of non-dimensional natural frequencies of tapered cantilever beam under different rotational speeds (Example 1)

λ	First mode			Second mode			Third mode		
	Present	Ref. [3]	Ref. [18]	Present	Ref. [3]	Ref. [18]	Present	Ref. [3]	Ref. [18]
0	3.8238	3.8238	3.8238	18.3173	18.3173	18.3173	47.2649	47.2648	47.2648
1	3.9866	3.9866	3.9866	18.4740	18.4740	18.4740	47.4173	47.4173	47.4173
2	4.4368	4.4368	4.4368	18.9366	18.9366	18.9366	47.8717	47.8716	47.8716
3	5.0927	5.0927	5.0927	19.6839	19.6839	19.6839	48.6190	48.6190	48.6190
4	5.8788	5.8788	5.8788	20.6852	20.6852	20.6852	49.6457	49.6456	49.6456
5	6.7434	6.7434	6.7345	21.9053	21.9053	21.9053	50.9338	50.9338	50.9338
6	7.6551	7.6551	7.6551	23.3093	23.3093	23.3093	52.4633	52.4633	52.4633
7	8.5956	8.5956	8.5956	24.8647	24.8647	24.8647	54.2125	54.2124	54.2124
8	9.5540	9.5540	9.5540	26.5437	26.5437	26.5437	56.1595	56.1595	56.1595
9	10.5239	10.5239	10.5239	28.3227	28.3227	28.3227	58.2834	58.2833	58.2833
10	11.5015	11.5015	11.5015	30.1827	30.1827	30.1827	60.5639	60.5639	60.5639
11	12.4845	12.4845	12.4845	32.1084	32.1085	32.1085	62.9829	62.9829	62.9829
12	13.4711	13.4711	13.4711	34.0877	34.0877	34.0877	65.5237	65.5237	65.5237

Wang and Wereley [3] used a single spectral finite element with the first 80 terms in the Frobenius power series for similar accuracy level while Hodges and Rutkowsky [18] used a variable order finite element with 15th order polynomials. The present analysis uses a single superelement with only 10 sine terms as interpolating functions to get comparable results.

Example 2 (Linear Mass, Linear Stiffness, Cantilevered Beam)

In the second example, the tapered beam is considered. For this particular problem, the taper is such that both the mass per unit length $m(x)$, and the bending flexural rigidity $EI(x)$ vary linearly along the length of the beam so that,

$$m(x) = m_0\,(1 - 0.8\,\xi) \tag{4.24}$$

and

$$EI(x) = EI_0\,(1 - 0.95\,\xi) \tag{4.25}$$

Taper parameters corresponding to Eqs. (4.20) and (4.21) stated earlier are

$$\alpha = 0.8, \qquad \beta_1 = 0.95, \qquad \beta_2 = \beta_3 = \beta_4 = 0$$

As shown in the preceding equation, when $\xi = 1$, the flexural stiffness drops to 5 % of the initial value of EI_0. The singularity is very close to $\xi = 1$, which results in a slower convergence of the results. Table 4.4 shows the comparison of the superelement results with the published works for the first three modes and Table 4.5 shows the comparison for fourth and fifth modes.

Table 4.4 Comparison of non-dimensional natural frequencies of tapered cantilever beam under different rotational speeds (Example 2)

λ	First mode			Second mode			Third mode		
	Present	Ref. [3]	Ref. [6]	Present	Ref. [3]	Ref. [6]	Present	Ref. [3]	Ref. [6]
0	5.2738	5.2738	5.2738	24.0041	24.0041	24.0041	59.9702	59.9708	59.9701
1	5.3903	5.3903	5.3903	24.1069	24.1069	24.1069	60.0697	60.0703	60.0696
2	5.7249	5.7249	5.7249	24.4130	24.4129	24.4130	60.3670	60.3676	60.3669
3	6.2402	6.2402	6.2402	24.9149	24.9148	24.9149	60.8591	60.8598	60.8590
4	6.8928	6.8928	6.8928	25.6013	25.6013	25.6013	61.5469	61.5420	61.5412
5	7.6443	7.6443	7.6443	26.4581	26.4581	26.4581	62.4070	62.4078	62.4069
6	8.4653	8.4653	8.4653	27.4693	27.4692	27.4693	63.4484	63.4494	63.4483
7	9.3347	9.3347	9.3347	28.6185	28.6184	28.6185	64.6567	64.6579	64.6566
8	10.2379	10.2379	10.2379	29.8894	29.8893	29.8894	66.0223	66.0238	66.0222
9	11.1650	11.1651	11.1650	31.2669	31.2667	31.2669	67.5352	67.5370	67.5351
10	12.1092	12.1092	12.1092	32.7369	32.7367	32.7369	69.1852	69.1875	69.1851
11	13.0657	13.0657	13.0657	34.2871	34.2868	34.2871	70.9623	70.9653	70.9622
12	14.0313	14.0313	14.0313	35.9064	35.9060	35.9064	72.8566	72.8604	72.8565

Table 4.5 Comparison of non-dimensional natural frequencies of tapered cantilever beam under different rotational speeds (Example 2)

λ	Fourth mode			Fifth mode		
	Present	Ref. [3]	Ref. [6]	Present	Ref. [3]	Ref. [6]
0	112.910	112.892	112.909	183.029	183.473	183.024
1	113.010	112.992	113.009	183.129	183.576	183.124
2	113.308	113.290	113.307	183.429	183.887	183.424
3	113.803	113.784	113.803	183.928	184.404	183.923
4	114.493	114.472	114.492	184.623	185.127	184.619
5	115.373	115.351	115.372	185.514	186.055	185.509
6	116.439	116.414	116.439	186.596	187.186	186.591
7	117.686	117.658	117.685	187.867	188.520	187.862
8	119.108	119.075	119.107	189.321	190.055	189.316
9	120.697	120.659	120.696	190.955	191.793	190.950
10	122.447	122.403	122.446	192.764	193.734	192.759
11	124.351	124.298	124.350	194.743	195.882	194.737
12	126.401	126.336	126.401	196.885	198.243	196.880

Wang and Wereley [3] used a single spectral finite element with as many as 350 terms in Frobenius power series. The present analysis uses one superelement with only 10 sine terms as the interpolating functions and the results compare very well with published work.

Example 3 (Linear Mass, Linear Stiffness, Hinged Beam)

In the third example, the tapered beam used by Wright et al. [6] with hinged boundary conditions is considered. The same superelement model can be used to obtain the natural frequencies of hinged tapered beams with different geometric boundary conditions.

For this particular problem, the taper is such that both the mass per unit length $m(x)$, and the bending flexural rigidity $EI(x)$ vary linearly along the length of the beam so that,

$$m(x) = m_0\,(1 - 0.8\,\xi) \tag{4.26}$$

and

$$EI(x) = EI_0\,(1 - 0.95\,\xi) \tag{4.27}$$

Taper parameters are thus identical to those considered in Example 2 and the only difference is in the boundary conditions. The taper parameters are given by

$$\alpha = 0.8, \qquad \beta_1 = 0.95, \qquad \beta_2 = \beta_3 = \beta_4 = 0$$

Earlier, Wright et al. [6], and Wang and Wereley [3] have discussed this type of tapered beam with hinged boundary conditions. Wang and Wereley [3] have used one spectral finite element with 350 terms in Frobenius functions to obtain the results. The method used by them is based on a similar principle of using power series as that

Table 4.6 Comparison of non-dimensional natural frequencies of tapered hinged beam under different rotational speeds (Example 3)

λ	First mode			Second mode			Third mode		
	Present	Ref. [3]	Ref. [6]	Present	Ref. [3]	Ref. [6]	Present	Ref. [3]	Ref. [6]
0	0.0	0.0	0.0	16.7328	16.7328	16.7328	48.4692	48.4696	48.4691
1	1.0	1.0	1.0	16.8711	16.8711	16.8711	48.5872	48.5876	48.5872
2	2.0	2.0	2.0	17.2794	17.2793	17.2794	48.9395	48.9399	48.9395
3	3.0	3.0	3.0	17.9390	17.9389	17.9388	49.5210	49.5214	49.5210
4	4.0	4.0	4.0	18.8233	18.8233	18.8233	50.3234	50.3239	50.3234
5	5.0	5.0	5.0	19.9023	19.9022	19.9022	51.3360	51.3366	51.3360
6	6.0	6.0	6.0	21.1457	21.1456	21.1456	52.5464	52.5471	52.5463
7	7.0	7.0	7.0	22.5260	22.5259	22.5260	53.9406	53.9415	53.9405
8	8.0	8.0	8.0	24.0195	24.0194	24.0195	55.5043	55.5054	55.5042
9	9.0	9.0	9.0	25.6061	25.6059	25.6060	57.2231	57.2245	57.2230
10	10.0	10.0	10.0	27.2692	27.2690	27.2692	59.0828	59.0847	59.0828
11	11.0	11.0	11.0	28.9956	28.9953	28.9956	61.0701	61.0727	61.0700
12	12.0	12.0	12.0	30.7745	30.7741	30.7745	63.1723	63.1758	63.1722

Table 4.7 Comparison of non-dimensional natural frequencies of tapered hinged beam under different rotational speeds (Example 3)

λ	Fourth mode			Fifth mode		
	Present	Ref. [3]	Ref. [6]	Present	Ref. [3]	Ref. [6]
0	97.1709	97.1599	97.1704	163.004	163.277	163.002
1	97.2828	97.2717	97.2823	163.114	163.388	163.111
2	97.6177	97.6062	97.6172	163.441	163.723	163.438
3	98.1732	98.1611	98.1727	163.985	164.281	163.983
4	98.9455	98.9324	98.9450	164.744	165.059	164.741
5	99.9292	99.9147	99.9287	165.714	166.055	165.771
6	101.118	101.102	101.117	166.892	167.268	166.889
7	102.504	102.485	102.504	168.272	168.695	168.269
8	104.080	104.058	104.079	169.851	170.333	169.847
9	105.836	105.809	105.835	171.621	172.180	171.618
10	107.762	107.730	107.762	173.577	174.236	173.573
11	109.851	109.810	109.850	175.711	176.500	175.708
12	112.091	112.040	112.090	178.019	178.978	178.105

of Wright et al. [6]. The results obtained using the present approach for first three modes are compared with those of published literature in Table 4.6. Table 4.7 presents comparison of fourth and fifth modes of same beam. Again, excellent agreement is obtained with the published results.

4.5 Summary

A Fourier-p superelement is used for finding the natural frequencies of rotating uniform and tapered beams, with cantilever and hinged boundary conditions. The shape functions used for modeling the finite element consist of a combination of product of polynomial functions and Fourier series, in addition to quintic polynomials. The classical h-FEM is therefore enhanced using the trigonometric functions to form the superelement. Since Fourier series are well behaved, the limitation of the higher order polynomial functions of being ill-conditioned, is removed. The results obtained from the current approach show an excellent match with the results obtained from different methods in the published literature for uniform and tapered rotating beams with cantilever and hinged boundary conditions. The superelement is easy to use for non-uniform mass and stiffness distributions which occur in helicopter and wind turbine blades. The stiffness matrix of even uniform rotating beams vary due to centrifugal effects and need to be calculated for each element in the conventional FEM formulation, leads to considerable analysis and assembly time, which is saved by the superelement.

References

1. Cai GP, Hong JZ, Yang SX (2004) Model study and active control of a rotating flexible cantilever beam. Int J Mech Sci 46(6):871–889
2. Thakkar D, Ganguli R (2006) Use of single crystal and soft piezoceramics for alleviation of flow separation induced vibration in a smart helicopter rotor. Smart Mater Struct 15(2):331–341
3. Wang G, Wereley NM (2004) Free vibration analysis of rotating blades with uniform tapers. AIAA J 42(12):2429–2437
4. Ganguli R, Chopra I, Weller WH (1998) Comparison of calculated vibratory rotor hub loads with experimental data. J Am Helicopter Soc 43(4):312–318
5. Banerjee JR (2000) Free vibration of centrifugally stiffened uniform and tapered beams using the dynamic stiffness method. J Sound Vib 233(5):857–875
6. Wright AD, Smith CE, Thresher RW, Wang JLC (1982) Vibration modes of centrifugally stiffened beams. J Appl Mech 49(2):197–202
7. Vinod KG, Gopalakrishnan S, Ganguli R (2006) Wave propogation characteristics of rotating uniform Euler–Bernoulli beams. CMES- Comput Model Eng Sci 16(3):197–208
8. Vinod KG, Gopalakrishnan S, Ganguli R, Free vibration and wave propogation analysis of uniform and tapered rotating beams using spectrally formulated finite elements. Int J Solids Struct (Accepted)
9. Ahmadian MT, Zangench MS (2002) Vibration analysis of orthotropic rectangular plates using superelements. Comput Methods Appl Mech Eng 191(19–20):2097–2103

10. Belyi MV (1993) Superelement method for transient dynamic analysis of structural systems. Int J Numer Methods Eng 36(13):2263–2286
11. Koko TS (1992) Vibration analysis of stiffened plates by super elements. J Sound Vib 158(1):149–167
12. West LJ, Bardell NS, Dunsdon JM, Loasby PM (1997) Some limitations associated with the use of K-orthogonal polynomials in hierarchial versions of the finite element method. In: The sixth international conference on recent advances in structural dynamics, Southhampton, UK, pp 217–227
13. Leung AYT, Chan JKW (1998) Fourier p-element for the analysis of beams and plates. J Sound Vib 212(1):179–185
14. Yongqiang L (2006) Free vibration analysis of plate using finite strip Fourier p-element. J Sound Vib 294(4–5):1051–1059
15. Kumar S, Roy N, Ganguli R (2007) Monitoring low cycle fatigue damage in turbine blade using vibration characteristics. Mech Syst Signal Process 21(1):480–501
16. Ganguli R (2001) A fuzzy logic system for ground based structural health monitoring of a helicopter rotor using modal data. J Intell Mater Syst Struct 12(6):397–407
17. Leung AYT, Zhu B (2004) Fourier p-elements for curved beam vibrations. Thin-Walled Struct 42:39–57
18. Hodges DJ, Rutkowsky MJ (1981) Free vibration analysis of rotating beams by a variable order finite element method. AIAA J 19(11):1459–1466

Chapter 5
Physics Based Basis Functions

We have seen from the previous chapters that the convergence of finite element methods can be improved if we assume basis functions which will closely resemble the displacement variation in the physical problem [3, 4, 10]. The stiffness matrix for dynamic analysis of any structural system is the same as that in static analysis. Hence a very good choice for basis functions is one which satisfies the static part of the governing partial differential equation. Several researchers [1, 5–7] have tried to simplify the governing equation such that an exact analytical solution can be found for the equation. It has been observed that these modified basis functions show improved convergence for the fundamental mode but converge slower than the cubic functions for the higher modes.

In order to obtain improved convergence for all the modes, the basis function must be able to capture the effects of the flexural stiffening as well as include the centrifugal stiffening effect. The centrifugal force varies along the length of the beam (maximum at the root and zero at the tip), hence the stiffness of the system varies along the length of the beam too. We try to split the static part of the differential equation into the flexural and the centrifugal part. This indirectly represents the analysis of a non-rotating beam and a rotating string separately. We assume a solution to the differential equation to be a linear combination of the two separate solutions. A method for solving differential equations is the collocation method [13]. In this method, we assume an approximate solution which is either defined in the global domain or in a piecewise manner and then use the residual in the differential equation to solve for the constants in the solution. We define an approximation in a given number of subintervals and then obtain a relation for the solution using the collocation method. This approximated solution is then used as a basis function in the finite element method.

Conventional finite element methods do not directly take the error into account. Gunda and Ganguli [6] tried to use the local form of differential equation for obtaining the shape functions and showed that improved convergence was not obtained in the

R. Ganguli, *Finite Element Analysis of Rotating Beams*,
Foundations of Engineering Mechanics, DOI 10.1007/978-981-10-1902-9_5

higher modes. The combination of the collocation and the finite element method is what gives the method discussed in this chapter an advantage over conventional finite element methods.

5.1 Basis Function

The governing differential equation of a Euler–Bernoulli rotating beam is given by [9].

$$\frac{\partial^2}{\partial x^2}(EI(x)\frac{\partial^2 w}{\partial x^2}) - \frac{\partial}{\partial x}(T(x)\frac{\partial w}{\partial x}) + m(x)\frac{\partial^2 w}{\partial t^2} = 0 \tag{5.1}$$

where $EI(x)$ is the flexural stiffness, $T(x)$ is the axial force due to centrifugal stiffening, $m(x)$ is the mass per unit length of the beam, w is the bending displacement and Ω is the rotation speed.

The axial force due to centrifugal stiffening, $T(x)$ is given as

$$T(x) = \int_x^L m(x)\Omega^2(R + x)\,dx + F \tag{5.2}$$

where R is the hub radius, L is the length of the beam and F is the axial force acting at the end of the beam. The static homogenous part of the governing differential equation for a rotating beam is obtained by ignoring the inertia term in Eq. (5.1) and is given by

$$\frac{d^2}{dx^2}\left(EI(x)\frac{d^2 w}{dx^2}\right) - \frac{d}{dx}\left(T(x)\frac{dw}{dx}\right) = 0 \tag{5.3}$$

Equation (5.3) does not have an exact closed form solution. In order to find an approximate solution, we assume $EI(x)$ and $m(x)$ to be constant. Furthermore, using $R = F = 0$, Eq. (5.3) can be simplified to

$$EI\frac{d^4 w}{dx^4} - \frac{m\Omega^2}{2}\frac{d}{dx}\left((L^2 - x^2)\frac{dw}{dx}\right) = 0 \tag{5.4}$$

The equation can be further simplified by introducing a dimensionless parameter $\lambda^2 = \frac{m\Omega^2 L^4}{EI}$

$$\frac{d^4 w}{dx^4} - \frac{\lambda^2}{2L^4}\frac{d}{dx}\left((L^2 - x^2)\frac{dw}{dx}\right) = 0 \tag{5.5}$$

We try to find an approximate solution to the static part of the governing differential equation by splitting Eq. (5.5) into two parts; the centrifugal part and the flexural part. For the equation to have an exact solution, we assume that a function w exists such that both the parts are satisfied i.e. both the operators must go to zero.

$$\frac{d^4 w}{dx^4} = 0 \tag{5.6}$$

$$\frac{d}{dx}\left((L^2 - x^2)\frac{dw}{dx}\right) = 0 \tag{5.7}$$

The cubic function satisfies Eq. (5.6)

$$\frac{d^4 w}{dx^4} = 0 \;\Rightarrow\; w^{(1)} = a_0 + a_1 x + a_2 x^2 + a_3 x^3 \tag{5.8}$$

Equation (5.8) represents the case of a non-rotating beam.

If we try to satisfy the centrifugal part given by Eq. (5.7) we get

$$\frac{d}{dx}\left((L^2 - x^2)\frac{dw}{dx}\right) = 0 \;\Rightarrow\; w^{(2)} = a \ln\left(\frac{L+x}{L-x}\right) \tag{5.9}$$

Equation (5.9) represents the case of a rotating string with no stiffness [8]. As the rotation speed of the beam increases, the centrifugal effects overwhelm the flexural stiffness and the rotating beam approaches the rotating string. It is reasonable to assume that a rotating beam solution should satisfy the extreme cases where rotation speed is zero and rotation speed is very high. Therefore, we consider a linear combination of Eqs. (5.8) and (5.9) as a solution to Eq. (5.3)

$$W_a = a_0 + a_1 x + a_2 x^2 + a_3 x^3 + a_4 \ln\left(\frac{L+x}{L-x}\right) \tag{5.10}$$

Any fourth order differential equation can have only four constants which can be found using the boundary conditions. Equation (5.10) contains an extra constant a_4; hence we eliminate a_4 from the solution by using an idea from the collocation method. The collocation method is an approach for solving differential equations by assuming a solution and finding the values of constants in the solution by forcing the differential equation to be satisfied at several 'collocation' points in the domain. The residual arising in the differential equation by assuming Eq. (5.10) to be a solution to Eq. (5.3) is given by

$$\varepsilon = EI\frac{d^4 W_a}{dx^4} - \frac{m\Omega^2}{2}\frac{d}{dx}\left(\left(R\,(L-x) + (L^2 - x^2)\right)\frac{dW_a}{dx}\right) \tag{5.11}$$

We set the error to zero at some points z within the domain

$$\varepsilon(z) = 0 \Rightarrow W_a = a_0 + a_1 x + a_2 x^2 + a_3 x^3 + f(a_0, a_1, a_2, a_3, z) \ln \left(\frac{L + x}{L - x} \right) \tag{5.12}$$

where z represents points within the domain $(0, L)$. We have considered the open interval because the natural and essential boundary conditions are imposed at the ends of the beam.

Until now, we have discussed on obtaining and approximate solution to the static part of the governing differential equation for a rotating beam. In order to use this solution as a basis function for the finite element method, the static part of the governing differential equation for the ith element is given by

$$\frac{d^2}{d\overline{x}^2} \left(EI(x_i + \overline{x}) \frac{d^2 w_i}{d\overline{x}^2} \right) - \frac{d}{d\overline{x}} \left(T(x_i + \overline{x}) \frac{d w_i}{d\overline{x}} \right) = 0 \tag{5.13}$$

We assume a solution of the form,

$$w = a_0 + a_1(x_i + \overline{x}) + a_2(x_i + \overline{x})^2 + a_3(x_i + \overline{x})^3 + a_4 \ln \left(\frac{L + x_i + \overline{x}}{L - x_i - \overline{x}} \right) \tag{5.14}$$

Equation (5.14) can be further simplified as

$$w = c_0 + c_1 \overline{x} + c_2 \overline{x}^2 + c_3 \overline{x}^3 + c_4 \ln \left(\frac{L + x_i + \overline{x}}{L - x_i - \overline{x}} \right) \tag{5.15}$$

We recall that the mass and flexural variation was assumed to be a constant in obtaining the solution given by Eq. (5.15). This assumption is validated by the fact that in the limit of mesh refinement the mass and flexural variation can be neglected. Hence, even though the solution was obtained for a uniform beam, it is used only for the finite element and is valid for all classes of tapered beams.

When we choose Eq. (5.15) as a basis function, it requires substitution at $x = l$ which will give rise to a discontinuity at the last element. Now, in order to simplify the finite element calculations, we approximate the equation as

$$w = c_0 + c_1 \overline{x} + c_2 \overline{x}^2 + c_3 \overline{x}^3 + c_4 \ln \left(\frac{L + \overline{x}}{L - \overline{x}} \right) \tag{5.16}$$

We can presume that the above approximation to the basis function will be negated by choosing a finer discretization of our beam, which is typical of finite element methods. Consider a two-noded four degree of freedom element shown in Fig. 5.1. For the sake of brevity, we give the solution for $R = 0$. The boundary conditions for the element are given by

$$w = c_0 + c_1 \overline{x} + c_2 \overline{x}^2 + c_3 \overline{x}^3 + c_4 \ln (\frac{L + \overline{x}}{L - \overline{x}}) \tag{5.17}$$

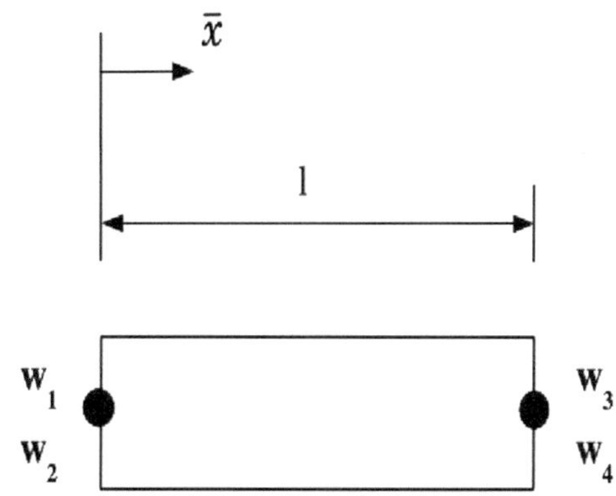

Fig. 5.1 Finite element model

$$w(0) = c_0 = w_1 \tag{5.18}$$

$$\frac{dw(0)}{d\overline{x}} = \frac{2\,c_4}{L} = w_2 \tag{5.19}$$

$$w(l) = c_0 + c_1\,l + c_2\,l^2 + c_3\,l^3 + c_4\,\ln(\frac{L+l}{L-l}) = w_3 \tag{5.20}$$

$$\frac{dw(l)}{d\overline{x}} = c_1 + 2c_2 l + 3c_3 L^2 + \frac{2\,c_4\,L}{(L^2 - l^2)} = w_4 \tag{5.21}$$

where

$$c_4 = \frac{(L^2 - z^2)^4\,\lambda^2\,(L^2 c_2 + (3c_3 L^2 - c_1)z - 3c_2 z^2 - 6c_3 z^3)}{48L^5 z(L^2 - z^2)} \tag{5.22}$$

Equation (5.22) is obtained by solving Eq. (5.12).

Since the basis function defined by Eq. (5.17) contains polynomial terms upto order three, it can represent a rigid body mode as well as a state of constant strain. Hence the above basis function will give rise to a complete interpolation polynomial [2].

We set the error to zero at a collocation point within the element. We choose $\overline{x} = l/2$ within each element so that for any general element this point will be $z = x_i + l/2$. Since we assume an uniform mesh $x_i = (i-1)l$, the zero error condition holds within each element(i) at

$$z = (i - 1/2)l \tag{5.23}$$

From Eqs. (5.22) and (5.23), the constants c_0, c_1, c_2, c_3 can be solved and the displacement function can be written in terms of the displacements and slopes at the nodes.

$$w = N_1 w_1 + N_2 w_2 + N_3 w_3 + N_4 w_4 \tag{5.24}$$

where N_1, N_2, N_3, N_4 are the interpolating functions and are given as

$$N_1 = \frac{R_1}{D},\ N_2 = \frac{R_2}{D},\ N_3 = \frac{R_3}{D},\ N_4 = \frac{R_4}{D}$$

where

$$\begin{aligned} R_1 =& l(12\,L\lambda^2(L^2 - z^2)^4(L^2 - l^2)(2L^2z - L^2l - 4z^3 + 3lz^2)(\ln(\frac{L+\overline{x}}{L-\overline{x}}) \\ &- \ln(\frac{L+l}{L-l})) - (l-\overline{x})(12\lambda^2(2L^2 - l(l-\overline{x}))z^{11} + l\lambda^2z^8(3L^2 \\ &- (2l+\overline{x})(l-\overline{x}))(26L^2 - 6z^2) - 2\lambda^2z^9(54L^4 - 2L^2(14l+\overline{x})(l-\overline{x}) \\ &+ (l+\overline{x})(l-\overline{x})l^2) + 8L^2\lambda^2z^7(24L^4 - (l-\overline{x})(13l+\overline{x})L^2 \\ &+ (l+\overline{x})(l-\overline{x})l^2) - 44L^4l(3L^2 - (2l+\overline{x})(l-\overline{x})) \\ &\lambda^2z^6 - 12L^4\lambda^2z^5(l^2(l+\overline{x})(l-\overline{x}) + 14L^4 - (4l+\overline{x}) \\ &(l-\overline{x})L^2 + 2(6+\lambda^2)l^2(l+\overline{x})(l-\overline{x})) + 14L^8l\lambda^2z^2 \\ &((2l+\overline{x})(l-\overline{x})) - 3L^2) - 2L^8z(6\lambda^2L^4 - 2(l-\overline{x}) \\ &(l-\overline{x})L^2) - 36L^6l\lambda^2z^4((2l+\overline{x})(l-\overline{x}) - 3L^2) + 4L^6z^3 \\ &(18\lambda^2L^4 - ((11l+4\overline{x})\lambda^2 + 24\overline{x} + 12l) \\ &((2l+\overline{x})\lambda^2 - 24\overline{x} - 12l)L^2 + l^2(l+\overline{x})(l-\overline{x}) \\ &\quad(-24+\lambda^2)) + 2L^{10}l\lambda^2(3L^2 - (\overline{x}+2l)(l-\overline{x})))) \end{aligned} \tag{5.25}$$

$$\begin{aligned} R_2 =& L((L^2 - z^2)^4\lambda^2(L^2 - l^2)\overline{x}((12l - 6\overline{x})z^3 - (9l^2 - 3\overline{x}^2) \\ &z^2 - ((6l - 3\overline{x})L^2 + 2l\overline{x}^2 - 3\overline{x}l^2)z + L^2(3l^2 - \overline{x}^2)) \\ &\ln(\frac{L+l}{L-l}) - l((L^2 - z^2)^4(6z^3 - 6lz^2 + (-3L^2 + l^2)z + L^2l)l\lambda^2 \\ &(L^2 - l^2)\ln(\frac{L+\overline{x}}{L-\overline{x}}) - (6\lambda^2z^{11} - 3\lambda^2(l+\overline{x})z^{10} \\ &- \lambda^2(27L^2 - \overline{x}l)z^9 + 13L^2\lambda^2(l+\overline{x})z^8 + 4L^2\lambda^2(12L^2 \\ &- \overline{x}l)z^7 + 22\lambda^2L^4(l+\overline{x})z^6 + 6\lambda^2L^4(7L^2 - \overline{x}l)z^5 \\ &- 18L^6\lambda^2(l+\overline{x})z^4 - 2L^6z(9\lambda^2L^4 - 2l(6l - 6\overline{x} + \overline{x} \\ &\lambda^2)L^2 + 12l^3(l-\overline{x}))L^4z^3 + 7L^8\lambda^2(l+\overline{x})z^2 + (3\lambda^2L^4 \\ &\quad - l(24l - \overline{x}\lambda^2 - 24\overline{x})L^2 + 24l^3(l-\overline{x})) - L^{10}\lambda^2(l+\overline{x}))2L\overline{x}(l-\overline{x}))) \end{aligned} \tag{5.26}$$

$$
\begin{aligned}
R_3 =\, & l(L\lambda^2(L^2-l^2)(L^2-z^2)^4\ln(\frac{L+\overline{x}}{L-\overline{x}})(12z^3-9lz^2+3L^2l-6L^2z)\\
&-2\overline{x}(-6\lambda^2(2L^2-(2l-\overline{x})l)z^{11}+3\lambda^2lz^{10}(3L^2\\
&z^{11}+3\lambda^2lz^{10}(3L^2-3l^2+\overline{x}^2)+\lambda^2(54L^4+(2\overline{x}^2\\
&-54l^2+24l\overline{x})L^2+\overline{x}l^2(3l-2\overline{x}))z^9+L^2\lambda^2\\
&lz^6(3L^2-l^2+\overline{x}^2)(11L^2-13z^2)+\overline{x}l^2(3l-2\overline{x})\\
&+4L^2\lambda^2(24L^4+(2\overline{x}^2+9l\overline{x}-24l^2)L^2)z^7\\
&+6L^4\lambda^2z^5(14L^4+(4l\overline{x}+2\overline{x}^2-14l^2)L^2\\
&+\overline{x}l^2(3l-2\overline{x}))+18L^6\lambda^2lz^4(\overline{x}^2-3l^2+3L^2)\\
&+2L^6z^3(18\lambda^2L^4+((4\overline{x}^2-18l^2+3l\overline{x})\\
&\lambda^2+24\overline{x}^2-36l\overline{x})L^2+2\overline{x}l^2(6+\lambda^2)(3l-2\overline{x}))\\
&+7L^8(\overline{x}^2-3l^2+3L^2)\lambda^2lz^2-L^8z(6\lambda^2L^4\\
&+((2\overline{x}^2-6l^2)\lambda^2-48\overline{x}^2+72l\overline{x})L^2+\overline{x}l^2\\
&(\lambda^2-24)(3l-2\overline{x}))-\lambda^2L^{10}l(\overline{x}^2-3l^2+3L^2))
\end{aligned}
\tag{5.27}
$$

$$
\begin{aligned}
R_4 =\, & (L^2-l^2)(L\lambda^2\overline{x}^2((-3z+\overline{x})L^2+(2z-\overline{x})3z^2)\\
&(L^2-z^2)^4\ln(\frac{L+l}{L-l})+l(L\lambda^2l((3z-l)L^2-3z^2\\
&(2z-l))(L^2-z^2)^4\ln(\frac{L+\overline{x}}{L-\overline{x}})-2(\lambda^2L^{10}(3z-l-\overline{x})\\
&-zL^8(\lambda^2(18z^2-7(l+\overline{x})z-l\overline{x})-24l\overline{x})\\
&+2((21z^2-9(l+\overline{x})z+2l\overline{x})\lambda^2+12l\overline{x})z^3L^6-2\lambda^2\\
&z^5L^4(24z^2-11(l+\overline{x})z+3l\overline{x})+\lambda^2z^7L^2(27z^2\\
&+4l\overline{x}-13(l+\overline{x})z)-\lambda^2(6z^2-3(l+\overline{x})z+l\overline{x})z^9)\overline{x}(l-\overline{x})))
\end{aligned}
\tag{5.28}
$$

$$
\begin{aligned}
D =\, & l((L^2-z^2)^4\lambda^2L((l-2z)L^2+(4z-3l)z^2)(L^2\\
&-l^2)\ln(\frac{L+l}{L-l})-2(3\lambda^2(l-2z)L^{12}+L^{10}\\
&((4l^2z-21lz^2-2l^3+36z^3)\lambda^2-24l^2z)-zL^8\\
&(\lambda^2(l^4+84z^4+22l^2z^2-14l^3z-54lz^3)\\
&-24l^4+24l^2z^2)+2z^3L^6(\lambda^2(l^4+24z^4\\
&+12l^2z^2-33lz^3-18l^3z)+6l^4)-\lambda^2z^5L^4(-44l^3z\\
&+54z^4+6l^4+52l^2z^2-39lz^3)-\lambda^2(-9lz^3+12z^4\\
&+4l^4-26l^3z+28l^2z^2)z^7L^2-l^2\lambda^2z^9(l^2-6zl+6z^2))l)
\end{aligned}
\tag{5.29}
$$

In the above equations $z = (i - 1/2)l, i = 1 \ldots n_e$ represents the position of the element in the beam and $L = ln_e$.

The interpolating functions obtained here are a function of the material properties (EI, m), geometric properties (L) and rotational speed (Ω) of the beam. They also depend on the position of the element in the beam (x_i). Though the new basis functions appear quite complicated, the behavior of the shape functions is smooth and continuous in the beam which can be observed by plotting them. Figures 5.2 and 5.3 show comparison of the shape functions of the first, middle and the last element using the hybrid basis function with the Hermite cubic shape functions at low and high rotation speeds, respectively, for a 10 element discretization ($n_e = 10$). At $\lambda = 12$, the shape functions are almost identical to the Hermite cubics showing that the effect of rotation is very small at low rotation speeds. The shape functions show increasing difference at the higher speed ($\lambda = 100$) and particularly near the root of the beam where the centrifugal effect is more (Eq. (5.2)). However, the effect of rotation is felt in the slope related shape functions N_2 and N_4 only and the displacement related functions show little change. The interpolating functions approach the Hermite cubic functions as the rotation speed tends to zero. The shape functions for the last element are almost similar to the Hermite cubic shape functions because the centrifugal force approaches zero. It is verified mathematically that

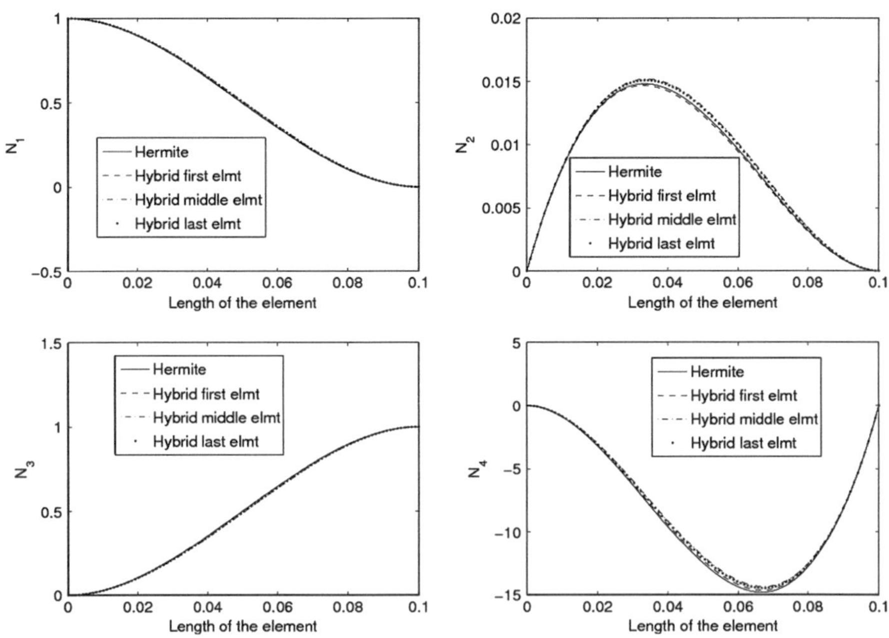

Fig. 5.2 Comparison of shape function values for $\lambda = 12$ for different elements in the beam

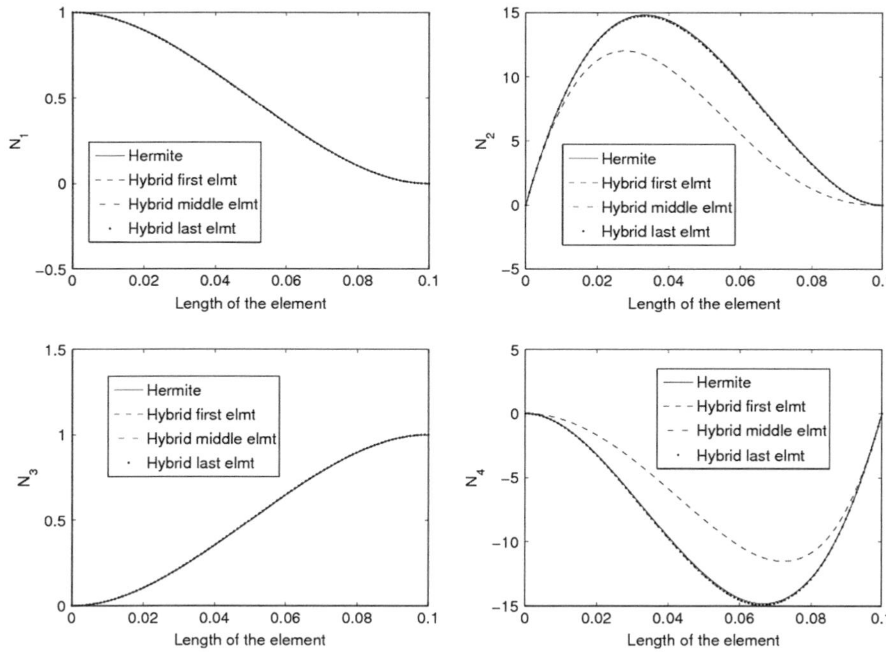

Fig. 5.3 Comparison of shape function values for $\lambda = 100$ for different elements in the beam

$$\lim_{\lambda \to 0} N_1 = \frac{l^3 - 3\overline{x}^2 l + 2\overline{x}^3}{l^3}$$

$$\lim_{\lambda \to 0} N_2 = \frac{\overline{x} l^2 - 2\overline{x}^2 l + \overline{x}^3}{l^2}$$

$$\lim_{\lambda \to 0} N_3 = \frac{3\overline{x}^2 l - 2\overline{x}^3}{l^3}$$

$$\lim_{\lambda \to 0} N_4 = \frac{-\overline{x}^2 l + \overline{x}^3}{l^2}$$

5.2 Finite Element Analysis

The mass and stiffness matrices can be obtained using the energy expressions. The kinetic energy for a rotating beam is given by Reddy [10]

$$T = \frac{1}{2} \int_0^L m(x)[\dot{w}(x,t)]^2 dx \tag{5.30}$$

where $\dot{w}(x,t)$ is the derivative of $w(x,t)$ with respect to time t.

The potential/strain energy is given by

$$U = \frac{1}{2}\int_0^L EI(x)[w''(x,t)]^2 dx + \frac{1}{2}\int_0^L T(x)[w'(x,t)]^2 dx \tag{5.31}$$

where $T(x)$ is defined in Eq. (5.2).

The mass and stiffness matrices (M_i and K_i) for a beam element can be obtained from the above energy expressions. The calculations for these matrices involve calculating the value of the following integrals.

$$M_i = \int_0^l m_i(\overline{x}) N^T N d\overline{x} \tag{5.32}$$

$$K_i = \int_0^l EI_i(\overline{x})(N'')^T N'' d\overline{x} + \int_0^l T_i(\overline{x})(N')^T N' d\overline{x} \tag{5.33}$$

where

$$T_i(\overline{x}) = \int_{x_i+\overline{x}}^L m_i(\overline{x})\Omega^2(\overline{x})\, d\overline{x} \tag{5.34}$$

No approximation is made in the variation of $EI(x)$ or $T(x)$ while calculating the mass and stiffness matrices. The assumptions have been made only to derive the new basis functions. The mass and stiffness matrices are calculated for each element and then assembled into the global mass and stiffness matrix. These matrices are then used to determine the natural frequencies and the mode shapes by solving the eigenvalue problem.

$$K\Phi = \omega^2 M\Phi \tag{5.35}$$

5.3 Numerical Results

The new hybrid basis function is used in the free vibration analysis of rotating uniform and tapered beams with different boundary conditions and the results obtained are compared with published literature.

5.3.1 *Uniform Beam*

In uniform rotating beams, $m(x)$ and $EI(x)$ are constant. Tables 5.1 and 5.2 show comparison of non-dimensional rotating frequencies of an uniform rotating cantilever

Table 5.1 Comparison of non-dimensional frequencies for cantilevered uniform rotating beam

Mode	Present FEM	Wang and Wereley [11]	Wright et al. [12]	Hodges and Rutkowski [9]
1	13.1702	13.1702	13.1702	13.1702
2	37.6031	37.6031	37.6031	37.6031
3	79.6145	79.6145	79.6145	79.6145
4	140.534	140.534	140.534	N/A
5	220.536	220.536	220.536	N/A

Table 5.2 Comparison of non-dimensional frequencies for hinged uniform rotating beam

Mode	Present FEM	Wang and Wereley [11]	Wright et al. [12]
1	12.0002	12.0000	1 2.0000
2	33.7603	33.7603	33.7603
3	70.8373	70.8373	70.8373
4	126.431	126.431	126.431
5	201.123	201.122	201.122

Table 5.3 Comparison of number of elements and degrees of freedom required for convergence of natural frequencies with accuracy of 1e-4 for cantilevered uniform rotating beam

	$\lambda = 12$		$\lambda = 100$	
Modes	Hybrid function	Cubic function	Hybrid function	Cubic function
1	9(18)	12(24)	33(66)	58(116)
2	15(30)	16(32)	48(96)	69(138)
3	24(48)	24(48)	54(108)	77(154)
4	35(70)	35(70)	60(120)	84(168)
5	47(94)	47(94)	61(122)	91(182)

and hinged beam with published literature [9, 11, 12]. Hodges and Rutkowski [9] used a variable order basis function in the finite element method while Wang and Wereley [11] and Wright et al. [12] used the Frobenius solution in the dynamic stiffness method to generate the natural frequencies of the rotating beam. We can see from Tables 5.1 and 5.2 that the current formulation gives identical results to the published literature.

Table 5.3 shows the comparison of the number of elements required for convergence of the first five modes upto four decimal places. The degrees of freedom is also specified in parentheses. The convergence of the formulation is compared with that of the Hermite cubic shape functions for two rotation speeds $\lambda = 12$ and $\lambda = 100$. It is seen that in the case of the cantilever beam, the hybrid function requires 47 elements for $\lambda = 12$ and 61 elements for $\lambda = 100$ to obtain converged values for

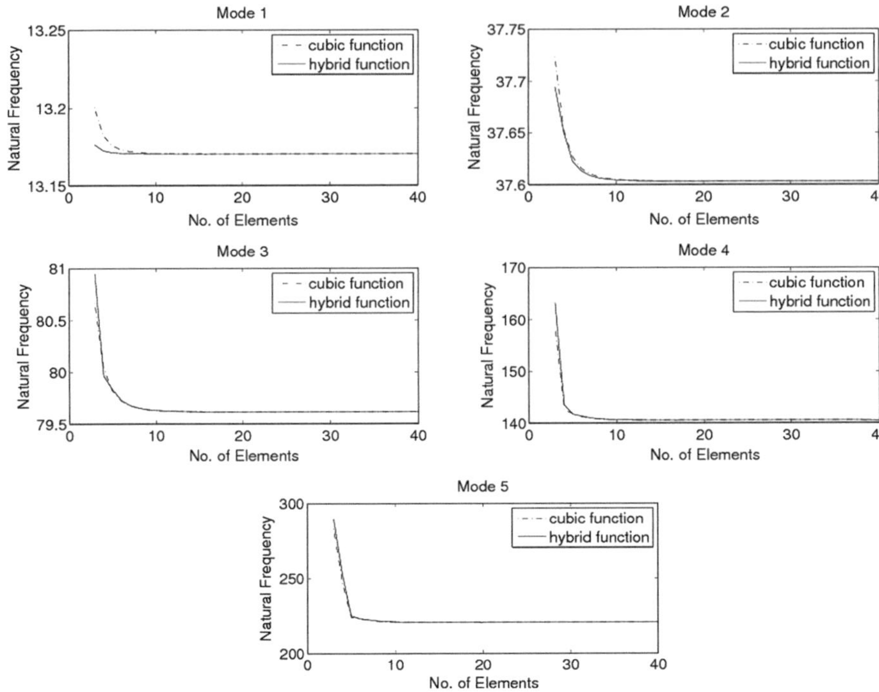

Fig. 5.4 Comparison of convergence of hybrid function and hermite cubic function for uniform cantilever beam with $\lambda = 12$

the first five modes. For $\lambda = 12$, the Hermite cubic functions also require only 47 elements but for $\lambda = 100$ it requires 91 of elements for converged values of the first five modes. We can see that the hybrid basis functions give some advantage in the fundamental mode convergence at $\lambda = 12$, however, this advantage becomes clear at $\lambda = 100$ for all the five modes. Figures 5.4 and 5.5 show the convergence of the natural frequencies with the hybrid and the cubic basis functions, at $\lambda = 12$ and $\lambda = 100$, respectively. In fact, with the new basis functions, a reduction of about 33 % is obtained at $\lambda = 100$ in the number of elements required for convergence when compared to the cubic basis functions. The number of degrees of freedom is reduced from 182 to 122 at $\lambda = 100$, which considerably reduces the eigenvalue problem size and makes the new basis functions attractive for the vibration analysis of high speed rotating beams.

The accuracy of the current hybrid function is observed by considering the difference in the value of natural frequency obtained for a very coarse discretization of the beam with the converged value of the natural frequency. This difference is a measure of the error. Table 5.4 shows a comparison of the difference in values of the non-dimensional natural frequency obtained by using the hybrid and the cubic function for a coarse discretization. At least a three element discretization is necessary to obtain the natural frequency for the first five modes. With a three element

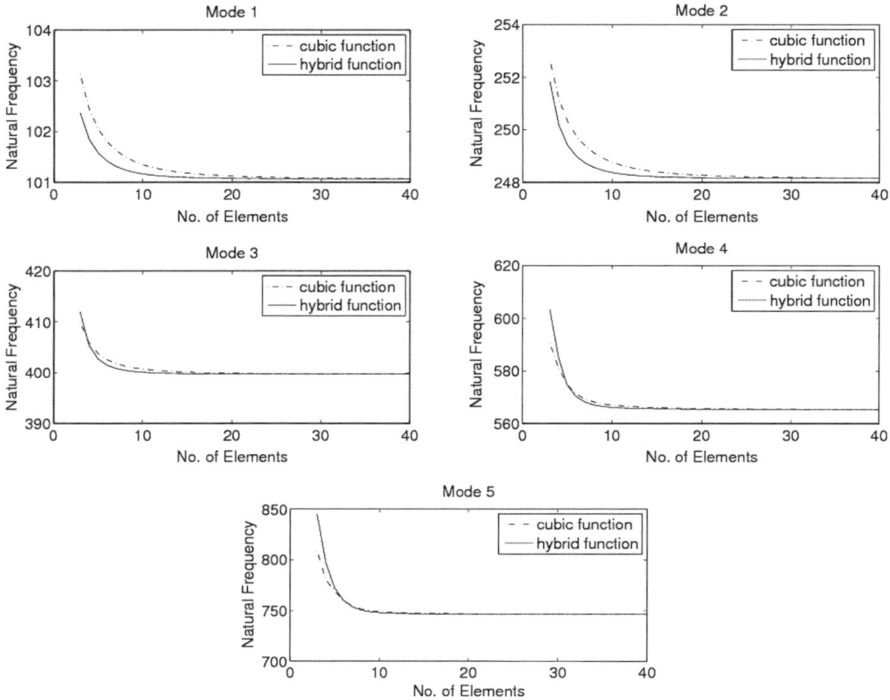

Fig. 5.5 Comparison of convergence of hybrid function and hermite cubic function for uniform cantilever beam with $\lambda = 100$

Table 5.4 Comparison of difference in the value of non-dimensional natural frequency obtained by using the hybrid and the cubic function for a low discretization in the FEM with the converged value for $\lambda = 100$

Mode	$n_e = 3$		$n_e = 7$		$n_e = 10$		$n_e = 20$	
	Hybrid function	Cubic function	Hybrid function	Cubic function	Hybrid function	Cubic function	Hybrid function	Cubic function
1	1.2961	2.0705	0.2417	0.5537	0.0941	0.2674	0.0088	0.0506
2	3.6876	4.5076	0.5615	1.2076	0.2125	0.6036	0.0193	0.1107
3	12.3151	10.2412	1.1423	2.1143	0.3928	1.0291	0.0355	0.1898
4	38.0513	25.3037	2.6878	3.8457	0.7370	1.6412	0.0669	0.3018
5	98.7069	62.4297	7.1107	7.9247	1.4760	2.5897	0.1298	0.4559

discretization, it is observed that the difference in the value obtained and the converged value is less for the first two modes when we use the current hybrid function. When the beam is discretized into seven elements, the current basis function shows better accuracy for all the five modes when compared to the cubic basis function. Thus we can construct low order models of the rotating beam which are often useful in control applications, using the new basis functions. Wang and Wereley [11] men-

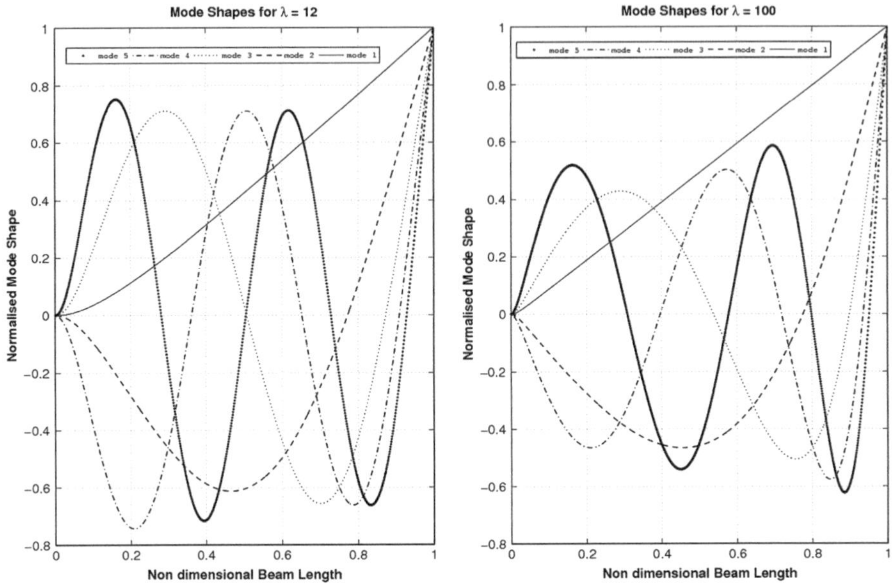

Fig. 5.6 Eigen mode shapes with the hybrid basis function for uniform cantilever beam with **a** $\lambda = 12$ **b** $\lambda = 100$

tion that the creation of low order models is critical for control of rotating beams when smart materials based actuation is used. For the 10 and 20 element cases, the new hybrid shape functions also perform better than the Hermite cubic's for all the five modes.

Figure 5.6 shows the Eigen Mode shapes obtained with the current hybrid function for $\lambda = 12$ and $\lambda = 100$. The mode shapes obtained by using the Hermite cubic shape functions is shown in Fig. 5.7. It is seen that the hybrid function is able to capture the mode shapes accurately especially for the fundamental mode which becomes linear as the rotation speed becomes considerably high. In such situations, the centrifugal force dominates the whole response of the system thereby making it behave similar to rotating string.

5.3.2 *Tapered Beam*

For a tapered beam, we assume that the variation of mass along the beam is given by

$$m = m_0(1 - \alpha\xi) \tag{5.36}$$

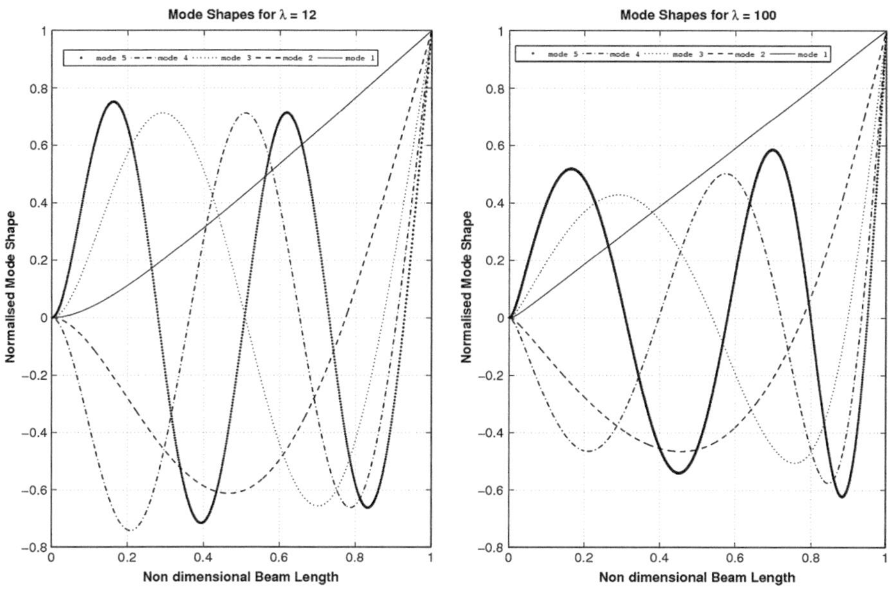

Fig. 5.7 Eigen mode shapes with the hermite cubic basis function for uniform cantilever beam with **a** $\lambda = 12$ **b** $\lambda = 100$

where m_0 corresponds to the value of mass per unit length at the thick end of the beam ($\xi = 0$), α is the taper parameter such that $0 < \alpha < 1$, ξ is the non-dimensional length of the beam: $\xi = x/L$.

The flexural stiffness variation along the tapered beam is given by

$$EI = EI_0(1 - \beta_1\xi - \beta_2\xi^2 - \beta_3\xi^3 - \beta_4\xi^4) \tag{5.37}$$

where EI_0 corresponds to the value of flexural rigidity at the thick end of the beam ($\xi = 0$), $\beta_i,\ i = 1\ldots4$ are taper parameters for the stiffness distribution. These parameters can be determined by using α for beams with a rectangular cross section and thickness varying along the length. However, it is not necessary that the taper and mass parameters be related and the only condition is that they should not give rise to a singularity at ($\xi = 1$). In order to validate the current formulation we consider two special cases of Eqs. 5.35 and 5.36 and verify the results with published literature.

Linear Mass and Linear Stiffness Beam

We assume $m(x) = m_0(1 - 0.8x)$ and $EI(x) = EI_0(1 - 0.95x)$. Tables 5.5 and 5.6 show comparison of non-dimensional rotating frequencies of a cantilever and hinged beam with published literature [11, 12]. The results compare very well with published literature.

Table 5.5 Comparison of non-dimensional frequencies for linear mass, linear stiffness cantilevered rotating beam

Mode	Present FEM	Wang and Wereley [11]	Wright et al. [12]
1	14.0313	14.0313	14.0313
2	35.9064	35.9060	35.9064
3	72.8565	72.8565	72.8565
4	126.401	126.336	126.401
5	198.880	198.243	198.880

Table 5.6 Comparison of non-dimensional frequencies for linear mass, linear stiffness hinged rotating beam

Mode	Present FEM	Wang and Wereley [11]	Wright et al. [12]
1	12.0000	12.0000	12.0000
2	30.7745	30.7741	30.7745
3	63.1722	63.1758	63.1722
4	112.090	112.040	112.090
5	178.016	178.978	178.105

Figures 5.8 and 5.9 show the comparison of the convergence rate of the current shape functions with the Hermite Cubic Shape functions. For the low speed case ($\lambda = 12$) the current basis function do not show improvement over the cubic polynomial. But in the higher speed case ($\lambda = 100$), the new function shows improved convergence.

Linear Mass and Cubic Stiffness Beam

We assume $m(x) = m_0(1 - 0.5x)$ and $EI(x) = EI_0(1 - 0.95x)^3$. Table 5.7 shows comparison of non-dimensional rotating frequencies of a cantilever beam with results from [11, 12]. Values for the first three modes are only available in published literature. Again we see that the results compare very well.

Figure 5.10 shows the comparison of the convergence rate of the current shape functions with the Hermite Cubic Shape functions. For the low speed case ($\lambda = 12$) the current basis function do not show an advantage compared to the cubic polynomial. However, in the higher speed case ($\lambda = 100$), the convergence is significantly improved.

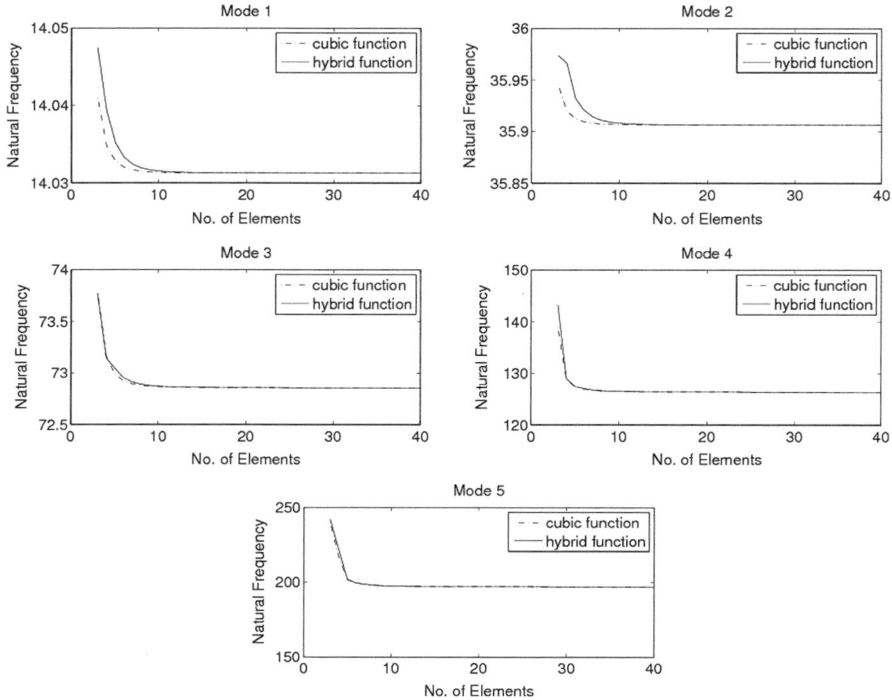

Fig. 5.8 Comparison of convergence of hybrid function and hermite cubic function for linearly tapered, linear flexural stiffness cantilever beam with $\lambda = 12$

5.3.3 Beams with Hub Offset

The new basis functions perform well for the test cases that have been used in published literature. Another advantage of the new basis function is that it is dependent of the hub offset. Even though the shape functions have been given for $R = 0$, hub offset can be included in the basis function. We expect the new basis function to perform better than the Hermite cubic function as hub offset increases the effect of the centrifugal force on the stiffness of the beam. Tables 5.8 and 5.9 show the natural frequencies for a cantilever rotating beam with linear mass and linear stiffness variation. Two different hub offset ratios R/L have been considered and the results have been validated with results from Wang and Wereley [11], Wright et al. wr24. It is seen that the values for the first three modes compare very well.

Figures 5.11, 5.12 and 5.13 show the convergence trend observed by using the new basis function to determine the natural frequency of a high speed rotating beam with hub offset. The convergence trends are shown for the first three modes of the rotating beam for two hub offset ratios $R/L = 1$ and 5. Furthermore, it is seen that, in the presence of hub offset, the new basis function shows a convergence trend which

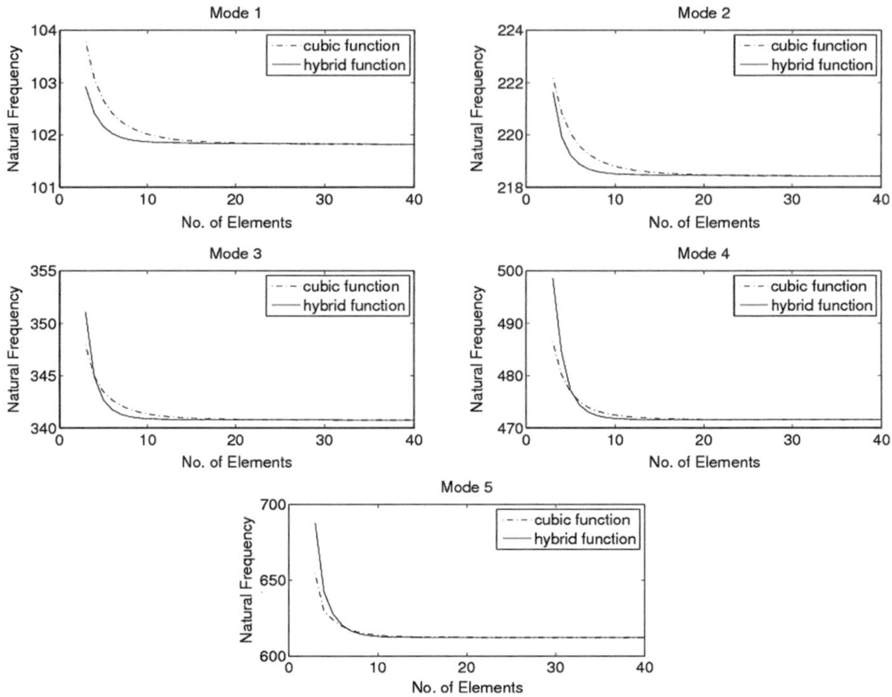

Fig. 5.9 Comparison of convergence of hybrid function and hermite cubic function for linearly tapered, linear flexural stiffness cantilever beam with $\lambda = 100$

Table 5.7 Comparison of non-dimensional frequencies for linear mass, cubic stiffness cantilevered rotating beam

Mode	Present FEM	Wang and Wereley [11]	Hodges and Rutkowsky [9]
1	13.4711	13.4711	13.4711
2	34.0877	34.0877	34.0877
3	65.5237	65.5237	65.5237

is almost independent of the geometric parameters of the rotating beam. This shows that there is a significant advantage in using the current basis function for high speed rotating beams.

5.4 Summary

New basis functions are created to solve the free vibration problem for rotating beams and improve the accuracy and convergence of the finite element method. The problem physics of rotating beams requires that the centrifugal stiffening effects in

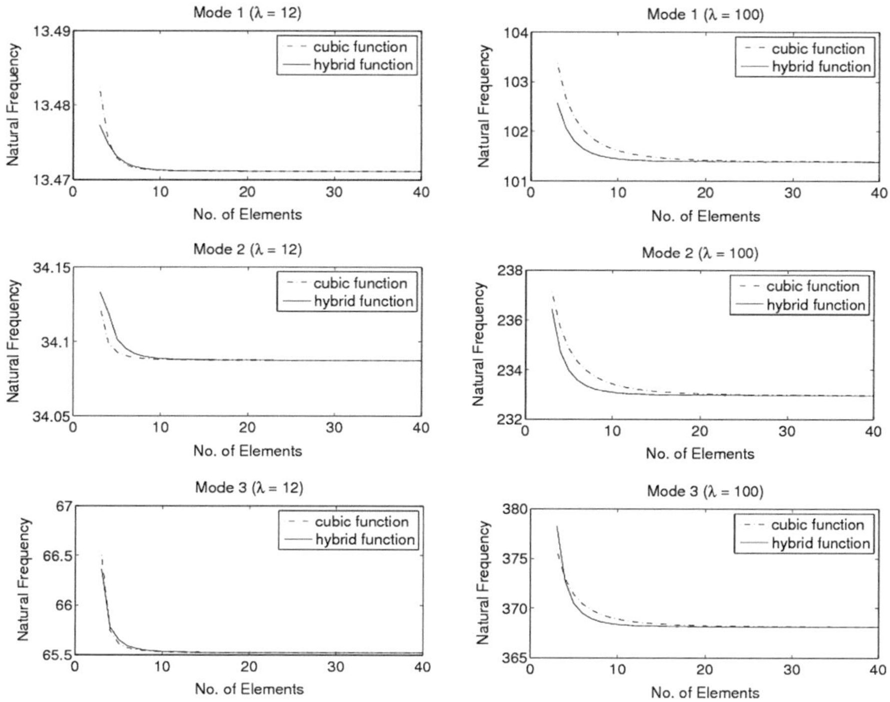

Fig. 5.10 Comparison of convergence of hybrid function and hermite cubic function for linearly tapered, cubic flexural stiffness cantilever beam with $\lambda = 12$ and $\lambda = 100$

Table 5.8 Comparison of non-dimensional frequencies for linear mass, linear stiffness cantilevered rotating beam for $\lambda = 5$ and $R/L = 1$

Mode	Present FEM	Wang and Wereley [11]	Wright et al. [12]
1	10.083	10.083	10.083
2	29.534	29.534	29.535
3	65.764	65.765	65.765

Table 5.9 Comparison of non-dimensional frequencies for linear mass, linear stiffness cantilevered rotating beam for $\lambda = 5$ and $R/L = 5$

Mode	Present FEM	Wang and Wereley [11]	Wright et al. [12]
1	16.512	16.512	16.512
2	39.412	39.411	39.413
3	77.602	77.610	77.602

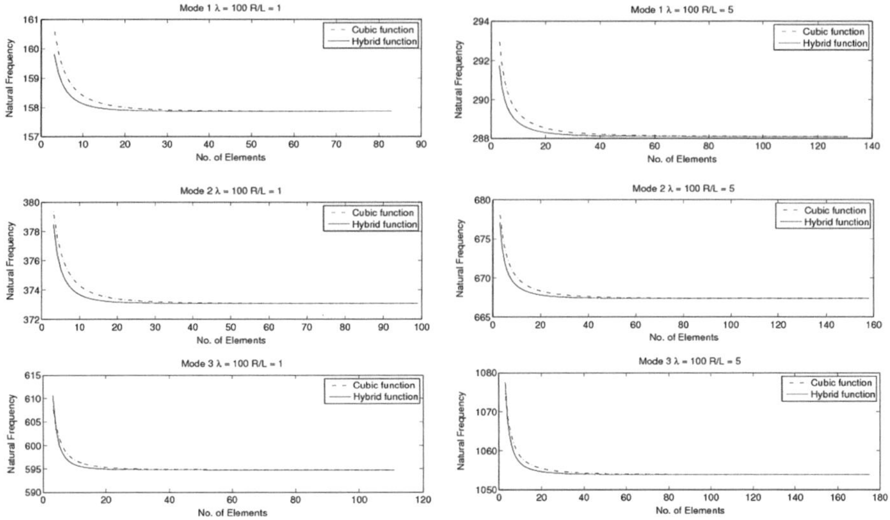

Fig. 5.11 Comparison of convergence of hybrid function and hermite cubic function for uniform cantilever beam with $\lambda = 100$ and $R/L = 1$ and 5

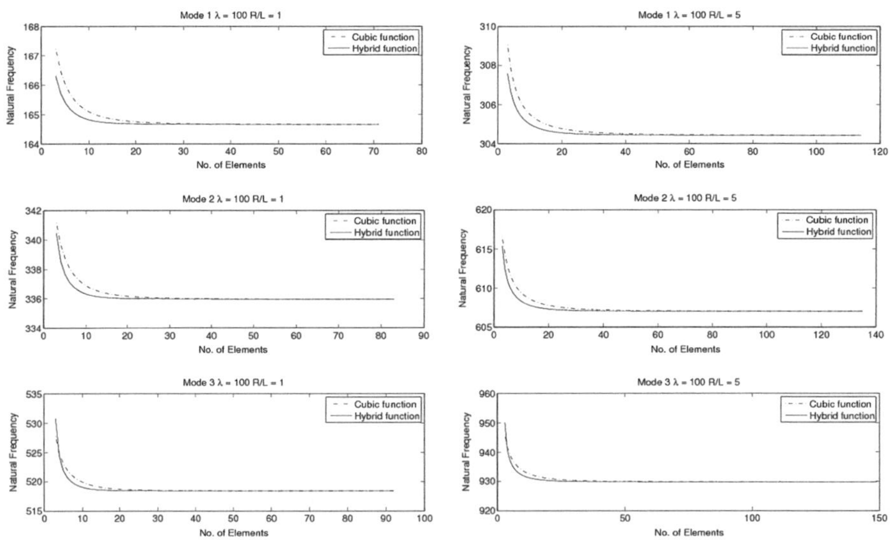

Fig. 5.12 Comparison of convergence of hybrid function and hermite cubic function for linearly tapered, linear flexural stiffness cantilever beam with $\lambda = 100$ and $R/L = 1$ and 5

the governing differential equation be properly captured by the basis functions. By assuming a linear combination of the solution to the two parts of the static part of the governing differential equation and forcing the error arising from the assumption to zero at collocation points within the domain, we generate new basis functions. The

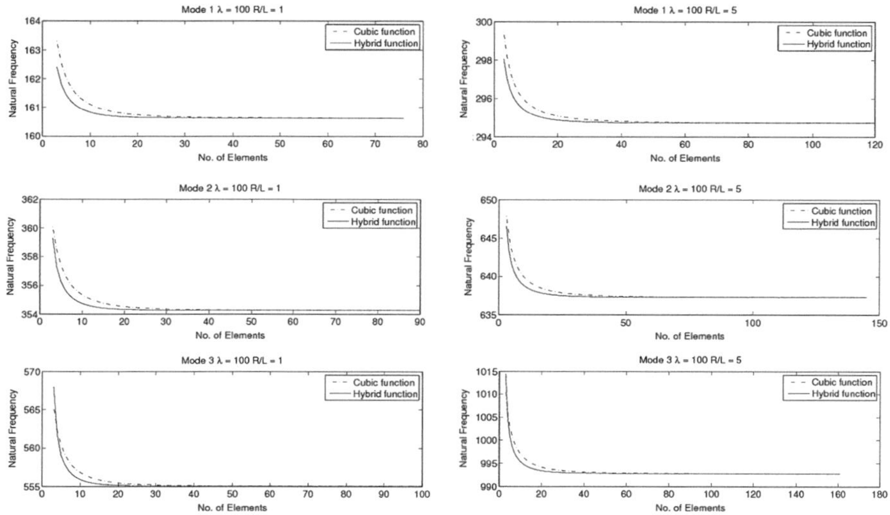

Fig. 5.13 Comparison of convergence of hybrid function and hermite cubic function for linearly tapered, cubic flexural stiffness cantilever beam with $\lambda = 100$ and $R/L = 1$ and 5

interpolation functions derived from this basis functions are a function of the element material and geometry parameters and of the speed of rotation. They also depend on the position of the element in the beam. They behave similar to the cubic interpolation functions for low rotation speeds and show better convergence rates as the rotation speed increases. The robustness of the functions is asserted by considering various cases of rotating beams and matching the obtained results with published literature. The new basis functions based on problem physics lead to a reduction in number of elements required for convergence by about 33 % for high speed rotating beams. They are recommended for use for high speed rotating beams such as turbine blades and for control applications where reasonably accurate low order models are required.

References

1. Al-Qaisia A, Al-Bedoor B (2005) Evaluation of different methods for the consideration of the effect of rotation on the stiffening of rotating beams. J Sound Vib 280(3–5):531–553
2. Bathe KJ (1996) Finite element procedures. Prentice Hall, Upper Saddle River
3. Chakraborty A, Gopalakrishnan S, Reddy JN (2003) A new beam finite element for the analysis of functionally graded materials. Int J Mech Sci 45(3):519–539
4. Cook RD, Malkus DS, Plesha ME, Witt RJ (2002) Concepts and applications of finite element analysis. Wiley, New York
5. Gunda JB, Ganguli R (2008a) New rational interpolation functions for finite element analysis of high speed rotating beams. Int J Mech Sci 50(3):578–588
6. Gunda JB, Ganguli R (2008b) Stiff-string basis functions for vibration analysis of high speed rotating beams. J Appl Mech 75(2):0245021–0245025

7. Gunda JB, Singh AP, Chhabra PS, Ganguli R (2007) Free vibration analysis of rotating tapered blades using fourier-p superelement. Struct Eng Mech 27227(2):243–257
8. Hildebrand FB (1965) Methods of applied mathematics. Prentice Hall, Upper Saddle River
9. Hodges DH, Rutkowski MJ (1981) Free-vibration analysis of rotating beams by a variable-order finite element method. AIAA J 19(11):1459–1466
10. Reddy JN (1993) An introduction to the finite element method, 2nd edn. McGraw-Hill, New York
11. Wang G, Wereley NM (2004) Free vibration analysis of rotating blades with uniform tapers. AIAA J 42(12):2429–2437
12. Wright AD, Smith CE, Thresher RW, Wang JLC (1982) Vibration modes of centrifugally stiffened beams. J Appl Mech 49(2):197–202
13. Wright K (2007) Adaptive methods for piecewise polynomial collocation for ordinary differential equations. BIT Numer Math 47:197–212

Chapter 6
Collocation Approach

In this chapter, we enforce satisfaction of the differential equation for the rotating beam at only some points in the domain, an idea which we borrow from the collocation method. In the collocation method, the governing differential equation is satisfied at several points along the whole domain. In this chapter, a new interpolating polynomial is developed which satisfies the static part of the homogeneous governing differential equation for rotating beams at certain *collocation points*. These points are selected within the element and lead to the creation of basis functions which are closer to the problem physics.

6.1 Governing Differential Equation

The governing partial differential equation for transverse bending vibration of an Euler–Bernoulli rotating beam, with varying geometry is given by

$$\frac{d^2}{dx^2}\left(EI(x)\frac{\partial^2 w}{\partial x^2}\right) + m(x)\frac{\partial^2 w}{\partial t^2} - \frac{d}{dx}\left(T(x)\frac{\partial w}{\partial x}\right) = 0 \tag{6.1}$$

where,

$$T(x) = \int_x^L m(x)\Omega^2(R+x)\,dx + F \tag{6.2}$$

is the centrifugal tensile load at a distance x from the axis of rotation. The static part of the partial differential Eq. 6.1 is given by

$$\frac{d^2}{dx^2}\left(EI(x)\frac{d^2 w}{dx^2}\right) - \frac{d}{dx}\left(T(x)\frac{dw}{dx}\right) = 0 \tag{6.3}$$

R. Ganguli, *Finite Element Analysis of Rotating Beams*,
Foundations of Engineering Mechanics, DOI 10.1007/978-981-10-1902-9_6

If a uniform beam is considered then $EI(x)$ and $m(x)$ are constant, and the homogeneous governing differential equation is given as,

$$EI\frac{d^4w}{dx^4} - T(x)\frac{d^2w}{dx^2} - \frac{dw}{dx}\frac{dT(x)}{dx} = 0 \tag{6.4}$$

Assuming that there is no offset (R) between the fixed end of the beam and the axis of rotation, and axial force (F) applied at the end of the beam to be zero, Eq. (2) for a uniform beam centrifugal tensile load reduces to,

$$T(x) = m\Omega^2 \int_x^L x dx = \frac{m\Omega^2}{2}\left(L^2 - x^2\right) \tag{6.5}$$

Now we differentiate the tensile load once which gives,

$$\frac{dT(x)}{dx} = -m\Omega^2 x \tag{6.6}$$

Substituting Eqs. 6.5 and 6.6 in Eq. 6.4, we obtain the homogeneous static governing differential equation as,

$$\frac{d^4w}{dx^4} - \frac{m\Omega^2(L^2 - x^2)}{2EI}\frac{d^2w}{dx^2} + \frac{m\Omega^2 x}{EI}\frac{dw}{dx} = 0 \tag{6.7}$$

which can be rewritten as,

$$\frac{d^4w}{dx^4} + \frac{m\Omega^2 x^2}{2EI}\frac{d^2w}{dx^2} - \frac{m\Omega^2 L^2}{2EI}\frac{d^2w}{dx^2} + \frac{m\Omega^2 x}{EI}\frac{dw}{dx} = 0 \tag{6.8}$$

Let $c = \frac{m\Omega^2}{EI}$, and $d = \frac{m\Omega^2 L^2}{2EI}$, and the Eq. 6.8 further simplifies to

$$\frac{d^4w}{dx^4} + \frac{cx^2}{2}\frac{d^2w}{dx^2} - d\frac{d^2w}{dx^2} + cx\frac{dw}{dx} = 0 \tag{6.9}$$

For ease of calculations we express rotating speed in non-dimensional form as $\lambda = \sqrt{\frac{m\Omega^2 L^4}{EI}}$. From Fig. 6.1, the relation between local coordinate $(\bar{x})$ and global coordinate (x) for an ith element along the beam, is given by, $x = x_i + \bar{x}$, where x_i is the location of the left edge of element i, given by, $x_i = (i-1)l$, assuming that all elements have the same length. Now, the simplified static homogeneous governing differential in terms of local coordinates is given as,

$$\frac{d^4w}{d\overline{x}^4} + \frac{c(x_i + \overline{x})^2}{2}\frac{d^2w}{d\overline{x}^2} - d\frac{d^2w}{d\overline{x}^2} + c(x_i + \overline{x})\frac{dw}{d\overline{x}} = 0 \tag{6.10}$$

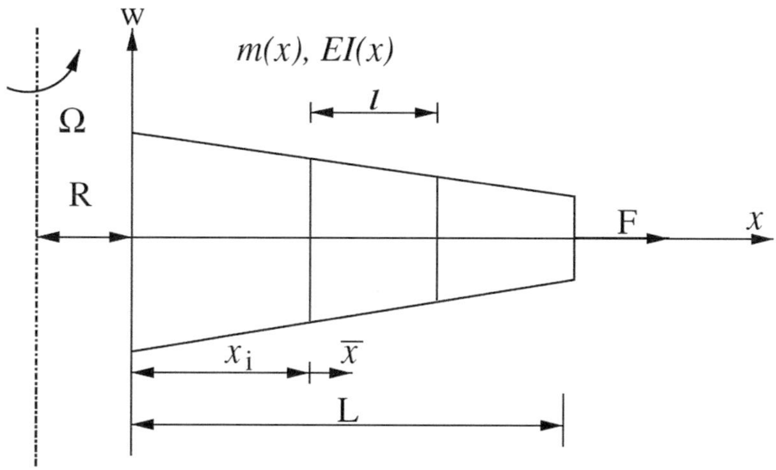

Fig. 6.1 Rotating beam geometry

Fig. 6.2 Beam element

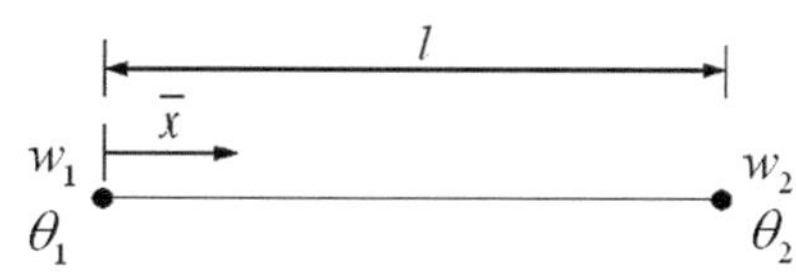

A two noded beam element with four degrees of freedom is shown in Fig. 6.2. Here w corresponds to the displacements and θ corresponds to slope degrees of freedom. Nodal degrees of freedom are given as, $w(0) = w_1$, $\frac{dw(0)}{d\bar{x}} = \theta_1$, $w(l) = w_2$, and $\frac{dw(l)}{d\bar{x}} = \theta_2$. The displacement field within the element using conventional finite element method is given by a cubic polynomial which can be written as, [3]

$$w = \alpha_0 + \alpha_1 \bar{x} + \alpha_2 \bar{x}^2 + \alpha_3 \bar{x}^3 \tag{6.11}$$

Then, α_0, α_1, α_2 and α_3 are solved in terms of nodal degrees of freedom and shape functions are obtained. Interpolating polynomial in terms of nodal degrees of freedom is given as,

$$w = w_1 N_1 + \theta_1 N_2 + w_2 N_3 + \theta_2 N_4 \tag{6.12}$$

Here, N_1, N_2, N_3 and N_4 are functions of $\bar{x}$ and are called as shape or basis functions. This interpolation polynomial satisfies the static part of homogeneous governing differential equation given by Euler Bernoulli beam theory for a non-rotating beam ($\frac{d^4w}{dx^4} = 0$). These shape functions are independent of the rotating speed and are given as

$$N_1 = 1 - 3\frac{\bar{x}^2}{l^2} + 2\frac{\bar{x}^3}{l^3} \tag{6.13}$$

$$N_2 = \bar{x} - 2\frac{\bar{x}^2}{l} + \frac{\bar{x}^3}{l^2} \tag{6.14}$$

$$N_3 = 3\frac{\bar{x}^2}{l^2} - 2\frac{\bar{x}^3}{l^3} \tag{6.15}$$

$$N_4 = -\frac{\bar{x}^2}{l} + \frac{\bar{x}^3}{l^2} \tag{6.16}$$

However, the Hermite cubics do not satisfy Eq. 6.10. Also, Eq. 6.10 does not have an exact solution. This leads us to use the idea of collocation methods for creating basis functions for rotating beams.

6.2 Point Collocation Approach

The collocation method is an approach to solve differential equations by assuming a solution and ensuring that it satisfies the differential equation at a number of collocation points in the domain [4, 7]. This results in simultaneous equations whose number is equal to the number of collocation points used. The collocation approach is an alternative to the finite element approach for solving differential equations. In this paper, we do not use the collocation method as an alternative to the finite element method. Instead, we apply the idea of collocation method to the generation of basis functions for the rotating beam problems. Thus, we force the basis function to satisfy Eq. 6.10 at one or more points in the domain. Hence extra information is available, which allows increasing the order of interpolation polynomial without increasing the degrees of freedom of the element. Using this new interpolating polynomial which is of higher order than the cubic polynomial, new shape functions are obtained which are used further. We next derive the basis functions for a general collocation point within the element.

6.2.1 Collocation Point at a Variable Location Within Beam Element

Collocation point is chosen at a point μ, whose value ranges from $0.1l$ to $0.9l$. The coordinate of this point is given by $x = x_i + \mu$. Order of the interpolation polynomial is increased from three to four which is given by,

$$w = \alpha_0 + \alpha_1\bar{x} + \alpha_2\bar{x}^2 + \alpha_3\bar{x}^3 + \alpha_4\bar{x}^4 \tag{6.17}$$

Now the interpolating polynomial at $\bar{x} = \mu$ should satisfy Eq. 6.10, which gives α_4 in terms of α_0, α_1, α_2 and α_3.

$$\alpha_4 = -\frac{(cx_i + c\mu)\alpha_1 + (cx_i^2 + 4cx_i\mu + 3c\mu^2 - 2d)\alpha_2}{2(12 + 3cx_i^2\mu^2 + 8cx_i\mu^3 + 5c\mu^4 - 6d\mu^2)} + \frac{(3cx_i^2\mu + 9cx_i\mu^2 + 6c\mu^3 - 6d\mu)\alpha_3}{2(12 + 3cx_i^2\mu^2 + 8cx_i\mu^3 + 5c\mu^4 - 6d\mu^2)} \tag{6.18}$$

Now, α_0, α_1, α_2 and α_3 are solved in terms of the nodal degrees of freedom. Displacement field in terms of shape functions is expressed as

$$w = N_1w_1 + N_2\theta_1 + N_3w_2 + N_4\theta_2 \tag{6.19}$$

and new shape functions are obtained as,

$$N_1 = 1 + U_1\overline{x}^2 + V_1\overline{x}^3 + W_1\overline{x}^4 \tag{6.20}$$

$$N_2 = x + U_2\overline{x}^2 + V_2\overline{x}^3 + W_2\overline{x}^4 \tag{6.21}$$

$$N_3 = -(U_1\overline{x}^2 + V_1\overline{x}^3 + W_1\overline{x}^4) \tag{6.22}$$

$$N_4 = U_4\overline{x}^2 + V_4\overline{x}^3 + W_4\overline{x}^4 \tag{6.23}$$

where the coefficients of the new shape functions are functions of position of collocation point, speed of rotation, material and geometric properties, length of the beam and position of element of the rotating beam as they involve the terms μ, x_i, c, d and l, whose value depends on the above mentioned factors. Coefficients of the shape functions in Eqs. 6.20–6.23 are given below.

$$\begin{aligned}P = & -36l^2 - 12dl^3\mu + 18dl^2\mu^2 + 12cl^3\mu^3 - 15cl^2\mu^4 + 18cl^3x_i\mu^2 \\ & - 24cl^2x_i\mu^3 + 6cl^3x_i^2\mu - 9cl^2x_i^2\mu^2\end{aligned}$$

$$\begin{aligned}Q = & -24l - 4dl^3\mu + 12dl\mu^2 + 6cl^3\mu^3 - 10cl\mu^4 + 8cl^3x_i\mu - 16clx_i\mu^3 \\ & + 2cl^3x_i^2 - 6clx_i^2\mu^2\end{aligned}$$

$$R = 24 - 12dl\mu^2 + 10c\mu^4 + 16cx_i\mu^3 + 6cx_i^2\mu^2$$

$$\begin{aligned}D = & -288l^4 + 24dl^6 - 144dl^5\mu + 288dl^4\mu^2 - 36cl^6\mu^2 - 12d^2l^6\mu^2 + 144cl^5\mu^3 \\ & + 72d^2l^5\mu^3 - 240cl^4\mu^4 - 72d^2l^4\mu^4 + 28cdl^6\mu^4 - 132cdl^5\mu^5 + 120cdl^4\mu^6 \\ & - 15c^2l^6\mu^6 + 60c^2l^5\mu^7 - 50c^2l^4\mu^8 - 48cl^6x_i\mu + 216cl^5x_i\mu^2 - 384cl^4x_i\mu^3 \\ & + 40cdl^6x_i\mu^3 - 204cdl^5x_i\mu^4 + 192cdl^4x_i\mu^5 - 44c^2l^6x_i\mu^5 + 192cdl^4x_i\mu^5 \\ & - 44c^2l^6x_i\mu^5 + 186c^2l^5x_i\mu^6 - 160c^2l^4x_i\mu^7 + 72cl^5x_i^2\mu - 12cl^6x_i^2 \\ & - 144cl^4x_i^2\mu^2 + 12cdl^6x_i^2\mu^2 - 72cdl^5x_i^2\mu^3 + 72cdl^4x_i^2\mu^4 - 46c^2l^6x_i^2\mu^4 \\ & + 210c^2l^5x_i^2\mu^5 - 188c^2l^4x_i^2\mu^6 - 20c^2l^6x_i\mu^3 + 102c^2l^5x_i^3\mu^4 \\ & - 96c^2l^4x_i^3\mu^5 - 3c^2l^6x_i^4\mu^2 + 18c^2l^5x_i^4\mu^3 - 18c^2l^4x_i^4\mu^4\end{aligned}$$

$$
\begin{aligned}
A_1 &= 576l + 96dl^3 - 576dl\mu^2 - 144cl^3\mu^2 - 48d^2l^3\mu^2 + 480cl\mu^4 + 144d^2l\mu^4 \\
&+ 112cdl^3\mu^4 - 240cdl\mu^6 - 60c^2l^3\mu^6 + 100c^2l\mu^8 - 192cl^3x_i\mu - 384cdlx_i\mu^5 \\
&- 176c^2l^3x_i\mu^5 + 320c^2lx_i\mu^7 - 48cl^3x_i^2 + 288clx_i^2\mu^2 + 48cdl^3x_i^2\mu^2 \\
&- 144cdlx_i^2\mu^4 - 184c^2l^3x_i^2\mu^4 + 376c^2lx_i^2\mu^6 - 80c^2l^3x_i^3\mu^3 + 192c^2lx_i^3\mu^5 \\
&- 12c^2l^3x_i^4\mu^2 + 36c^2lx_i^4\mu^4
\end{aligned}
$$

$$
\begin{aligned}
A_2 &= 288l^2 + 72dl^4 + 24cl^5\mu - 288d^2l\mu^2 - 108cl^4\mu^2 - 36d^2l^4\mu^2 - 12cdl^5\mu^3 \\
&+ 240cl^2\mu^4 + 72d^2l^2\mu^4 + 84cdl^4\mu^4 + 10c^2l^5\mu^5 - 120cdl^2\mu^6 - 45c^2l^4\mu^6 \\
&+ 50c^2l^2\mu^8 + 24cl^5x_i - 144cl^4x_i\mu - 12cdl^5x_i\mu^2 + 384cl^2x_i\mu^3 \\
&- 132c^2l^4x_i\mu^5 + 160c^2l^2x_i\mu^7 + 120cdl^4x_i\mu^3 + 26c^2l^5x_i\mu^4 - 192cdl^2x_i\mu^5 \\
&- 36cl^4x_i^2 + 144cl^2x_i^2\mu^2 + 36cdl^4x_i^2\mu^2 + 22c^2l^5x_i^2\mu^3 - 72cdl^2x_i^2\mu^4 \\
&- 138c^2l^4x_i^2\mu^4 + 188c^2l^2x_i^2\mu^6 + 6c^2l^5x_i^3\mu^2 - 60c^2l^4x_i^3\mu^3 + 96c^2l^2x_i^3\mu^5 \\
&- 9c^2l^4x_i^4\mu^2 + 18c^2l^2x_i^4\mu^4
\end{aligned}
$$

$$
\begin{aligned}
A_4 &= 288l^2 + 24dl^4 - 288d^2l\mu^2 - 36cl^4\mu^2 - 12d^2l^4\mu^2 + 240cl^2\mu^4 + 72d^2l^2\mu^4 \\
&+ 28cdl^4\mu^4 - 120cdl^2\mu^6 - 15c^2l^4\mu^6 + 50c^2l^2\mu^8 - 48cl^4x_i\mu + 384cl^2x_i\mu^3 \\
&+ 40cdl^4x_i\mu^3 - 192cdl^2x_i\mu^5 - 44c^2l^4x_i\mu^5 + 160c^2l^2x_i\mu^7 - 12cl^4x_i^2 \\
&+ 144cl^2x_i^2\mu^2 + 12cdl^4x_i^2\mu^2 - 72cdl^2x_i^2\mu^4 - 46c^2l^4x_i^2\mu^4 + 188c^2l^2x_i^2\mu^6 \\
&- 20c^2l^4x_i^3\mu^3 + 96c^2l^2x_i^3\mu^5 - 3c^2l^4x_i^4\mu^2 + 18c^2l^2x_i^4\mu^4
\end{aligned}
$$

$$
\begin{aligned}
U_1 &= \frac{A_1 P}{DQ} \\
V_1 &= -\frac{A_1}{D} \\
W_1 &= -(-2d + 3c\mu^2 + 4cx_i\mu + cx_i^2)\frac{U_1}{R} - (6d\mu - 6c\mu^2 - 9cx_i\mu^2 - 3cx_i^2\mu)\frac{V_1}{R} \\
U_2 &= \frac{12 - 2cl^3\mu - 6d\mu^2 + 5c\mu^4 - 2cl^3x_i + 8cx_i\mu^3 + 3cx_i^2\mu^2}{Q} + \frac{A_2 P}{DQ} \\
V_2 &= -\frac{A_2}{D}
\end{aligned}
$$

$$
\begin{aligned}
W_2 = &-\frac{c(x_i + \mu)}{R} + (2d - 3c\mu^2 - 4cx_i\mu - cx_i^2)\frac{U_2}{R} \\
&- (6d\mu - 6c\mu^2 - 9cx_i\mu^2 - 3cx_i^2\mu)\frac{V_2}{R} \\
U_4 = &\frac{-12 + 6dl\mu^2 - 5c\mu^4 - 16cx_i\mu^3 - 3cx_i^2\mu^2}{Q} + \frac{A_4 P}{DQ}
\end{aligned}
$$

$$V_4 = -\frac{A_4}{D}$$

$$W_4 = -(-2d + 3c\mu^2 + 4cx_i\mu + cx_i^2)\frac{U_4}{R} - (6d\mu - 6c\mu^2 - 9cx_i\mu^2 - 3cx_i^2\mu)\frac{V_4}{R}$$

At $\lambda = 0$, i.e., when beam is not rotating, all the $\overline{x}^4$ terms disappear and the shape functions reduce to

$$N_1 = 1 - 3\frac{\bar{x}^2}{l^2} + 2\frac{\bar{x}^3}{l^3}$$

$$N_2 = \bar{x} - 2\frac{\bar{x}^2}{l} + \frac{\bar{x}^3}{l^2}$$

$$N_3 = 3\frac{\bar{x}^2}{l^2} - 2\frac{\bar{x}^3}{l^3}$$

$$N_4 = -\frac{\bar{x}^2}{l} + \frac{\bar{x}^3}{l^2}$$

These shape functions are the Hermite cubic interpolating polynomials. Now these shape functions are studied near three points, two near the nodes and one at the midpoint of beam by giving suitable value of μ.

6.2.2 Collocation Point Near the Left Node of Beam Element

In the first case, a single collocation point is chosen near the left node of the beam element which is given by $\mu = 0.1l$. Now, α_4 is expressed in terms of $\alpha_0, \alpha_1, \alpha_2$ and α_3 by inserting $\mu = 0.1l$ in (18) which after simplification yields,

$$\alpha_4 = \frac{(5cx_i + 0.5cl)\alpha_1 + (5cx_i^2 + 2cx_il + 0.15cl^2 - 10d)\alpha_2}{-0.5(240 + 0.6cx_i^2l^2 + 0.16cx_il^3 + 0.01cl^4 - 1.2dl^2)} + \frac{(1.5cx_i^2l + 0.45cx_il^2 + 0.03cl^3 - 3dl)\alpha_3}{-0.5(240 + 0.6cx_i^2l^2 + 0.16cx_il^3 + 0.01cl^4 - 1.2dl^2)} \quad (6.24)$$

Shape functions are once again given by Eqs. 6.20, 6.21, 6.22 and 6.23. Coefficients are obtained given below.

$$U_1 = \frac{(-720 - 20.4dl^2 + 0.21cl^4 + 3.12cl^3x_i + 10.2cl^2x_i^2)}{l^2D}$$

$$V_1 = \frac{(480 + 77.6dl^2 - 1.18cl^4 - 15.68cl^3x_i - 38.8cl^2x_i^2)}{l^3D}$$

$$W_1 = -\left(\frac{U_1 A}{S} + \frac{V_1 B}{S}\right)$$

$$U_2 = \frac{(-480l - 15.624dl^3 - 8.4cl^5 - 7.62cl^4 x_i + 7.8cl^3 x_i^2)}{l^2 D}$$

$$V_2 = \frac{(240l + 58.8dl^3 + 1.11cl^5 + 8.16cl^4 x_i - 29.4cl^3 x_i^2)}{l^3 D}$$

$$W_2 = -\left(\frac{(0.5cl + 5cx_i)}{S} + \frac{U_1 A}{S} + \frac{V_1 B}{S}\right)$$

$$U_4 = \frac{(-240l - 4.8dl^3 + 0.05cl^5 + 0.74cl^4 x_i + 2.4cl^3 x_i^2)}{l^2 D}$$

$$V_4 = \frac{(240l + 18.8dl^3 - 0.29cl^5 - 3.84cl^4 x_i - 9.4cl^3 x_i^2)}{l^3 D}$$

$$W_4 = -\left(\frac{U_4 A}{S} + \frac{V_4 B}{S}\right)$$

$$A = (0.15cl^2 - 10d + 2cx_i l + 5cx_i^2)$$

$$B = (0.03cl^2 - 3d + 0.45cx_i l + 1.5cx_i^2)$$

$$D = (240 - 9.2dl^2 + 0.19cl^4 + 2.36cl^3 x_i + 4.6cl^2 x_i^2)$$

$$S = 0.5(240 - 1.2dl^2 + 0.01cl^4 + 0.16cl^3 x_i + 0.6cl^2 x_i^2).$$

6.2.3 *Collocation Point at the Midpoint of Beam Element*

In this case, the collocation point is chosen at the center of each element which is given by $\mu = 0.5l$. Now, the error is forced to become zero at the mid point of the beam element. Using this condition, α_4 in terms of α_0, α_1, α_2 and α_3 is given as,

$$\alpha_4 = \frac{(4cx_i + 2cl)\alpha_1 + (4cx_i^2 + 8cx_i l + 3cl^2 - 8d)\alpha_2}{-0.5(192 + 12cx_i^2 l^2 + 16cx_i l^3 + 5cl^4 - 24dl^2)} + \frac{(6cx_i^2 l + 9cx_i l^2 + 3cl^3 - 12dl)\alpha_3}{-0.5(192 + 12cx_i^2 l^2 + 16cx_i l^3 + 5cl^4 - 24dl^2)} \tag{6.25}$$

Shape functions are once again given by Eqs. 6.20, 6.21, 6.22 and 6.23 but coefficients change and are given below.

$$U_1 = \frac{(576 + 24dl^2 - 9cl^4 - 24cl^3x_i - 12cl^2x_i^2)}{l^2 D}$$

$$V_1 = \frac{(-384 - 16dl^2 + 14cl^4 + 32cl^3x_i + 8cl^2x_i^2)}{l^3 D}$$

$$W_1 = -\left(\frac{U_1 A}{S} + \frac{V_1 B}{S}\right)$$

$$U_2 = \frac{(384l + 24dl^3 - 4cl^5 - 14cl^4x_i - 12cl^3x_i^2)}{l^2 D}$$

$$V_2 = \frac{(-192l - 24dl^3 + 5cl^5 + 16cl^4x_i + 12cl^3x_i^2)}{l^3 D}$$

$$W_2 = -\left(\frac{(2cl + 4cx_i)}{S} + \frac{U_1 A}{S} + \frac{V_1 B}{S}\right)$$

$$U_4 = \frac{(192l - cl^5 - 2cl^4x_i - 12cl^3x_i^2)}{l^2 D}$$

$$V_4 = \frac{(-192l + 8dl^3 + cl^5 - 4cl^3x_i^2)}{l^3 D}$$

$$W_4 = -\left(\frac{U_4 A}{S} + \frac{V_4 B}{S}\right)$$

$$A = (3cl^2 - 8d + 8cx_i l + 4cx_i^2)$$

$$B = (3cl^2 - 12d + 9cx_i l + 6cx_i^2)$$

$$D = (-192 - 8dl^2 + cl^4 + 4cl^3x_i + 4cl^2x_i^2)$$

$$S = 0.5(192 - 24dl^2 + 5cl^4 + 16cl^3x_i + 12cl^2x_i^2).$$

6.2.4 Collocation Point Near the Right Node of Beam Element

In the third case, the collocation point is chosen near the right of each element which is given by $\mu = 0.9l$. Now α_4 is obtained in terms of α_0, α_1, α_2 and α_3 and is given as,

$$\alpha_4 = \frac{(5cx_i + 4.5cl)\alpha_1 + (5cx_i^2 + 1.8cx_il + 12cl^2 - 10d)\alpha_2}{-1.5(80 + 16.2cx_i^2l^2 + 38.88cx_il^3 + 21.87cl^4 - 32.4dl^2)} + \frac{(13.5cx_i^2l + 36.45cx_il^2 + 21cl^3 - 27dl)\alpha_3}{-1.5(80 + 16.2cx_i^2l^2 + 38.88cx_il^3 + 21.87cl^4 - 32.4dl^2)} \tag{6.26}$$

Shape functions are given by Eqs. 6.20, 6.21, 6.22 and 6.23, and the coefficients are given below.

$$U_1 = \frac{(-720 + 75.6dl^2 - 21.87cl^4 - 58.32cl^3x_i - 37.8cl^2x_i^2)}{l^2D}$$

$$V_1 = \frac{(480 - 114.4dl^2 + 34.02cl^4 + 89.28cl^3x_i + 57.2cl^2x_i^2)}{l^3D}$$

$$W_1 = -\left(\frac{U_1A}{S} + \frac{V_1B}{S}\right)$$

$$U_2 = \frac{(-480l + 32.4dl^3 - 9cl^5 - 24.58cl^4x_i - 16.2cl^3x_i^2)}{l^2D}$$

$$V_2 = \frac{(240l - 37.2dl^3 + 10.71cl^5 + 28.64cl^4x_i + 18.6cl^3x_i^2)}{l^3D}$$

$$W_2 = -\left(\frac{(4.5cl + 5cx_i)}{S} + \frac{U_1A}{S} + \frac{V_1B}{S}\right)$$

$$U_4 = \frac{(-240l + 43.2dl^3 - 21.87cl^5 - 43.74cl^4x_i - 21.6cl^3x_i^2)}{l^2D}$$

$$V_4 = \frac{(240l - 77.2dl^3 + 41.31cl^5 + 80.64cl^4x_i + 38.6cl^3x_i^2)}{l^3D}$$

$$W_4 = -\left(\frac{U_4A}{S} + \frac{V_4B}{S}\right)$$

$$A = (12.15cl^2 - 10d + 18cx_i l + 5cx_i^2)$$

$$B = (21.87cl^2 - 27d + 36.45cx_i l + 13.5cx_i^2)$$

$$D = (240 - 9.2dl^2 + 2.43cl^4 + 6.84cl^3 x_i + 4.6cl^2 x_i^2)$$

$$S = 3(80 - 32.4dl^2 + 21.87cl^4 + 38.8cl^3 x_i + 16.2cl^2 x_i^2).$$

6.2.5 Two Point Collocation

It is natural to think that increasing the number of collocation points within the element will lead to better shape functions. Therefore, we increase the number of collocation points from one to two. So, we satisfy the governing differential equation at two points within the domain. For simplicity of the resulting expressions, instead of choosing variable points, two collocation points are chosen at $x = x_i + l/3$ and $x = x_i + 2l/3$. The order of interpolating polynomial is increased from four to five. The displacement field is given as,

$$w = \alpha_0 + \alpha_1 \bar{x} + \alpha_2 \bar{x}^2 + \alpha_3 \bar{x}^3 + \alpha_4 \bar{x}^4 + \alpha_5 \bar{x}^5 \tag{6.27}$$

Now, Eq. 6.10 is satisfied using Eq. 6.27 at $\overline{x} = l/3$ and $\overline{x} = 2l/3$. Which gives α_5 and α_4 in terms of α_0, α_1, α_2 and α_3. Then, α_0, α_1, α_2 and α_3 are once again solved in terms of nodal degrees of freedom and shape functions are obtained. Interpolating polynomial in terms of nodal degrees of freedom is given by Eq. (19). The new shape functions are obtained as,

$$N_1 = 1 + U_1 \overline{x}^2 + V_1 \overline{x}^3 + W_1 \overline{x}^4 + Z_1 \overline{x}^5 \tag{6.28}$$

$$N_2 = x + U_2 \overline{x}^2 + V_2 \overline{x}^3 + W_2 \overline{x}^4 + Z_2 \overline{x}^5 \tag{6.29}$$

$$N_3 = -(U_1 \overline{x}^2 + V_1 \overline{x}^3 + W_1 \overline{x}^4 + Z_1 \overline{x}^5) \tag{6.30}$$

$$N_4 = U_4 \overline{x}^2 + V_4 \overline{x}^3 + W_4 \overline{x}^4 + Z_4 \overline{x}^5 \tag{6.31}$$

The coefficients of shape functions are given below.

$$
\begin{aligned}
D = & -3967185807360l^4 - 117546246144dl^6 + 125709179904cl^8 - 15033403008d^2l^8 \\
& + 11250487872cdl^{10} - 744489792d^3l^{10} - 1251943776c^2l^{12} + 477185904cd^2l^{12} \\
& - 9237888d^4l^{12} - 89103240c^2dl^{14} + 6492960cd^3l^{14} + 4454352c^3l^{16} \\
& - 1519920c^2d^2l^{16} + 132288c^3dl^{18} - 3520c^4l^{20} + 224072531712cl^7x_i \\
& + 27073730304cdl^9x_i - 7152014880c^2l^{11}x_i + 1406783376cd^2l^{11}x_i \\
& - 556250328c^2dl^{13}x_i + 20785248cd^3l^{13}x_i + 46306080c^3l^{15}x_i \\
& - 10275984c^2d^2l^{15}x_i + 1461888c^3dl^{17}x_i - 57024c^4l^{19}x_i + 58773123072cl^6x_i^2 \\
& + 15033403008cdl^8x_i^2 - 15235586784c^2l^{10}x_i^2 + 1116734688cd^2l^{10}x_i^2 \\
& - 1323677376c^2dl^{12}x_i^2 + 18475776cd^3l^{12}x_i^2 + 196716276c^3l^{14}x_i^2 \\
& - 26636688c^2d^2l^{14}x_i^2 + 6666192c^3dl^{16}x_i^2 - 396000c^4l^{18}x_i^2 - 13536865152c^2l^9x_i^3 \\
& - 1406783376c^2dl^{11}x_i^3 + 438878412c^3l^{13}x_i^3 - 31177872c^2d^2l^{13}x_i^3 \\
& + 16107984c^3dl^{15}x_i^3 - 1539648c^4l^{17}x_i^3 - 3758350752c^2l^8x_i^4 - 558367344c^2d^{10}x_i^4 \\
& + 542542212c^3l^{12}x_i^4 - 13856832c^2d^2l^{12}x_i^4 + 21766968c^3dl^{14}x_i^4 - 3667356c^4l^{16}x_i^4 \\
& + 351695844c^3l^{11}x_i^5 + 15588936c^3dl^{13}x_i^5 - 5484996c^4l^{15}x_i^5 + 93061224c^3l^{10}x_i^6 \\
& + 4618944c^3dl^{12}x_i^6 - 5035932c^4l^{14}x_i^6 - 2598156c^4l^{13}x_i^7 - 577368c^4l^{12}x_i^8
\end{aligned}
$$

$$
\begin{aligned}
P = & -1259712l - 186624dl^3 + 39204cl^5 - 4212d^2l^5 + 1578cdl^7 - 268c^2l^9 \\
& + 145800cl^4x_i + 5508cdl^6x_i - 1584c^2l^8x_i + 93312cl^3x_i^2 - 4212cdl^5x_i^2 \\
& - 3141c^2l^7x_i^2 - 2754c^2l^6x_i^3 - 1053c^2l^5x_i^4
\end{aligned}
$$

$$
\begin{aligned}
Q = & -1889568l^2 - 122472dl^4 + 28188cl^6 - 1944d^2l^6 + 828cdl^8 - 96c^2l^{10} \\
& + 87480cl^5x_i + 2592cdl^7x_i - 612c^2l^9x_i + 61236cl^4x_i^2 + 1944cdl^6x_i^2 \\
& - 1350c^2l^8x_i^2 - 1296c^2l^7x_i^3 - 486c^2l^6x_i^4
\end{aligned}
$$

$$
\begin{aligned}
R = & \ 157464 - 2916dl^2 - 3807cl^4 + 648d^2l^4 - 258cdl^6 + 20c^2l^8 - 5103cl^3x_i \\
& - 810cdl^5x_i + 144c^2l^7x_i + 1458cl^2x_i^2 - 648cdl^4x_i^2 + 369c^2l^6x_i^2 + 405c^2l^5x_i^3 \\
& + 162c^2l^4x_i^4
\end{aligned}
$$

$$
S = 80l(81 - 6dl^2 + 2cl^4 + 5cl^3x_i + 3cl^2x_i^2)
$$

$$
\begin{aligned}
A_1 = {} & -1190155742208 0l^2 - 550998028800dl^4 + 465287224320cl^6 - 46936869120d^2l^6 \\
& + 40077737280cdl^8 - 2947726080d^3l^8 - 6408784800c^2l^{10} + 2193998400cd^2l^{10} \\
& - 50388480d^4l^{10} - 503768160c^2dl^{12} + 41523840cd^3l^{12} + 37169280c^3l^{14} \\
& - 12588480c^2d^2l^{14} + 1653120c^3dl^{16} - 76800c^4l^{18} + 936696648960cl^5x_i \\
& + 92343188160cdl^7x_i - 33813819360c^2l^9x_i + 6330052800cd^2l^9x_i \\
& - 3013860960c^2dl^{11}x_i + 130170240cd^3l^{11}x_i + 345137760c^3l^{13}x_i - 80637120c^2d^2l^{13}x_i \\
& + 16269120c^3dl^{15}x_i - 1042560c^4l^{17}x_i + 275499014400cl^4x_i^2 + 46936869120cdl^6x_i^2 \\
& - 61930591200c^2l^8x_i^2 + 4421589120cd^2l^8x_i^2 - 6623985600c^2dl^{10}x_i^2 \\
& + 100776960cd^3l^{10}x_i^2 + 1293829200c^3l^{12}x_i^2 - 189190080c^2d^2l^{10}x_i^2 \\
& + 64955520c^3dl^{14}x_i^2 - 6022080c^4l^{16}x_i^2 - 46171594080c^2l^7x_i^3 - 6330052800c^2dl^9x_i^3 \\
& + 2537794800c^3l^{11}x_i^3 - 195255360c^2d^2l^{11}x_i^3 + 135846720c^3dl^{13}x_i^3 \\
& - 19401120c^4l^{15}x_i^3 - 11734217280c^2l^6x_i^4 - 2210794560c^2dl^8x_i^4 \\
& + 2763493200c^3l^{10}x_i^4 - 75582720c^2d^2l^{10}x_i^4 + 158047200c^3dl^{12}x_i^4 - 38316240c^4l^{14}x_i^4 \\
& + 1582513200c^3l^9x_i^5 + 97627680c^3dl^{11}x_i^5 - 47764080c^4l^{13}x_i^5 + 368465760c^3l^8x_i^6 \\
& + 25194240c^3dl^{10}x_i^6 - 36916560c^4l^{12}x_i^6 - 16271280c^4l^{11}x_i^7 - 36916560c^4l^{10}x_i^8
\end{aligned}
$$

$$U_1 = -\frac{A_1}{D}$$

$$V_1 = \frac{P}{Q}A_1$$

$$
\begin{aligned}
W_1 = {} & (11664d + 3888cl - 1512d^2l^2 + 702cdl^4 - 84c^2l^6 + 2214cdl^3x_i - 540c^2l^5x_i \\
& - 5832cx_i^2 + 1512cdl^2 - 1191c^2l^4x_i^2 - 1107c^2l^3x_i^3 - 378c^2l^2x_i^4)\frac{-9U_1}{8R} \\
& - (-7776cl^3 + 1296d^2l^3 - 540cdl^5 + 48c^2l^7 - 11664cl^2x_i - 1692cdl^4x_i \\
& + 332c^2l^6x_i - 1296cdl^3x_i^2 + 810c^2l^5x_i^2 + 846c^2l^4x_i^3 + 324c^2l^3x_i^4)\frac{9V_1}{8R}
\end{aligned}
$$

$$
\begin{aligned}
Z_1 = {} & -(162d - 108cl^2 - 216clx_i - 81cx_i^2)\frac{U_1}{S} \\
& + (324dl - 144cl^3 - 324cl^2x_i - 162clx_{i^2})\frac{V_1}{S} \\
& - (2187 - 486dl^2 + 180cl^4 + 432cl^3x_i + 243cl^2x_i^2)\frac{8W_1}{9S}
\end{aligned}
$$

$$
\begin{aligned}
A_2 = {} & -7934371614720l^3 - 499571546112dl^5 + 273458280960cl^7 - 31291246080d^2l^7 \\
& + 30527860608cdl^9 - 2463996672d^3l^9 - 3161457216c^2l^{11} + 1588636800cd^2l^{11} \\
& - 43670016d^4l^{11} - 294939792c^2dl^{13} + 31415040cd^3l^{13} + 14723208c^3l^{15} \\
& - 7674336c^2d^2l^{15} + 728112c^3dl^{17} - 22880c^4l^{19} + 547324708608cl^6x_i \\
& + 67514264640cdl^8x_i - 18374159232c^2l^{10}x_i + 4657785120cd^2l^{10}x_i \\
& - 1838987064c^2dl^{12}x_i + 100357056cd^3l^{12}x_i + 152568036c^3l^{14}x_i - 51457680c^2d^2l^{14}x_i \\
& + 7843608c^3dl^{16}x_i - 349776c^4l^{18}x_i + 249785773056cl^5x_i^2 + 31291246080cdl^7x_i^2 \\
& - 39365370144c^2l^9x_i^2 + 3695995008cd^2l^9x_i^2 - 4378986360c^2dl^{11}x_i^2 \\
& + 87340032cd^3l^{11}x_i^2 + 647430732c^3l^{13}x_i^2 - 131223888c^2d^2l^{11}x_i^2 \\
& + 34801272c^3dl^{15}x_i^2 - 2301624c^4l^{17}x_i^2 - 33757132320c^2l^8x_i^3 - 4657785120c^2dl^9x_i^3 \\
& + 1446175620c^3l^{12}x_i^3 - 150535584c^2d^2l^{12}x_i^3 + 81803520c^3dl^{14}x_i^3 \\
& - 8523468c^4l^{16}x_i^3 - 7822811520c^2l^7x_i^4 - 1847997504c^2dl^9x_i^4 \\
& + 1792333980c^3l^{11}x_i^4 - 65505024c^2d^2l^{12}x_i^4 + 107662608c^3dl^{13}x_i^4 - 19450692c^4l^{15}x_i^4 \\
& + 1164446280c^3l^{10}x_i^5 + 75267792c^3dl^{12}x_i^5 - 28037340c^4l^{14}x_i^5 + 307999584c^3l^8x_i^6 \\
& + 21835008c^3dl^{11}x_i^6 - 24952212c^4l^{13}x_i^6 - 12544632c^4l^{12}x_i^7 - 2729376c^4l^{11}x_i^8
\end{aligned}
$$

$$
\begin{aligned}
B_2 = {} & (-629856 + 11664dl^2 + 41472cl^4 - 2592d^2l^4 + 60cdl^6 - 350c^2l^8 + 125388cl^3x_i \\
& + 2754cdl^5x_i - 1737c^2l^7x_i - 5832cl^2x_i^2 + 2592cdl^4x_i^2 - 2448c^2l^6x_i^2 - 1377c^2l^5x_i^3 \\
& - 648c^2l^4x_i^4)
\end{aligned}
$$

$$
U_2 = -\frac{A_2}{D}
$$

$$
V_2 = \frac{-B_4}{Q} + \frac{PA_2}{Q}
$$

$$
\begin{aligned}
W_2 = {} & (-216cdl^3 + 90c^2l^5 + 5832cx_i - 756cdl^2x_i + 489c^2l^4x_i + 783c^2l^3x_i^3 + 378c^2l^2x_i^3) \\
& - (11664d + 3888cl^2 - 1512d^2l^2 + 702cdl^4 - 84c^2l^6 + 2214cdl^3x_i - 540c^2l^5x_i \\
& - 5832cx_i^2 + 1512cdl^2x_i^2 - 1191c^2l^4x_i^2 - 1107c^2l^3x_i^3 - 378c^2l^2x_i^4)\frac{9U_2}{8R} \\
& - (-7776cl^3 + 1296d^2l^3 - 540cdl^5 + 48c^2l^7 - 11664cl^2x_i - 1692cdl^4x_i \\
& + 332c^2l^6x_i - 1296cdl^3x_i^2 + 810c^2l^5x_i^2 + 846c^2l^4x_i^3 + 324c^2l^3x_i^4)\frac{9V_2}{8R}
\end{aligned}
$$

$$
\begin{aligned}
Z_2 = {} & -(54cl + 81cx_i)\frac{1}{S} - (162d - 108cl^2 - 216clx_i - 81cx_i^2)\frac{U_2}{S} \\
& + (324dl - 144cl^3 - 324cl^2x_i - 162clx_{i^2})\frac{V_2}{S} \\
& - (2187 - 486dl^2 + 180cl^4 + 432cl^3x_i + 243cl^2x_i^2)\frac{8W_2}{9S}
\end{aligned}
$$

$$
\begin{aligned}
A_4 = {} & -3967185807360l^3 - 51426482688dl^5 + 125709179904cl^7 - 15645623040d^2l^7 \\
& + 9549876672cdl^9 - 483729408d^3l^9 - 1274618592c^2l^{11} + 355938624cd^2l^{11} \\
& - 6718464d^4l^{11} - 77818320c^2dl^{13} + 5069952cd^3l^{13} + 5002560c^3l^{15} \\
& - 1368288c^2d^2l^{15} + 156480c^3dl^{17} - 6400c^4l^{19} + 224072531712cl^6x_i \\
& + 24522813504cdl^8x_i - 7242714144c^2l^{10}x_i + 1054378944cd^2l^{10}x_i \\
& - 480930048c^2dl^{12}x_i + 15956352cd^3l^{12}x_i + 149233744c^3l^{14}x_i - 8895744c^2d^2l^{14}x_i \\
& + 1581984c^3dl^{16}x_i - 89280c^4l^{18}x_i + 25713241344cl^5x_i^2 + 15645623040cdl^7x_i^2 \\
& - 14453305632c^2l^9x_i^2 + 725594112cd^2l^9x_i^2 - 1087166448c^2dl^{11}x_i^2 \\
& + 13436928cd^3l^{11}x_i^2 + 195657768c^3l^{13}x_i^2 - 21749472c^2d^2l^{11}x_i^2 \\
& + 6563808c^3dl^{15}x_i^2 - 534240c^4l^{17}x_i^2 - 12261406752c^2l^8x_i^3 - 1054378944c^2dl^9x_i^3 \\
& + 403422768c^3l^{12}x_i^3 - 23934528c^2d^2l^{12}x_i^3 + 14455584c^3dl^{14}x_i^3 \\
& - 1800144c^4l^{16}x_i^3 - 3911405760c^2l^7x_i^4 - 362797056c^2dl^9x_i^4 \\
& + 454598568c^3l^{11}x_i^4 - 10077696c^2d^2l^{12}x_i^4 + 17947008c^3dl^{13}x_i^4 - 3757752c^4l^{15}x_i^4 \\
& + 263594736c^3l^{10}x_i^5 + 11967264c^3dl^{12}x_i^5 - 5003856c^4l^{14}x_i^5 + 60466176c^3l^8x_i^6 \\
& + 3359232c^3dl^{11}x_i^6 - 4169880c^4l^{13}x_i^6 - 1994544c^4l^{12}x_i^7 - 419904c^4l^{11}x_i^8
\end{aligned}
$$

$$
\begin{aligned}
B_4 = {} & (629856 - 11664dl^2 - 15228cl^4 + 2592d^2l^4 - 1032cdl^6 + 80c^2l^8 - 20412cl^3x_i \\
& - 3240cdl^5x_i + 576c^2l^7x_i + 5832cl^2x_i^2 - 2592cdl^4x_i^2 + 1476c^2l^6x_i^2 + 1620c^2l^5x_i^3 \\
& + 648c^2l^4x_i^4)
\end{aligned}
$$

$$
U_4 = -\frac{A_4}{D}
$$

$$
V_4 = \frac{-B_4}{Q} + \frac{PA_4}{Q}
$$

$$
\begin{aligned}
W_4 = {} & (11664d + 3888cl^2 - 1512d^2l^2 + 702cdl^4 - 84c^2l^6 + 2214cdl^3x_i - 540c^2l^5x_i \\
& - 5832cx_i^2 + 1512cdl^2x_i^2 - 1191c^2l^4x_i^2 - 1107c^2l^3x_i^3 - 378c^2l^2x_i^4)\frac{-9U_4}{8R} \\
& - (-7776cl^3 + 1296d^2l^3 - 540cdl^5 + 48c^2l^7 - 11664cl^2x_i - 1692cdl^4x_i \\
& + 332c^2l^6x_i - 1296cdl^3x_i^2 + 810c^2l^5x_i^2 + 846c^2l^4x_i^3 + 324c^2l^3x_i^4)\frac{9V_2}{8R}
\end{aligned}
$$

$$
\begin{aligned}
Z_4 = {} & -(162d - 108cl^2 - 216clx_i - 81cx_i^2)\frac{U_4}{S} \\
& + (324dl - 144cl^3 - 324cl^2x_i - 162clx_{i^2})\frac{V_2}{S} \\
& - (2187 - 486dl^2 + 180cl^4 + 432cl^3x_i + 243cl^2x_i^2)\frac{8W_4}{9S}
\end{aligned}
$$

It can be seen that two collocation points greatly complicates the shape functions.

6.2.6 Analysis of Shape Functions

We can see that the shape functions obtained using a collocation point vary from one another when the position of the collocation point changes. Compared to the Hermite cubic, these shape functions clearly shows the influence of rotation effects which come in through the governing differential equation.

The shape functions obtained using two collocation points and one collocation point at three different positions within the beam element, are compared with the Hermite cubics in Figs. 6.3 and 6.4 for $\lambda = 12$ and $\lambda = 100$. These results are obtained for a single element. We can see that using two collocation points and collocation point at midpoint of the beam element modify the shape functions for both lower and higher speeds. However for the case of one collocation point near the nodes, the shape function corresponding to the slope of other node is modified only at high speeds. This shows the dependence of shape functions on terms c and d. At lower speeds, shape functions corresponding to a single collocation point at the midpoint of the beam element and shape functions corresponding to two collocation points almost coincide with each other.

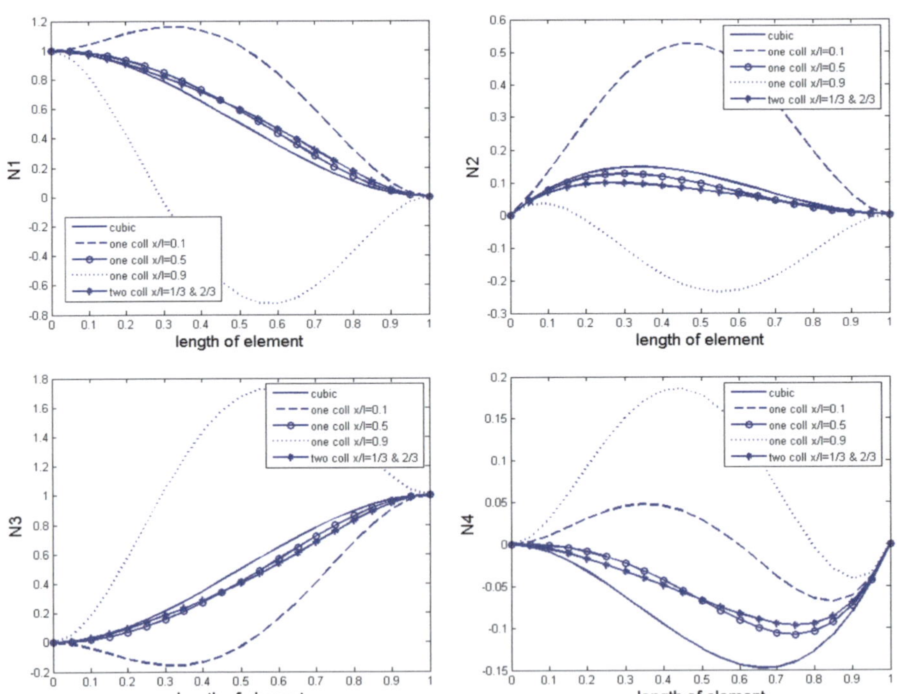

Fig. 6.3 Variation of shape functions along the element using different basis functions at $\lambda = 12$ for a single element

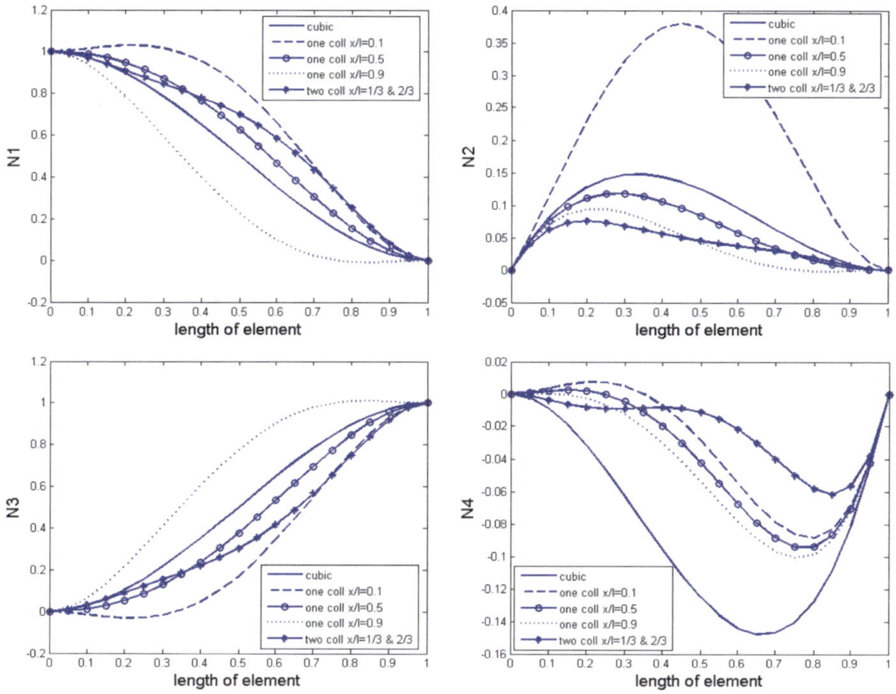

Fig. 6.4 Variation of shape functions along the element using different basis functions at $\lambda = 100$ for a single element

In this case, increasing the number of collocation points does not have much effect on the shape functions. But at higher speeds, the collocation points effects the performance of shape functions. Since term d not only depends on rotating speed but also on the length of beam, shape functions vary not only with the rotating speed but also with the length of the beam. To exemplify this, shape functions are once again plotted for $\lambda = 12$ and $\lambda = 100$ for the first element, with $L = 10$ and $l = 1$ which are shown in Figs. 6.5 and 6.6, respectively. We can see that with increase in length which increases the magnitude of d, shape functions corresponding to that of lower speeds are modified not only for two point collocation and collocation point at midpoint but also for one collocation point near the nodes.

Inducing a collocation point modifies the shape functions, when compared to those given by the approximate solutions which do not satisfy the governing differential equation of the problem. But they also increase the complexity of shape functions enormously, which limits the increase in number of collocation points that can be used and hence the order of the interpolating polynomial. However, the use of symbolic algebra software which also includes numerical computation capabilities such as Mathematica [9] and MATLAB [6] can reduce this problem to a large extent. We used

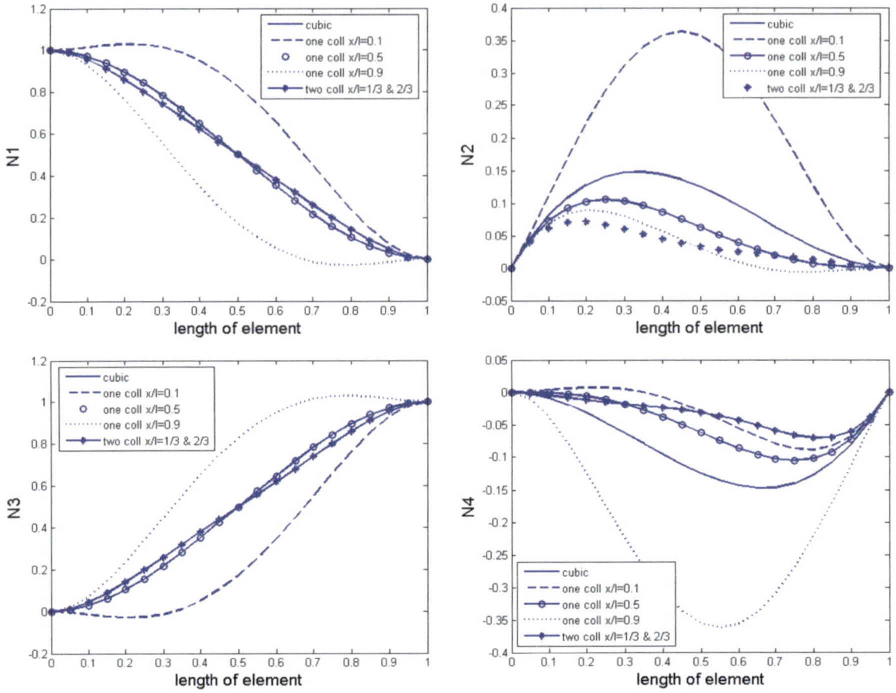

Fig. 6.5 Variation of shape functions along the element using different basis functions at $\lambda = 12$ for the first element ($x_i = 0$) of ten elements used with $l = 1$

Mathematica for the generation of shape functions and MATLAB for the solution of the eigenvalue problems in finite element method.

6.3 Finite Element Formulation

The refined basis functions obtained through collocation approach are used to obtain the stiffness and mass matrices. The kinetic energy of the beam is given by [3],

$$T = \frac{1}{2}\int_0^l m(x)\dot{w}(x,t)^2 dx \tag{6.32}$$

and strain energy is given by,

$$V = \frac{1}{2}\int_0^l EI(x)\left(\frac{\partial^2 w(x,t)}{\partial x^2}\right)^2 dx + \frac{1}{2}\int_0^l T(x)\left(\frac{\partial w(x,t)}{\partial x}\right)^2 dx \tag{6.33}$$

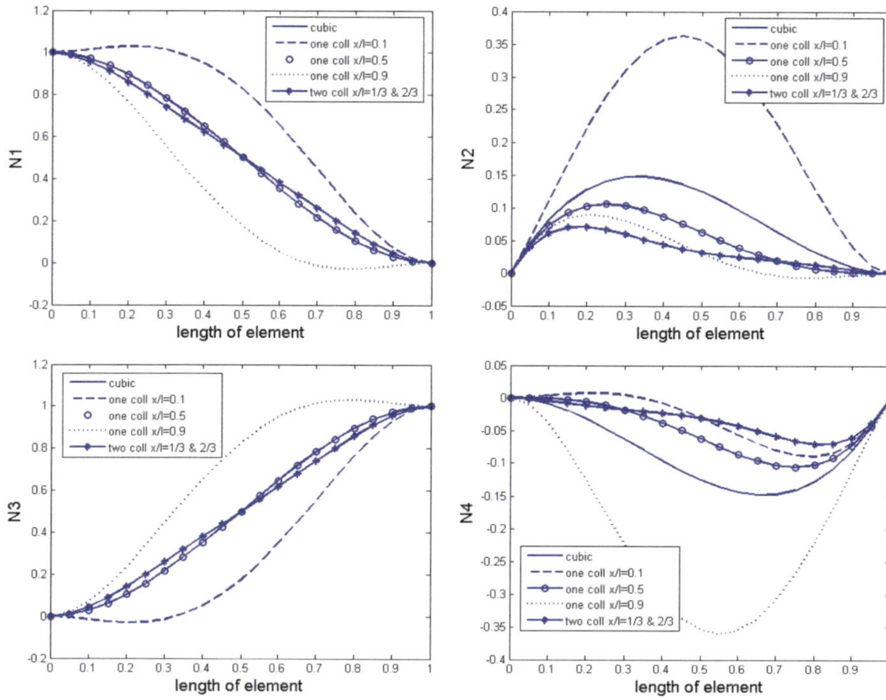

Fig. 6.6 Variation of shape functions along the element using different basis functions at $\lambda = 100$ for the first element ($x_i = 0$) of ten elements used with $l = 1$

Lagrangian is given as,

$$L = T - V \tag{6.34}$$

From Hamilton's principle, integrating Lagrangian over a small time interval and taking variation [2],

$$\delta \int_{t_1}^{t_2} L dt = 0 \tag{6.35}$$

$$\delta \int_{t_1}^{t_2} \int_0^l \left(m(x) \left(\frac{\partial w}{\partial t} \right)^2 - EI(x) \left(\frac{\partial^2 w}{\partial x^2} \right)^2 - T(x) \left(\frac{\partial w}{\partial x} \right)^2 \right) dx dt = 0 \tag{6.36}$$

Integrating this by parts and setting coefficients of δw to zero, governing differential equation is obtained as

$$\frac{d^2}{dx^2} \left(EI(x) \frac{\partial^2 w}{\partial x^2} \right) + m(x) \frac{\partial^2 w}{\partial t^2} - \frac{d}{dx} \left(T(x) \frac{\partial w}{\partial x} \right) = 0 \tag{6.37}$$

Assuming a periodic motion, the weak form is given by

$$\int_{x_a}^{x_b}\left(EI(x)\frac{d^2v}{dx^2}\frac{d^2w}{dx^2}+T(x)\frac{dv}{dx}\frac{dw}{dx}-\omega^2m(x)vw\right)dx+$$
$$\left[v\left\{\frac{d}{dx}\left(EI(x)\frac{d^2w}{dx^2}\right)-T(x)\frac{dw}{dx}\right\}\right]_{x_a}^{x_b}-\left[\frac{dv}{dx}EI(x)\frac{d^2w}{dx^2}\right]_{x_a}^{x_b}=0 \tag{6.38}$$

where v is a weight function. Using finite element approximation, $w=[N]\{w_e\}$, where N and w_e are shape functions and nodal displacements respectively, the mass and stiffness matrices for an element are obtained as,

$$M_i=\int_0^l m_i(\bar{x})N^TNd\bar{x} \tag{6.39}$$

$$K_i=\int_0^l EI_i(\bar{x})N''^TN''d\bar{x}+\int_0^l T_i(\bar{x})N'^TN'd\bar{x} \tag{6.40}$$

where,

$$T_i(\bar{x})=\sum_{j=i}^{N}\int_{x_j}^{x_{j+1}}m_j(\bar{x})\Omega^2\bar{x}dx-\int_{x_i}^{x_{i+1}}m_i(\bar{x})\Omega^2\bar{x}dx \tag{6.41}$$

x_i is given by $(i-1)l$. After obtaining the element stiffness and mass matrices, they are assembled according to nodal connectivity and finally boundary conditions are applied. Frequencies are obtained by solving the eigenvalue problem,

$$K\Phi=\omega^2M\Phi. \tag{6.42}$$

6.4 Numerical Results

6.4.1 Uniform Rotating Beam

The frequencies of a rotating beam at various speeds are calculated using the new basis functions obtained through two point collocation and one collocation point at various locations of beam and compared with those obtained conventionally, using Hermite cubic interpolating polynomials. Tables 6.1 and 6.2 show that the current formulation converges to the well known results from the published literature.

Table 6.3 shows that the number of degrees of freedom required for convergence by the new element is considerably less than that required using Hermite cubic shape functions at high speeds. We can see that the difference in degrees of freedom required is not much for lower speeds, when compared to that of cubic. However, at higher speeds, the new formulation gives convergence of the first mode with almost half the

Table 6.1 Comparison of non-dimensional natural frequencies of a uniform cantilever beam at various rotating speeds

λ	First natural frequency			Second natural frequency		
	Present	Ref. [5]	Ref. [1]	Present	Ref. [5]	Ref. [1]
0	3.5160	3.5160	3.5160	22.0345	22.0345	22.0345
1	3.6816	3.6816	3.6816	22.1810	22.1810	22.1810
2	4.1373	4.1373	4.1373	22.6149	22.6149	22.6149
3	4.7973	4.7973	4.7973	23.3203	23.3203	23.3203
4	5.5850	5.5850	5.5850	24.2733	24.2733	24.2733
5	6.4495	6.4495	6.4495	25.4461	25.4461	25.4461
6	7.3604	7.3604	7.3604	26.8091	26.8091	26.8091
7	8.2996	8.2996	8.2996	28.3341	28.3341	28.3341
8	9.2568	9.2568	9.2568	29.9954	29.9954	29.9954
9	10.2257	10.2257	10.2257	31.7705	31.7705	31.7705
10	11.2023	11.2023	11.2023	33.6404	33.6404	33.6404

Table 6.2 Comparison of non-dimensional natural frequencies of a uniform cantilever beam at ($\lambda = 12$)

Mode	Cubic	One Coll point	Two Coll points	Wang and Wereley [8]	Gunda and Ganguli [3]
1	13.1702	13.1702	13.1702	13.1702	13.1702
2	37.6031	37.6032	37.6031	37.6031	37.6031
3	79.6145	79.6145	79.6145	79.6145	79.6145
4	140.534	140.534	140.534	140.534	140.534

Table 6.3 Comparison of degrees of freedom required using the four types of basis functions to obtain natural or fundamental frequency at various λ with an error <0.00001

Mode	Cubic	One Coll point $\overline{x} = 0.1l$	One Coll point $\overline{x} = 0.5l$	One Coll point $\overline{x} = 0.9l$	Two Coll points $\overline{x} = l/3$ and $2l/3$
$\lambda = 12$					
1	32	26	26	28	28
2	58	66	66	66	66
$\lambda = 50$					
1	152	100	82	96	82
2	164	142	140	144	140
$\lambda = 100$					
1	366	218	168	208	162
2	298	230	216	228	214

degrees of freedom compared to those used in cubic formulation. There is not much advantage gained by using the two point collocation and the one point collocation at the element center is sufficient for getting good results. At higher rotation speeds, Hermite cubic shape functions require more degrees of freedom for convergence of fundamental frequency than those required for higher modes. This is clearly evident from the number of elements required for convergence of modes at $\lambda = 100$ given in Table 6.3. Hence as angular speed increases, convergence of fundamental mode becomes more difficult with cubic shape functions, and the collocation method based shape functions become extremely useful. However, at lower speeds, cubic shape functions are as good as collocation based shape functions.

Convergence of natural frequencies, with number of elements at $\lambda = 12$ and $\lambda = 100$ is shown in Figs. 6.7 and 6.8, respectively. It can be seen that the new basis functions obtained by introducing the collocation point at the center of the beam and two collocation points at $l/3$ and $2l/3$, show superior convergence for the lower modes. At low speeds, the convergence obtained using collocation based shape functions is very close to that of Hermite cubic shape functions, but with increase in rotation speed we obtain superior convergence. Collocation point near the nodes of the beam element, gives poor performance initially but rate of convergence improves fast with number of elements and rotation speed, and even these functions give better

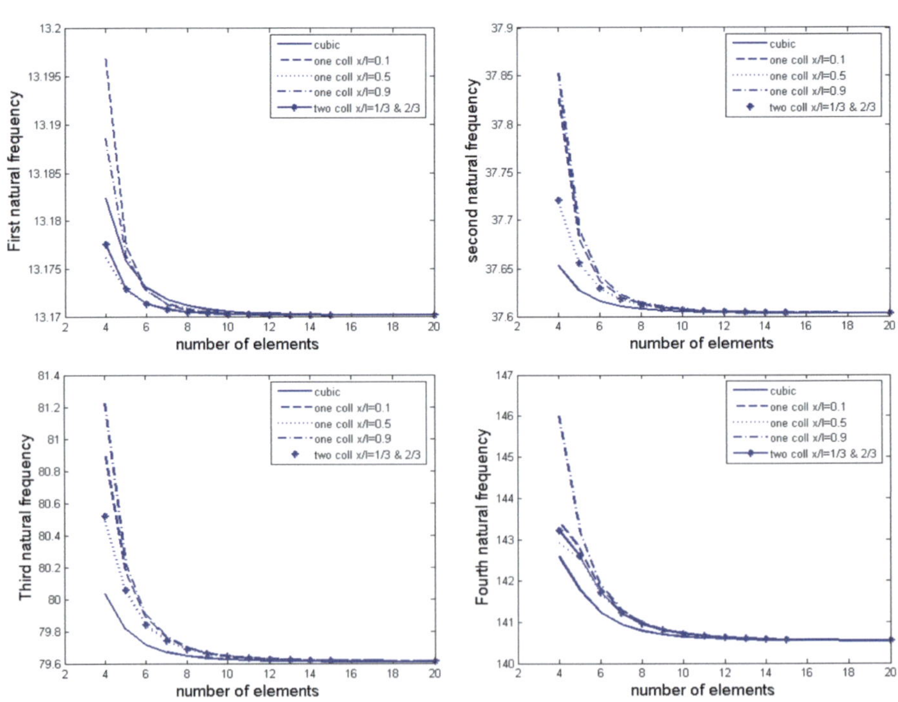

Fig. 6.7 Convergence of natural frequencies at $\lambda = 12$

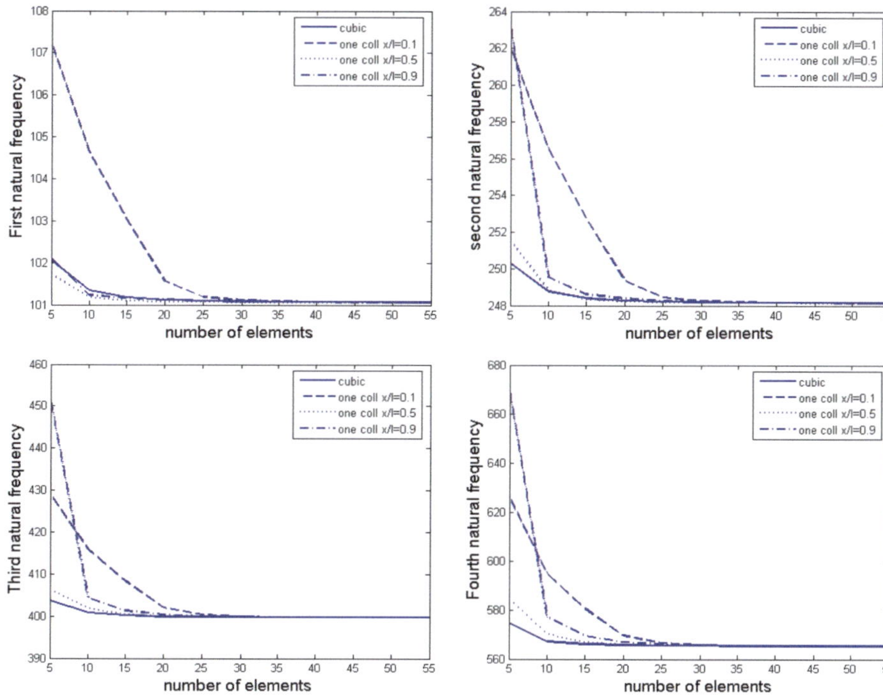

Fig. 6.8 Convergence of natural frequencies at $\lambda = 100$

convergence than the Hermite cubic shape functions for lower modes. Even though the convergence of second mode is initially poor, with increase in rotation speed the new shape functions give better convergence.

In the classical collocation method, the location of the collocation point is known to be important. The number of degrees of freedom required for convergence of first and second mode with the position of collocation point in a beam element at non dimensional rotating speeds $\lambda = 50$ and $\lambda = 100$ is shown in Fig. 6.9. Here convergence of the frequencies to an error less than 0.00001 is considered. We see that the minimum number of degrees of freedom is required when collocation point is located at midpoint of the beam element. The curve is almost symmetric in nature about the element midpoint. The increase in degrees of freedom in comparison with collocation point at midpoint is almost same as the collocation point moves away from midpoint of the beam element in either direction. It can also be observed that slope of curve increases with increase in rotation speed. Hence at higher rotation speeds, position of collocation point is more important as it effects the degrees of freedom for convergence to a greater extent than at lower. Slope of the curve changes even with modes and it is steeper for the fundamental mode and it becomes flatter for second mode. The numerical results verify the intutive selection of the collocation point at the element center.

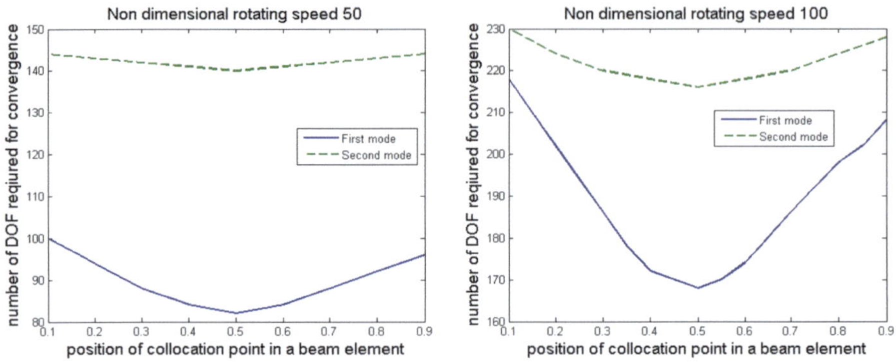

Fig. 6.9 Number of degrees of freedom required for convergence with position of collocation point at $\lambda = 50$ and $\lambda = 100$

6.4.2 *Tapered Rotating Beam*

Two types of linearly tapered beams are used based on the works of Wang and Wereley [8] and Gunda and Ganguli [3]. The variation of mass along the beam is defined as

$$m(\xi) = m_0(1 - \alpha\xi) \tag{6.43}$$

where m_0 is the value of mass per unit length at the thick end of beam ($\xi = 0$), and α is the taper parameter whose value varies from zero to unity. Flexural stiffness distribution is given as

$$EI(\xi) = EI_0(1 - \beta_1\xi - \beta_2\xi^2 - \beta_3\xi^3 - \beta_4\xi^4) \tag{6.44}$$

where EI_0 is the flexural stiffness at thicker end ($\xi = 0$). β_i, $i = 1, 4$, are taper parameters. Two examples are mentioned below.

Example 1 In this example, the mass $m(\xi)$ and stiffness $EI(\xi)$ distribution are given as,

$$m(\xi) = m_0(1 - 0.5\xi) \tag{6.45}$$

and

$$EI(\xi) = EI_0(1 - 0.5\xi)^3 \tag{6.46}$$

where $\alpha = 0.5, \beta_1 = 3\alpha = 1.5, \beta_2 = -3\alpha^2 = -0.75, \beta_3 = \alpha^3 = 0.125$ and $\beta_4 = 0$. The frequencies obtained are compared with Wang and Wereley [8] and Gunda and Ganguli [3] at various rotating speeds and Table 6.4 shows that they confirm well.

Table 6.4 Comparison of non-dimensional natural frequencies of a uniform cantilever beam at various λ (Example 1)

λ	First natural frequency			Second natural frequency		
	Present	Ref. [8]	Ref. [3]	Present	Ref. [8]	Ref. [3]
0	3.8238	3.8238	3.8239	18.3173	18.3173	18.3173
1	3.9866	3.9866	3.9866	18.4740	18.4740	18.4740
2	4.4368	4.4368	4.4368	18.9366	18.9366	18.9366
3	5.0927	5.0927	5.0927	19.6839	19.6839	19.6839
4	5.8788	5.8788	5.8788	20.6852	20.6852	20.6852
5	6.7434	6.7434	6.7434	21.9053	21.9053	21.9053
6	7.6551	7.6551	7.6551	23.3093	23.3093	23.3093
7	8.5956	8.5956	8.5956	24.8647	24.8647	24.8647
8	9.5540	9.5540	9.5540	26.5437	26.5437	26.5437
9	10.5239	10.5239	10.5239	28.3227	28.3227	28.3227
10	11.5015	11.5015	11.5016	30.1827	30.1827	30.1827

Table 6.5 Comparison of non-dimensional natural frequencies of a uniform cantilever beam at various λ (Example 2)

λ	First natural frequency			Second natural frequency		
	Present	Ref. [8]	Ref. [3]	Present	Ref. [8]	Ref. [3]
0	5.2738	5.2738	5.2738	24.0041	24.0041	24.0041
1	5.3903	5.3903	5.3903	24.1069	24.1069	24.1069
2	5.7249	5.7249	5.7249	24.4129	24.4129	24.4130
3	6.2402	6.2402	6.2402	24.9148	24.9148	24.9149
4	6.8928	6.8928	6.8928	25.6013	25.6013	25.6013
5	7.6443	7.6443	7.6443	26.4580	26.4580	26.4581
6	8.4653	8.4653	8.4653	27.4692	27.4692	27.4693
7	9.3347	9.3347	9.3347	28.6184	28.6184	28.6185
8	10.2379	10.2379	10.2379	29.8894	29.8894	29.8894
9	11.1650	11.1651	11.1650	31.2669	31.2669	31.2669
10	12.1092	12.1092	12.1092	32.7367	32.7367	32.7367

Example 2 In this example, the mass $m(\xi)$ and stiffness $EI(\xi)$ distribution are given as,

$$m(\xi) = m_0(1 - 0.8\xi) \tag{6.47}$$

and

$$EI(\xi) = EI_0(1 - 0.95\xi) \tag{6.48}$$

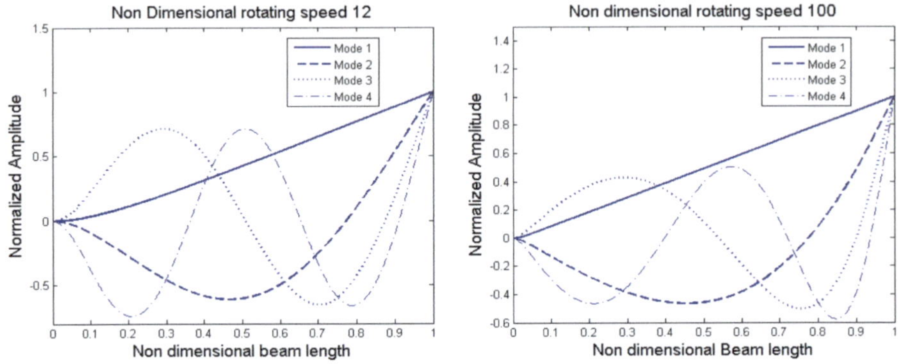

Fig. 6.10 Mode shapes of cantilever beam at $\lambda = 12$ and $\lambda = 100$

where $\alpha = 0.8$, $\beta_1 = 0.95$, $\beta_2 = 0$, $\beta_3 = 0$ and $\beta_4 = 0$. Table 6.5 shows the frequencies obtained for the above mass and stiffness distribution and they compare well with the published literature.

Until, now, we have considered the natural frequency of the beam. It is important to check the mode shapes predicted using the new basis functions. The first four mode shapes of the beam at two non-dimensional rotating speeds $\lambda = 12$ and $\lambda = 100$ are plotted in Fig. 6.10. We can see that amplitude of the beam along the length of the beam reduces with increase in speed which clearly shows the centrifugal stiffening effect due to beam rotation. These results obtained for the different collocation based basis functions match exactly with the mode shapes predicted using the Hermite cubics.

It is conceptually possible to increase the number of collocation points further in the domain. However, the shape functions become very complicated and the numerical results do not show much improvement over the single point collocation approach. For the rotating beam problem, the center point collocation is a good way to accelerate the convergence of finite element analysis at high rotation speeds. The idea of enhancing the basis functions by satisfying the governing differential equations can also be used for other computational science problems.

In this chapter, we have considered a beam undergoing only flapwise bending motion as given in the classical formulation by Hodges and Rutkowski [5]. However, other more sophisticated forms of the rotating beam equations which model axial and edgewise deformations along with nonlinear terms have been proposed by researchers such as Chung and Yoo [1]. Extension of the proposed collocation approach for such coupled nonlinear analysis is a fertile subject for future research.

6.5 Summary

New shape functions are derived by satisfying the static differential equation for a rotating beam at one variable point whose value can be changed within the domain of the element and at two points. These basis functions are functions of rotating speed, material and geometric properties, position of collocation point and location of the element, which affect the centrifugal load acting on the beam element. These refined basis functions show superior convergence for lower modes which are dominated by centrifugal stiffening. For higher modes, this formulation is as good as the Hermite cubic shape functions as high modes are dominated by flexure bending which can be effectively solved using Hermite cubic shape functions. We can see from the number of elements required for achieving desired accuracy in natural frequencies at high speeds, that the new refined basis functions require approximately half the number of elements used in conventional methods. However, collocation point near the nodes when a single collocation point is used, is not very desirable as they give poor performance when compared to the collocation points at the mid point of beam, but are still better than that of cubic for lower modes. Performance of two point collocation for a rotating beam problem is almost as good as collocation point at the midpoint of the beam. However, the two point approach increases the complexity of shape functions and is not very useful when compared to one point collocation. This method of increasing the order of interpolating polynomial without increasing the degrees of freedom of element is quite simple to implement as it only requires changes in shape function in the existing codes. The approach of deriving the basis functions by satisfying the static homogeneous differential equation at one or more collocation points in the domain can be applied not only to a rotating beam problem but also to solve various other problems in computational mechanics which are governed by differential equations which do not have exact solutions.

References

1. Chung J, Yoo HH (2002) Dynamic analysis of a rotating cantilever beam by using the finite element method. J Sound Vib 249(1):147–164
2. Cook RD, Malkus DS, Plesha ME (1989) Concept and Application of Finite Element Analysis, 3rd edn. Wiley, New York
3. Gunda JB, Gupta RK, Ganguli R (2009) Hybrid stiff-string-polynomial basis functions for vibration analysis of high speed rotating beams. Comput Struct 87(3–4):254–265
4. Herrera I, Yates R, Rubio E (2007) Collocation methods: more efficient procedures for applying collocation. Adv Eng Softw 38(10):657–667
5. Hodges DJ, Rutkowsky MJ (1981) Free vibration analysis of rotating beams by a variable order finite element method. AIAA J 19(11):1459–1466
6. Pratap R (2006) Getting Stated with MATLAB 7, 1st edn. Oxford University Press, New Delhi
7. Reddy JN (2004) An Introduction to Finite Element Methods, 3rd edn. McGraw-Hill, New York
8. Wang G, Wereley NM (2004) Free vibration analysis of rotating blades with uniform tapers. AIAA J 42(12):2429–2437
9. Wolfram S (1999) The Mathematica Book, 4th edn. Cambridge University Press, New York

Chapter 7
Rotor Blade Finite Element

Previous chapters have focused on rotating beams with only the out-of-plane motion. In this chapter, we study rotating blades which have in-plane bending, torsion and axial degrees of freedom. Rotor blades are essential and critical components of helicopters, turbines, compressors and other rotatory machinery [1–6]. Blade failure is a severe accident causing the entire machine to shut down. Among the various factors leading to a blade failure, resonance is regarded as the most important [7]. Therefore, determining the free vibration characteristics of the rotating blades is essential toward their reliable design and avoidance of their catastrophic failure.

The accurate prediction of natural frequencies of rotating blades using the finite element methods is an active area of research [8, 9]. In earlier works, as reported in the literature, some researchers have used two and three dimensional finite element models for accurate prediction of the vibration characteristics. Leissa [10] used shell finite elements to study the vibrations of turbine engine blades. Ramamurti and Sreenivasamoorthy [11] used a three dimensional 20 noded isoparametric finite element for the analysis of the stresses in a rotating blade. Bogomolov et al. [12] used a superparametric shell finite element to predict the vibrational characteristics of a real blade. Bucco and Mazumdar [13] and Repetskii [14] extended their works with the shell finite elements to analyze different blade configurations. Jiang et al. [15] studied the non-linear vibrations of rotating beams by using the tetrahedral and the solid finite elements. Qin et al. [16] used a three dimensional finite element model with 8-noded brick element to model a turbine blade. Typically, detailed 3D and shell finite element formulations are useful for the stress analysis.

The two and three dimensional analysis of the whole structure are computationally more expensive and not practical for dynamic analysis and control applications. To reduce the computation effort and computation time for the analysis, some researchers have used one-dimensional beam models for the finite element analysis of rotating blades. Such models are especially useful for vibration control applications where the model size must be kept small [17]. Most works on low order FEM modeling have concentrated on beams undergoing out-of-plane bending motion only. Very

R. Ganguli, *Finite Element Analysis of Rotating Beams*,
Foundations of Engineering Mechanics, DOI 10.1007/978-981-10-1902-9_7

few works have addressed the beam undergoing out-of-plane bending, in-plane bending, torsion and axial deformations. Durocher and Kane [18] investigated the blade deflections with a beam finite element. They included the effects of shear stress, axial-torsion coupling and torsional stiffness in their finite element formulation. Ormiston and Hodges [19] included the effects of pre-cone, variable elastic coupling, pitch-lag coupling and the aerodynamics of induced flow in their formulation and studied the stability characteristics of the rotor blade flap-lag oscillations in hover. In an extension of one-dimensional finite element analysis, Qin and Mao [20] used a shaft finite element model with ten degrees of freedom for the coupled torsional-flexural vibrations of the rotor systems. Similar models have also been developed for rotary shaft systems by Ruhl and Booker [21] and Thorkildson [22]. Belo and Marques [23] used the finite element analysis to analyze a helicopter blade undergoing the coupled motions of flapping, lead-lagging, axial stretching and torsion. They also included the pretwist angle and the offsets between the mass and the elastic axes in their blade model.

Some researchers have also used non-conventional modeling techniques for the vibration analysis of rotating structures. Putter and Manor [24] demonstrated the use of assumed mode approximation method for the modal analysis of a rotating beam. Ramamurti and Kielb [25] employed a plate theory to calculate the natural frequencies of rotating pre-twisted blades. Yoo et al. [26] used a modeling method which employs hybrid deformation variables for the vibration analysis of pre-twisted blades. Banerjee [27, 28] used Dynamic Stiffness Method (DSM) for the free vibration analysis of centrifugally stiffened uniform and tapered beams. Hashemi and Richard [29] combined the well-known weighted residual method, as used in the conventional FEM and some interesting features of DSM, and developed a Dynamic Finite Element (DFE) for vibrational analysis of the rotating structures. Kuang and Hsu [30] used generalized differential quadrature method to calculate the eigensolutions of grouped turbo blades.

A set of governing equations of a rotating blade was developed by Hodges and Dowell [31]. A fifteen degree of freedom FEM model was used by Bir et al. [32] to develop an aeroelastic analysis based on the equations derived in [31]. This FE model included the effects of the coupled motions in flapwise bending, lead-lag bending, axial deformation and torsion. They considered four degrees of freedom in each of flapwise, lead-lag bending and axial deformation, and three degrees of freedom in torsion. In their model, all the shape functions are polynomials. Also, almost all the works on rotating blade finite element analysis use polynomial shape functions, which may not have good convergence properties.

In the present chapter, we seek to develop new shape functions using the exact solutions of the governing static differential equations of a rotating blade coupled in flapwise bending, lead-lag bending, axial deformation and torsion. Current research shows that such an approach leads to an element with superior convergence properties [33–35]. For example, in Reddy's superconvergent Timoshenko beam element [35], the transverse deflection is modeled using a cubic polynomial and the rotation field by a quadratic polynomial. These polynomials are the exact solutions of the homogenous form of the equation from Timoshenko beam theory. The resulting element is free from shear locking and gives exact results at the nodes because the basis from the exact solution is also used to represent the finite element solution. Generally, locking may be avoided if inter-dependence of the approximated fields is not neglected; for example, by considering exact solutions. Following Reddy's work, Gopalakrishnan and his co-workers [33, 34, 36] applied the idea to more complex structures. In general, these studies consider a uniform structure when developing the shape functions which can then be used for non-uniform finite element analysis.

For rotating blade problems, the displacement field resulting from such an approach captures the effect of centrifugal stiffening more accurately. Typically, the shape functions derived by using this approach tend to be the functions of the element cross-sectional properties. However, the research done till now in this area has considered only simple problems where the exact solution could be easily obtained. In this chapter, we consider a much more complex problem and show that by capturing the primary effect of the centrifugal stiffening, shape functions can be derived which have considerably superior convergence characteristics as compared to polynomials.

7.1 Energy Expressions

The energy expressions for a rotating blade including the effects of moderate rotations were derived by Hodges and Dowell [31] and are used in UMARC [32]. This approach assumes moderate deflection in deriving the strain displacement relations and isotropic material properties for the stress-strain relations. The rotor blade is idealized as a beam undergoing flap bending, lag bending, torsion and axial deformation and the blade is pitched at an angle θ_0. For the free vibration analysis of a rotating blade, the variation of strain energy expression is given as:

$$\frac{\delta U_b}{m_0 \Omega^2 L^3} = \int_0^1 \big(U_{u_e'} \delta u_e' + U_{v'} \delta v' + U_{w'} \delta w' + U_{v''} \delta v'' + U_{w''} \delta w'' + U_{\phi} \delta \phi + U_{\phi'} \delta \phi' + U_{\phi''} \delta \phi''\big)\, dx \tag{7.1}$$

In Eq. (7.1), $U_{w'}$ is obtained by collecting together the coefficients of the term $\delta w'$. Other terms are similarly obtained.

Where,

$$U_{u_e'} = EA \left[u_e' + K_A^2 \theta_0' \left(\phi' + w'v''\right) + {K_A}^2 \frac{\phi'^2}{2} \right] - EAe_A \left[v'' \left(\cos\theta_0 - \phi \sin\theta_0\right) + w'' \left(\sin\theta_0 + \phi \cos\theta_0\right) \right] \tag{7.2}$$

$$U_{v'} = 0 \tag{7.3}$$

$$
\begin{aligned}
U_{v''} = {} & v''\left(EI_z\cos^2\theta_0 + EI_y\sin^2\theta_0\right) + w''\left(EI_z - EI_y\right)\cos\theta_0\sin\theta_0 \\
& - EAe_A u_e'\left(\cos\theta_0 - \phi\sin\theta_0\right) - \phi'\mathrm{EB}_2\theta_0'\cos\theta_0 \\
& + w''\phi\left(EI_z - EI_y\right)\cos 2\theta_0 - v''\phi\left(EI_z - EI_y\right)\sin 2\theta_0 \\
& + \left(GJ + \mathrm{EB}_1{\theta'}_0^2\right)\phi' w' + EAK_A^2\theta_0' w' u_e'
\end{aligned}
\tag{7.4}
$$

$$
U_{w'} = \left(GJ + \mathrm{EB}_1\theta_0'^2\right)\phi' v'' + EAK_A^2\theta_0' v'' u_e'
\tag{7.5}
$$

$$
\begin{aligned}
U_{w''} = {} & w''\left(EI_y\cos^2\theta_0 + EI_z\sin^2\theta_0\right) + v''\left(EI_z - EI_y\right)\cos\theta_0\sin\theta_0 \\
& - EAe_A u_e'\left(\sin\theta_0 + \phi\cos\theta_0\right) - \phi'\mathrm{EB}_2\theta_0\sin\theta_0 \\
& + w''\phi\left(EI_z - EI_y\right)\sin 2\theta_0 + v''\phi\left(EI_z - EI_y\right)\cos 2\theta_0 \\
U_\phi = {} & w''^2\left(EI_z - EI_y\right)\sin\theta_0\cos\theta_0 + v''w''\left(EI_{z-EI_y}\right)\cos 2\theta_0 \\
& - v''^2\left(EI_z - EI_y\right)\sin\theta_0\cos\theta_0 \\
U_\phi' = {} & GJ\left(\phi' + w'v''\right) + \mathrm{EB}_1\theta_0'^2\phi' + EAK_A^2\left(\theta_0' + \phi'\right)u_e' \\
& - \mathrm{EB}_2\theta_0'\left(v''\cos\theta_0 + w''\sin\theta_0\right) \\
U_\phi'' = {} & \mathrm{EC}_1\phi'' + \mathrm{EC}_2\left(w''\cos\theta_0 - v''\sin\theta_0\right)
\end{aligned}
\tag{7.6}
$$

Here, E and G are non-dimensionalized with $m_0\Omega^2$. All the dimensional quantities having dimension of length are non-dimensionalized with blade radius L. An antisymmetric warping function, λ_T, is assumed ($\lambda_T \propto \eta\zeta$). Here EC_1 and EC_2 are related to the restraint of warping displacements. Therefore, they are generally more important for the open section beams.

The variation of kinetic energy expression for the blade is given as:

$$
\frac{\delta T_b}{m_0\Omega^2 L^3} = \int_0^1 m\left(T_{u_e}\delta u_e + T_v\delta v + T_w\delta w + T_{v'}\delta v' + T_{w'}\delta w' + T_\phi\delta\phi + T_F\right)dx
\tag{7.7}
$$

Where,

$$
T_{u_e} = x + u_e + 2\dot{v} - \ddot{u}_e
\tag{7.8}
$$

$$
\begin{aligned}
T_v = {} & e_g\left(\cos\theta_0 + \ddot{\theta}_0\sin\theta_0\right) + v - \phi e_g\sin\theta_0 + 2\dot{w}\beta_p + 2\dot{v}' e_g\cos\theta_0 \\
& + 2\dot{w}' e_g\sin\theta_0 - \ddot{v} + \ddot{\phi} e_g\sin\theta_0 - 2\dot{u}_e + 2\int_0^x\left(v'\dot{v}' + w'\dot{w}'\right)d\zeta
\end{aligned}
\tag{7.9}
$$

$$
T_{v'} = -e_g\left(x\cos\theta_0 - \phi x\sin\theta_0 + 2\dot{v}\cos\theta_0\right)
\tag{7.10}
$$

$$
T_w = -x\beta_p - \ddot{\theta}_0 e_g\cos\theta_0 - 2\dot{v}\beta_p - \ddot{w} - \ddot{\phi} e_g\cos\theta_0
\tag{7.11}
$$

$$
T_{w'} = -e_g\left(x\sin\theta_0 + \phi x\cos\theta_0 + 2\dot{v}\sin\theta_0\right)
\tag{7.12}
$$

$$
\begin{aligned}
T_\phi = {} & -K_m^2\ddot{\phi} - \left(K_{m_2}^2 - K_{m_1}^2\right)\cos\theta_0\sin\theta_0 - x\beta_p e_g\cos\theta_0 - v e_g\sin\theta_0 - \ddot{w} e_g\cos\theta_0 \\
& + v' x e_g\sin\theta_0 - w' x e_g\cos\theta_0 + \ddot{v} e_g\sin\theta_0 - \phi\left(K_{m_2}^2 - K_{m_1}^2\right)\cos 2\theta_0 - K_m^2\ddot{\theta}_0
\end{aligned}
\tag{7.13}
$$

$$
T_F = -\left(x + 2\dot{v}\right)\int_0^x\left(v'\delta v' + w'\delta w'\right)d\xi
\tag{7.14}
$$

As indicated by the foreshortening term, T_F, the variation of kinetic energy, δT_b, results in the following double integral expression.

$$-\int_0^1 mT_F dx = \int_0^1 m\,(x + 2\dot{v}) \left[\int_0^x \left(v'\delta v' + w'\delta w' \right) d\xi \right] dx \tag{7.15}$$

Integrating Eq. (7.15) by parts yields,

$$\begin{aligned} -\int_0^1 mT_F dx &= \int_0^1 \left(v'\delta v' + w'\delta w' \right) \left[\int_x^1 m\,(x + 2\dot{v})\, d\xi \right] dx \\ &= \int_0^1 F_A \left(v'\delta v' + w'\delta w' \right) dx \\ &\quad + \int_0^1 \left(v'\delta v' + w'\delta w' \right) \left[\int_x^1 2m\dot{v} d\xi \right] dx \end{aligned} \tag{7.16}$$

with the axial centrifugal force F_A, defined as,

$$F_A(x) = \int_x^1 mx d\xi \tag{7.17}$$

Here, F_A reflects the "centrifugal stiffening" effect on the flap and lag equations.

7.2 Governing Differential Equations

The rotor blades are modeled as Euler–Bernoulli beams undergoing axial, flap, lag and torsion deformations. The chordwise offset of mass centroid (center of gravity) and area centroid (tension axis) from elastic axis, initial precone and pretwist angles are considered in the formulation. Hamilton's variational principle is used to derive the blade equations of motion. In the absence of any external loads, the Hamilton's principle can be written as,

$$\delta \Pi_b = \int_{t_1}^{t_2} (\delta U - \delta T)\, dt = 0 \tag{7.18}$$

where δU is virtual strain energy and δT is virtual kinetic energy.

By using Eqs. (7.1), (7.7) and (7.18), the governing differential equations of a rotating blade are obtained. By solving these differential equations, we can find the displacements u_e, v, w and ϕ. Here u_e and ϕ have continuous second derivatives and v and w have continuous fourth derivatives.

$$m\left(x+u_e+2\dot{v}-\ddot{u}_e\right)+EAu_e''+EAK_A^2\left(\theta_0''\phi'+\theta_0'\phi''\right)-EAe_Av'''\cos\theta_0$$
$$+EAe_Av''\sin\theta_0\theta_0'-EAe_A\left(w''\cos\theta_0\theta_0'+w'''\sin\theta_0\right)=0 \tag{7.19}$$

$$2me_g\cos\theta_0+m\ddot{\theta}_0\sin\theta_0+mv-2m\phi e_g\sin\theta_0+2m\dot{w}\beta_p-m\ddot{v}+4m\dot{v}'$$
$$e_g\cos\theta_0+2m\dot{w}'e_g\sin\theta_0+m\ddot{\phi}e_g\sin\theta_0-2m\dot{u}_e-mxe_g\sin\theta_0\theta_0'-me_g$$
$$\phi'x\sin\theta_0-me_g\phi x\cos\theta_0\theta_0'-2me_g\dot{v}\sin\theta_0\theta_0'-v^{\mathrm{IV}}\left(EI_z\cos^2\theta_0+EI_y\sin^2\theta_0\right)$$
$$+2EI_zv'''\sin 2\theta_0\theta_0'-2EI_yv'''\sin 2\theta_0\theta_0'+2EI_zv''\cos 2\theta_0\theta_0'^2+EI_zv''$$
$$\sin 2\theta_0\theta_0''-2EI_y\cos 2\theta_0v''\theta_0'^2-EI_y\sin 2\theta_0v''\theta_0''-w^{\mathrm{IV}}\left(EI_z-EI_y\right)$$
$$\frac{\sin 2\theta_0}{2}-2w'''\left(EI_z-EI_y\right)\cos 2\theta_0\theta_0'+2w''\left(EI_z-EI_y\right)\sin 2\theta_0\theta_0'^2$$
$$-3\phi'\mathrm{EB}_2\sin\theta_0\theta_0'\theta_0''+EAe_Au_e'''\cos\theta_0-2EAe_Au_e''\sin\theta_0\theta_0'$$
$$-EAe_Au_e'\cos\theta_0\theta_0'^2-EAe_Au_e'\sin\theta_0\theta_0''+\phi'''\mathrm{EB}_2\theta_0'\cos\theta_0$$
$$+2\phi''\mathrm{EB}_2\theta_0''\cos\theta_0-2\phi''\mathrm{EB}_2\sin\theta_0\theta_0'^2-\phi'\mathrm{EB}_2\cos\theta_0\theta_0'^3$$
$$-w''\left(EI_z-EI_y\right)\cos 2\theta_0\theta_0''+\phi'\mathrm{EB}_2\theta_0'''\cos\theta_0+\left[v'\int_x^1 mxd\xi\right]'=0 \tag{7.20}$$

$$-mx\beta_p-m\ddot{\theta}_0e_g\cos\theta_0-2m\dot{v}\beta_p-m\ddot{w}-m\ddot{\phi}e_g\cos\theta_0+me_g\sin\theta_0+me_gx$$
$$\cos\theta_0\theta_0'+me_g\phi'x\cos\theta_0+me_g\phi\cos\theta_0-me_g\phi x\sin\theta_0\theta_0'+2me_g\dot{v}'\sin\theta_0+2me_g\dot{v}$$
$$\cos\theta_0\theta_0'-w^{\mathrm{IV}}\left(EI_y\cos^2\theta_0+EI_z\sin^2\theta_0\right)-2w'''\left(EI_z\sin 2\theta_0-EI_y\sin 2\theta_0\right)$$
$$\theta_0'-2w''EI_z\cos 2\theta_0\theta_0'^2-w''EI_z\sin 2\theta_0\theta_0''+2w''EI_y\cos 2\theta_0\theta_0'^2+2EI_yw''\sin 2\theta_0\theta_0''$$
$$-v^{\mathrm{IV}}\left(EI_z-EI_y\right)\frac{\sin 2\theta_0}{2}-2v'''\left(EI_z-EI_y\right)\cos 2\theta_0\theta_0'+EAe_Au_e'''\sin\theta_0$$
$$+2v''\left(EI_z-EI_y\right)\sin 2\theta_0\theta_0'^2-v''\left(EI_z-EI_y\right)\cos 2\theta_0\theta_0''+2EAe_Au_e''\cos\theta_0\theta_0'$$
$$+EAe_Au_e'\cos\theta_0\theta_0''+EAe_Au_e'\sin\theta_0\theta_0'^2+\phi'''\mathrm{EB}_2\theta_0'\sin\theta_0+2\phi''\mathrm{EB}_2\theta_0''\sin\theta_0$$
$$+2\phi''\mathrm{EB}_2\cos\theta_0\theta_0'^2-\phi'\mathrm{EB}_2\sin\theta_0\theta_0'^3+3\phi'\mathrm{EB}_2\cos\theta_0\theta_0'\theta_0''$$
$$+\phi\mathrm{EB}_2\sin\theta_0\theta_0'''+\left[w'\int_x^1 mxd\xi\right]'=0 \tag{7.21}$$

$$-mK_m^2\ddot{\phi}-m\left(K_{m_2}^2-K_{m_1}^2\right)\frac{\sin 2\theta_0}{2}-mx\beta_pe_g\cos\theta_0-mve_g\sin\theta_0+mv'xe_g\sin\theta_0$$
$$-mw'xe_g\cos\theta_0-m\phi\left(K_{m_2}^2-K_{m_1}^2\right)\cos 2\theta_0+m\ddot{v}e_g\sin\theta_0-m\ddot{w}e_g\cos\theta_0-mK_m^2\ddot{\theta}_0$$
$$+GJ\phi''+\mathrm{EB}_1\theta_0'^2\phi''+2\mathrm{EB}_1\theta_0'\theta_0''\phi'+EAK_A^2\theta_0''u_e'+EAK_A^2\theta_0'u_e''-\mathrm{EB}_2\theta_0'v'''\cos\theta_0$$
$$-\mathrm{EB}_2\theta_0'w'''\sin\theta_0-\mathrm{EB}_2\theta_0''\left(v''\cos\theta_0+w''\sin\theta_0\right)+\mathrm{EB}_2v''\sin\theta_0\theta_0'^2-\mathrm{EB}_2w''$$
$$\cos\theta_0\theta_0'^2-\mathrm{EC}_1\phi^{\mathrm{IV}}-\mathrm{EC}_2w^{\mathrm{IV}}\cos\theta_0+2\mathrm{EC}_2w'''\sin\theta_0\theta_0'+\mathrm{EC}_2w''\cos\theta_0\theta_0'^2$$
$$+\mathrm{EC}_2w''\sin\theta_0\theta_0''+\mathrm{EC}_2v^{\mathrm{IV}}\sin\theta_0+2\mathrm{EC}_2v'''\cos\theta_0\theta_0'$$
$$+\mathrm{EC}_2v''\cos\theta_0\theta_0''-\mathrm{EC}_2v''\sin\theta_0\theta_0'^2=0 \tag{7.22}$$

7.3 Derivation of the Shape Functions

Ideally, the shape functions could be derived using exact solutions of the static part of the differential equations given by Eqs. (7.19)–(7.22). However, these equations are complicated due to coupling terms. For example, the presence of twist θ_0 strongly couples the lag equation (7.20) and flap equation (7.21). Our aim is to capture the effect of centrifugal stiffening in the shape functions, as it is the primary effect of rotation. For simplifying the derivation of the shape functions, we set a number of variables to zero. These variables are $\theta_0,\ \theta_0',\ \theta_0'',\ \dot{\theta}_0,\ \ddot{\theta}_0,\ e_g,\ \beta_p$ and e_A. We also set $\mathrm{EC}_1,\ \mathrm{EC}_2,\ \mathrm{EB}_1$ and EB_2 equal to zero, as they are higher order effects relating to warping.

By making these assumptions, the non-dimensional differential equations (7.19)–(7.22) reduce to,

$$EAu_e'' + m\left(x + u_e + 2\dot{v} - \ddot{u}_e\right) = 0 \tag{7.23}$$

$$EI_z v^{\mathrm{IV}} - mv + m\ddot{v} + 2m\dot{u}_e - \left[v' \int_x^1 mx d\xi\right]' = 0 \tag{7.24}$$

$$EI_y w^{\mathrm{IV}} + m\ddot{w} - \left[w' \int_x^1 mx d\xi\right]' = 0 \tag{7.25}$$

$$GJ\phi'' - mK_m^2\ddot{\phi} - m\phi\left(K_{m_2}^2 - K_{m_1}^2\right) = 0 \tag{7.26}$$

To derive the shape functions based on the logic discussed in the introduction section, we remove the inertial and velocity terms to obtain the static part of the governing homogenous differential equations. These equations in dimensional form are given as:

$$EAu_e'' + m\Omega^2 u_e = 0 \tag{7.27}$$

$$EI_z v^{\mathrm{IV}} - m\Omega^2 v - \left[v' \int_x^L m\Omega^2 x d\xi\right]' = 0 \tag{7.28}$$

$$EI_y w^{\mathrm{IV}} - \left[w' \int_x^L m\Omega^2 x d\xi\right]' = 0 \tag{7.29}$$

$$GJ\phi'' - m\Omega^2\phi\left(K_{m_2}^2 - K_{m_1}^2\right) = 0 \tag{7.30}$$

where,

$$T = \int_x^L m\Omega^2 x d\xi \tag{7.31}$$

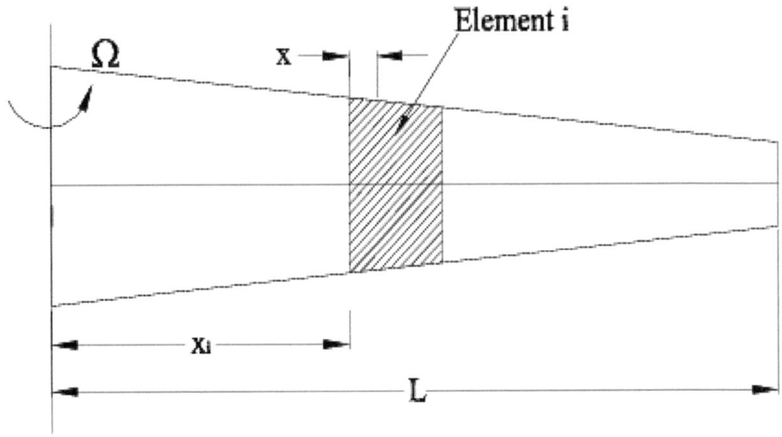

Fig. 7.1 Element geometry of the rotating blade

The axial and torsion equations are second order equations and can be solved for constant coefficients. Therefore they 'strongly' satisfy the requirements for a superconvergent element. But the exact solutions for the flap and the lag equations are not possible, even for a uniform blade. In these bending cases, we assume T as a constant value. The dimensionalized differential equations (7.27)–(7.31) are solved using *Maple*^{TM} [39] by assuming constant tension T_i prevailing in the ith finite element. For the full blade, the approximation of the constant tension is not appropriate. But, if we discretize the blade into N number of finite elements, then the approximation of constant tension for the finite element becomes realistic. We can say that the bending shape functions satisfy the superconvergence requirements in a 'weak' sense.

The constant tension T_i for the element is approximated by taking the average centrifugal tension in the element. The centrifugal tension T_i for the ith element as shown in Fig. 7.1 can be expressed as:

$$T_i = \int_{x_i}^{L} m_i \Omega^2 x dx \tag{7.32}$$

Here x_i is the location of the left edge of the finite element. The solutions of the differential equations (7.27)–(7.31) are obtained as:

$$u_e = C_1 \sin\left(\sqrt{\frac{m\Omega^2}{EA}} x\right) + C_2 \cos\left(\sqrt{\frac{m\Omega^2}{EA}} x\right) \tag{7.33}$$

$$v = C_3 exp\left[-\sqrt{\frac{T - \sqrt{T^2 + 4EI_z m\Omega^2}}{2EI_z}} x\right] + C_4 exp\left[\sqrt{\frac{T - \sqrt{T^2 + 4EI_z m\Omega^2}}{2EI_z}} x\right]$$

$$+ C_5 exp\left[-\sqrt{\frac{T+\sqrt{T^2+4EI_z m\Omega^2}}{2EI_z}}x\right]$$

$$+ C_6 exp\left[\sqrt{\frac{T+\sqrt{T^2+4EI_z m\Omega^2}}{2EI_z}}x\right] \tag{7.34}$$

$$w = C_7 + C_8 x + C_9 exp\left[\sqrt{\frac{T}{EI_y}}x\right] + C_{10} exp\left[-\sqrt{\frac{T}{EI_y}}x\right] \tag{7.35}$$

$$\phi = C_{11} exp\left[\sqrt{\frac{m\Omega^2\left(K_{m_2}^2 - K_{m_1}^2\right)}{GJ}}x\right]$$

$$+ C_{12} exp\left[-\sqrt{\frac{m\Omega^2\left(K_{m_2}^2 - K_{m_1}^2\right)}{GJ}}x\right] \tag{7.36}$$

These solutions are used to derive the new shape functions. The new shape functions so derived satisfy the static homogenous part of the governing differential equations at the elemental level. For this reason, these new shape functions have better convergence properties, as discussed in literature [33, 34] and proved later by numerical studies. These new shape functions are functions of the rotational speed, blade length, element mass, element stiffness and the location of the element from the root. So, they capture the effect of the rotation speed and the element location from the rotation axis on the element displacements, which is not possible with Hermite cubic or linear shape functions used conventionally for the finite element analysis of the rotating structures. The new shape functions for each type of motion are studied in detail in the next few sections.

7.3.1 Shape Functions for Flapwise Bending

Consider a finite element for flapwise bending with displacement and slope degrees of freedom at the two ends, as shown in Fig. 7.2. Let

$$\lambda_1 = \sqrt{\frac{T_i}{EI_y}} \tag{7.37}$$

Writing the solution of the differential equation (7.35) for flapwise bending in terms of λ_1, we get,

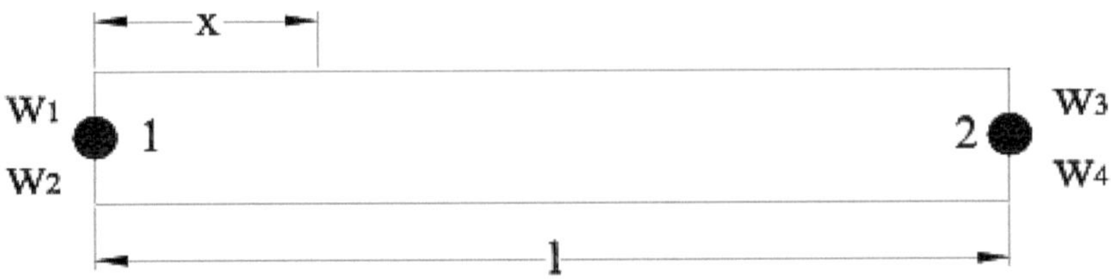

Fig. 7.2 Flapwise degrees of freedom of a rotating blade

$$w = C_7 + C_8 x + C_9 e^{\lambda_1 x} + C_{10} e^{-\lambda_1 x} \tag{7.38}$$

Boundary conditions are: $w(0) = w_1$, $\frac{dw(0)}{dx} = w'(0) = w_2$, $w(l) = w_3$ and $\frac{dw(l)}{dx} = w'(l) = w_4$. Putting these boundary conditions into Eq. (7.38), we get,

$$w_1 = C_7 + C_9 + C_{10} \tag{7.39}$$

$$w_2 = C_8 + \lambda_1 C_9 - \lambda_1 C_{10} \tag{7.40}$$

$$w_3 = C_7 + C_8 l + C_9 e^{\lambda_1 l} + C_{10} e^{-\lambda_1 l} \tag{7.41}$$

$$w_4 = C_8 + \lambda_1 C_9 e^{\lambda_1 l} - \lambda_1 C_{10} e^{-\lambda_1 l} \tag{7.42}$$

From these equations, the values of C_7, C_8, C_9 and C_{10} can be found out. So the shape functions are derived as,

$$\begin{aligned} H_{w_1} = \alpha \big[& -\lambda_1^2 l e^{\lambda_1 l} - 2\lambda_1 + \lambda_1 e^{\lambda_1 l} + \lambda_1 e^{-\lambda_1 l} - \lambda_1 e^{-\lambda_1 x} + \lambda_1 e^{(\lambda_1 x - \lambda_1 l)} \\ & - \lambda_1 e^{\lambda_1 x} - \lambda_1^2 x e^{-\lambda_1 l} + \lambda_1^2 x e^{\lambda_1 l} + \lambda_1 e^{(-\lambda_1 x + \lambda_1 l)} + \lambda_1^2 l e^{-\lambda_1 l} \big] \end{aligned} \tag{7.43}$$

$$\begin{aligned} H_{w_2} = \alpha \big[& -e^{\lambda_1 x} - e^{-\lambda_1 l} + e^{\lambda_1 l} - 2\lambda_1 x + e^{-\lambda_1 x} - e^{(-\lambda_1 x + \lambda_1 l)} - \lambda_1 l e^{\lambda_1 l} \\ & + \lambda_1 l e^{(\lambda_1 x - \lambda_1 l)} + e^{(\lambda_1 x - \lambda_1 l)} 1 l + \lambda_1 x e^{-\lambda_1} + \lambda_1 l e^{(-\lambda_1 x + \lambda_1 l)} \\ & - \lambda_1 l e^{-\lambda_1 l} + \lambda_1 x e^{\lambda_1 l} \big] \end{aligned} \tag{7.44}$$

$$\begin{aligned} H_{w_3} = \alpha \big[& -2\lambda_1 + \lambda_1 e^{\lambda_1 l} + \lambda_1 e^{-\lambda_1 l} + \lambda_1 e^{\lambda_1 x} + \lambda_1 e^{-\lambda_1 x} - \lambda_1 e^{(\lambda_1 x - \lambda_1 l)} \\ & + \lambda_1^2 x e^{-\lambda_1 l} - \lambda_1^2 x e^{\lambda_1 l} - \lambda_1 e^{(-\lambda_1 x + \lambda_1 l)} \big] \end{aligned} \tag{7.45}$$

$$\begin{aligned} H_{w_4} = \alpha \big[& e^{-\lambda_1 l} + e^{(-\lambda_1 x + \lambda_1 l)} + 2\lambda_1 l - 2\lambda_1 x - e^{\lambda_1 l} + e^{\lambda_1 x} - e^{-\lambda_1 x} \\ & - e^{(\lambda_1 x - \lambda_1 l)} + \lambda_1 x e^{\lambda_1 l} + \lambda_1 x e^{-\lambda_1 l} - \lambda_1 l e^{\lambda_1 x} - \lambda_1 l e^{-\lambda_1 x} \big] \end{aligned} \tag{7.46}$$

where

$$\alpha = \frac{1}{\lambda_1 \left\{ \lambda_1 l e^{-\lambda_1 l} - 4 + 2e^{-\lambda_1 l} - \lambda_1 l e^{\lambda_1 l} + 2e^{\lambda_1 l} \right\}} \tag{7.47}$$

Thus, the flapwise bending shape functions are given as:

$$\mathbf{H}_w = \left[H_{w_1} \; H_{w_2} \; H_{w_3} \; H_{w_4} \right] \tag{7.48}$$

These shape functions become Hermite cubics as the analytical limit of the rotation speed tends to zero.

$$\lim_{\Omega \to 0} H_{w_1} = \frac{2x^3 - 3x^2 l + l^3}{l^3}, \qquad \lim_{\Omega \to 0} H_{w_2} = \frac{x^3 - 2x^2 l + x l^2}{l^2} \tag{7.49}$$

$$\lim_{\Omega \to 0} H_{w_3} = \frac{-2x^3 + 3x^2 l}{l^3}, \qquad \lim_{\Omega \to 0} H_{w_4} = \frac{-x^2 l + x^3}{l^2} \tag{7.50}$$

As the rotation speed tends to infinity, the shape functions H_{w_1} and H_{w_3} become linear and H_{w_2} and H_{w_4} vanish, which are the solutions for a stiff string.

$$\lim_{\Omega \to \infty} H_{w_1} = 1 - \frac{x}{l}, \qquad \lim_{\Omega \to \infty} H_{w_2} = 0 \tag{7.51}$$

$$\lim_{\Omega \to \infty} H_{w_3} = \frac{x}{l}, \qquad \lim_{\Omega \to \infty} H_{w_4} = 0 \tag{7.52}$$

The asymptotic behavior show the beauty of the new shape functions as they capture the effects of both zero and very high rotation speeds.

7.3.2 *Shape Functions for Lead-Lag Bending*

The lead-lag degrees of freedom are shown in Fig. 7.3. Again, displacement and slope are taken at both the ends.

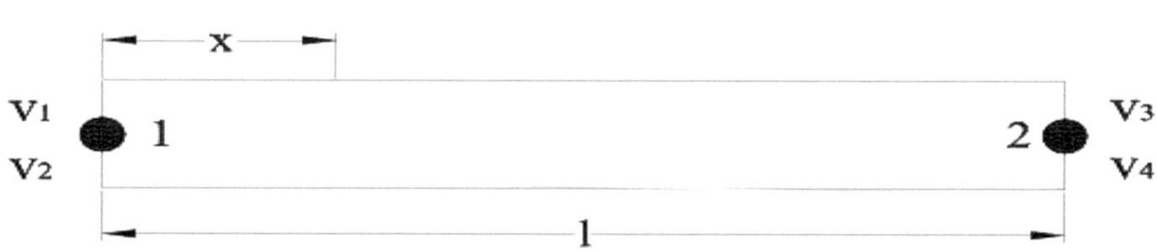

Fig. 7.3 Lead-lag degrees of freedom of a rotating blade

Let

$$\lambda_2 = \sqrt{\frac{T_i - \sqrt{T_i^2 + 4EI_z m\Omega^2}}{2EI_z}}, \qquad \lambda_3 = \sqrt{\frac{T_i + \sqrt{T_i^2 + 4EI_z m\Omega^2}}{2EI_z}} \tag{7.53}$$

Writing the solution of the differential equation (7.53) for lead-lag bending in terms of λ_2 and λ_3, we get,

$$v = C_3 e^{-\lambda_2 x} + C_4 e^{\lambda_2 x} + C_5 e^{-\lambda_3 x} + C_6 e^{\lambda_3 x} \tag{7.54}$$

Boundary conditions are: $v(0) = v_1$, $\frac{dv(0)}{dx} = v'(0) = v_2, v(l) = v_3$ and $\frac{dv(l)}{dx} = v'(l) = v_4$ Putting these boundary conditions into Eq. (7.54), we get,

$$v_1 = C_3 + C_4 + C_5 + C_6 \tag{7.55}$$
$$v_2 = -\lambda_2 C_3 + \lambda_2 C_4 - \lambda_3 C_5 + \lambda_3 C_6 \tag{7.56}$$
$$v_3 = C_3 e^{-\lambda_2 l} + C_4 e^{\lambda_2 l} + C_5 e^{-\lambda_3 l} + C_6 e^{-\lambda_3 l} \tag{7.57}$$
$$v_4 = -\lambda_2 e^{-\lambda_2 l} C_3 + \lambda_2 e^{\lambda_2 l} C_4 - \lambda_3 e^{-\lambda_3 l} C_5 + C_6 \lambda_3 e^{\lambda_3 l} \tag{7.58}$$

The values of C_3, C_4, C_5 and C_6 can be found out and the shape functions are derived as,

$$\begin{aligned} H_{v_1} = \frac{1}{\beta}\Big[&-2\lambda_2\lambda_3 e^{-\lambda_2 x} + \lambda_2\lambda_3 e^{(\lambda_3 x - \lambda_2 l - \lambda_3 l)} + \lambda_2\lambda_3 e^{(\lambda_2 x - \lambda_2 l + \lambda_3 l)} \\ &+ \lambda_2\lambda_3 e^{(-\lambda_3 x + \lambda_2 l + \lambda_3 l)} + \lambda_2\lambda_3 e^{(-\lambda_2 x + \lambda_2 l + \lambda_3 l)} + \lambda_2\lambda_3 e^{(-\lambda_2 x + \lambda_2 l - \lambda_3 l)} \\ &\lambda_2\lambda_3 e^{(\lambda_2 x - \lambda_2 l - \lambda_3 l)} + \lambda_2\lambda_3 e^{(\lambda_3 x + \lambda_2 l - \lambda_3 l)} - \lambda_2^2 e^{(\lambda_3 x - \lambda_2 l - \lambda_3 l)} \\ &+ \lambda_2\lambda_3 e^{(-\lambda_3 x - \lambda_2 l + \lambda_3 l)} + \lambda_2^2 e^{(-\lambda_3 x - \lambda_2 l + \lambda_3 l)} - \lambda_3^2 e^{(-\lambda_2 x + \lambda_2 l + \lambda_3 l)} \\ &+ \lambda_3^2 e^{(-\lambda_2 x + \lambda_2 l - \lambda_3 l)} - \lambda_3^2 e^{(\lambda_2 x - \lambda_2 l - \lambda_3 l)} + \lambda_2^2 e^{(\lambda_3 x + \lambda_2 l - \lambda_3 l)} \\ &+ \lambda_3^2 e^{(\lambda_2 x - \lambda_2 l + \lambda_3 l)} - \lambda_2^2 e^{(-\lambda_3 x + \lambda_2 l + \lambda_3 l)} \\ &- 2\lambda_2\lambda_3 e^{\lambda_3 x} - 2\lambda_2\lambda_3 e^{\lambda_2 x} - 2\lambda_2\lambda_3 e^{-\lambda_3 x}\Big] \end{aligned} \tag{7.59}$$

$$\begin{aligned} H_{v_2} = \frac{1}{\beta}\Big[&-2\lambda_3 e^{\lambda_2 x} + 2\lambda_2 e^{-\lambda_3 x} - 2\lambda_2 e^{\lambda_3 x} - \lambda_2 e^{(-\lambda_2 x + \lambda_2 l - \lambda_3 l)} + 2\lambda_3 e^{-\lambda_2 x} \\ &+ \lambda_3 e^{(\lambda_2 x - \lambda_2 l + \lambda_3 l)} + \lambda_2 e^{(\lambda_3 x - \lambda_2 l - \lambda_3 l)} + \lambda_3 e^{(-\lambda_3 x + \lambda_2 l + \lambda_3 l)} \\ &- \lambda_3 e^{(-\lambda_2 x + \lambda_2 l - \lambda_3 l)} - \lambda_3 e^{(\lambda_3 x - \lambda_2 l - \lambda_3 l)} + \lambda_3 e^{(\lambda_3 x + \lambda_2 l - \lambda_3 l)} \\ &- \lambda_2 e^{(-\lambda_3 x - \lambda_2 l + \lambda_3 l)} - \lambda_3 e^{(-\lambda_3 x - \lambda_2 l + \lambda_3 l)} - \lambda_2 e^{(-\lambda_3 x + \lambda_2 l + \lambda_3 l)} \\ &- \lambda_3 e^{(-\lambda_2 x + \lambda_2 l + \lambda_3 l)} + \lambda_3 e^{(\lambda_2 x - \lambda_2 l - \lambda_3 l)} + \lambda_2 e^{(\lambda_2 x - \lambda_2 l + \lambda_3 l)} \\ &+ \lambda_2 e^{(\lambda_3 x + \lambda_2 l - \lambda_3 l)} + \lambda_2 e^{(-\lambda_2 x + \lambda_2 l + \lambda_3 l)} - \lambda_2 e^{(\lambda_2 x - \lambda_2 l - \lambda_3 l)}\Big] \end{aligned} \tag{7.60}$$

$$
\begin{aligned}
H_{v_3} = \frac{1}{\beta}\Big[&\lambda_2\lambda_3 e^{(-\lambda_3 x-\lambda_2 l)} + \lambda_3^2 e^{(\lambda_2 x-\lambda_3 l)} + \lambda_2\lambda_3 e^{(\lambda_2 x+\lambda_3 l)} + \lambda_2^2 e^{(-\lambda_3 x+\lambda_2 l)} \\
&- \lambda_3^2 e^{(-\lambda_2 x-\lambda_3 l)} - 2\lambda_2\lambda_3 e^{(-\lambda_2 x+\lambda_2 l)} - \lambda_3^2 e^{(\lambda_2 x+\lambda_3 l)} + \lambda_3^2 e^{(-\lambda_2 x+\lambda_3 l)} \\
&- \lambda_2^2 e^{(\lambda_3 x+\lambda_2 l)} - 2\lambda_2\lambda_3 e^{(\lambda_3 x-\lambda_3 l)} - 2\lambda_2\lambda_3 e^{(\lambda_2 x-\lambda_2 l)} - 2\lambda_2\lambda_3 e^{(-\lambda_3 x+\lambda_3 l)} \\
&+ \lambda_2\lambda_3 e^{(\lambda_3 x-\lambda_2 l)} + \lambda_2\lambda_3 e^{(-\lambda_2 x-\lambda_3 l)} + \lambda_2\lambda_3 e^{(\lambda_2 x-\lambda_3 l)} + \lambda_2\lambda_3 e^{(-\lambda_3 x+\lambda_2 l)} \\
&+ \lambda_2^2 e^{(\lambda_3 x-\lambda_2 l)} + \lambda_2\lambda_3 e^{(-\lambda_2 x+\lambda_3 l)} - \lambda_2^2 e^{(-\lambda_3 x-\lambda_2 l)} + \lambda_2\lambda_3 e^{(\lambda_3 x+\lambda_2 l)}\Big] \quad (7.61)
\end{aligned}
$$

$$
\begin{aligned}
H_{v_4} = \frac{1}{\beta}\Big[&-\lambda_3 e^{(-\lambda_2 x+\lambda_3 l)} + 2\lambda_3 e^{(-\lambda_2 x+\lambda_2 l)} + \lambda_2 e^{(-\lambda_2 x-\lambda_3 l)} - \lambda_3 e^{(-\lambda_3 x+\lambda_2 l)} \\
&- \lambda_3 e^{(-\lambda_2 x-\lambda_3 l)} + \lambda_3 e^{(\lambda_3 x-\lambda_2 l)} - \lambda_2 e^{(-\lambda_2 x+\lambda_3 l)} - \lambda_2 e^{(-\lambda_3 x+\lambda_2 l)} \\
&+ \lambda_3 e^{(\lambda_2 x+\lambda_3 l)} - 2\lambda_2 e^{(\lambda_3 x-\lambda_3 l)} + \lambda_3 e^{(\lambda_2 x-\lambda_3 l)} + \lambda_2 e^{(\lambda_3 x+\lambda_2 l)} \\
&- \lambda_3 e^{(\lambda_3 x+\lambda_2 l)} + \lambda_2 e^{(\lambda_3 x-\lambda_2 l)} + \lambda_2 e^{(\lambda_2 x-\lambda_3 l)} - 2\lambda_3 e^{(\lambda_2 x-\lambda_2 l)} \\
&- \lambda_2 e^{(-\lambda_3 x-\lambda_2 l)} + 2\lambda_2 e^{(-\lambda_3 x+\lambda_3 l)} - \lambda_2 e^{(\lambda_2 x+\lambda_3 l)} + \lambda_3 e^{(-\lambda_3 x-\lambda_2 l)}\Big] \quad (7.62)
\end{aligned}
$$

where

$$
\begin{aligned}
\beta = {} & 2\lambda_2\lambda_3 e^{(\lambda_2-\lambda_3)l} - \lambda_3^2 e^{(\lambda_2+\lambda_3)l} - \lambda_2^2 e^{(\lambda_2+\lambda_3)l} + \lambda_3^2 e^{(\lambda_2-\lambda_3)l} - 8\lambda_2\lambda_3 \\
&+ \lambda_2^2 e^{(\lambda_2-\lambda_3)l} + 2\lambda_2\lambda_3 e^{(\lambda_2+\lambda_3)l} + \lambda_3^2 e^{(\lambda_3-\lambda_2)l} + \lambda_2^2 e^{(\lambda_3-\lambda_2)l} \\
&- \lambda_3^2 e^{-(\lambda_2+\lambda_3)l} - \lambda_2^2 e^{-(\lambda_2+\lambda_3)l} + 2\lambda_2\lambda_3 e^{-(\lambda_2+\lambda_3)l} + 2\lambda_2\lambda_3 e^{(\lambda_3-\lambda_2)l} \quad (7.63)
\end{aligned}
$$

Thus, the lead-lag bending shape functions are given as:

$$\mathbf{H}_v = \left[H_{v_1}\ H_{v_2}\ H_{v_3}\ H_{v_4}\right] \quad (7.64)$$

These shape functions become Hermite cubics as the analytical limit of the rotation speed tends to zero as in Eqs. (7.59) and (7.60).

7.3.3 Shape Functions for Axial Deflection

For axial deflection, two end nodes with displacement degrees of freedom are used as shown in Fig. 7.4.

Let

$$\lambda_4 = \sqrt{\frac{m\Omega^2}{EA}} \quad (7.65)$$

Writing the solution of the differential equation (7.65) for axial deformation in terms of λ_4, we get,

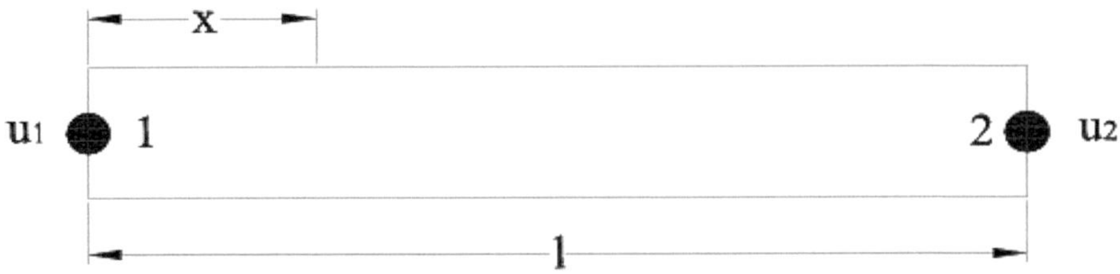

Fig. 7.4 Axial degrees of freedom of a rotating blade

$$u_e = C_1 \sin(\lambda_4 x) + C_2 \cos(\lambda_4 x) \tag{7.66}$$

Boundary conditions are:

$$u(0) = u_1 \qquad \textit{and} \qquad u(l) = u_2 \tag{7.67}$$

Putting these boundary conditions into Eq. (7.66), we get,

$$u_1 = C_2 \tag{7.68}$$

$$u_2 = C_1 \sin \lambda_4 l + C_2 \cos \lambda_4 l \tag{7.69}$$

From these equations, the values of C_1 and C_2 can be calculated. So the shape functions are derived as,

$$H_{u_1} = \cos \lambda_4 x - \cot \lambda_4 l \; \sin \lambda_4 x \tag{7.70}$$

$$H_{u_2} = \frac{\sin \lambda_4 x}{\sin \lambda_4 l} \tag{7.71}$$

Thus, the axial deformation shape functions are given as:

$$\mathbf{H}_u = \left[H_{u_1} \; H_{u_2}\right] \tag{7.72}$$

These shape functions become linear as the analytical limit of the rotation speed tends to zero.

$$\lim_{\Omega \to 0} H_{u_1} = 1 - \frac{x}{l}, \lim_{\Omega \to 0} H_{u_2} = \frac{x}{l} \tag{7.73}$$

7.3.4 Shape Functions for Torsion

Two end nodes with elastic twist degrees of freedom are considered for the torsion motion as shown in Fig. 7.5.

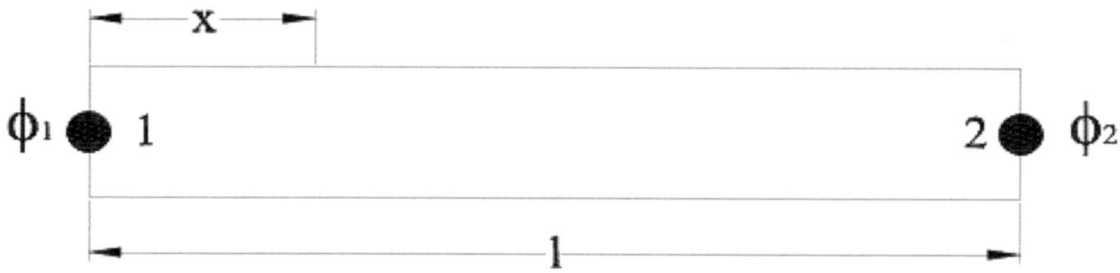

Fig. 7.5 Torsion degrees of freedom of a rotating blade

Let

$$\lambda_5 = \sqrt{\frac{m\Omega^2\left(K_{m_2}^2 - K_{m_1}^2\right)}{GJ}} \tag{7.74}$$

Writing the solution of the differential equation (7.74) for torsion in terms of λ_5, we get,

$$\phi = C_{11}e^{\lambda_5 x} + C_{12}e^{-\lambda_5 x} \tag{7.75}$$

Boundary conditions are:

$$\phi(0) = \phi_1 \quad \textit{and} \quad \phi(l) = \phi_2 \tag{7.76}$$

Putting these boundary conditions into Eq. (7.75), we get,

$$\phi_1 = C_{11} + C_{12} \tag{7.77}$$
$$\phi_2 = C_{11}e^{\lambda_5 l} + C_{12}e^{-\lambda_5 l} \tag{7.78}$$

From these equations, the values of C_{11} and C_{12} can be calculated. So the shape functions are derived as,

$$H_{\phi_1} = \frac{e^{\lambda_5(x-l)} - e^{\lambda_5(l-x)}}{e^{-\lambda_5 l} - e^{\lambda_5 l}} \tag{7.79}$$

$$H_{\phi_2} = \frac{e^{-\lambda_5 x} - e^{\lambda_5 x}}{e^{-\lambda_5 l} - e^{\lambda_5 l}} \tag{7.80}$$

Thus, the torsion shape functions are given as:

$$\mathbf{H}_\phi = \left[H_{\phi_1}\ H_{\phi_2}\right] \tag{7.81}$$

These shape functions become linear as the analytical limit of the rotation speed tends to zero as in Eq. (7.73). At infinite rotation speed, they vanish.

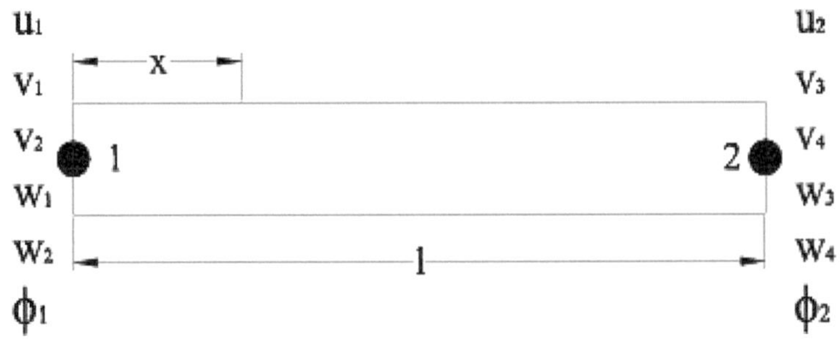

Fig. 7.6 Twelve degrees of freedom finite element model of a rotating blade

7.4 Finite Element Method

The shape functions, derived in the previous section, are used to create a new finite element. The potential energy expression, kinetic energy expression and Hamilton's principle, given by Eqs. (7.1), (7.7) and (7.18) are used. The new finite element is a twelve degrees of freedom model, having four degrees of freedom in each of flapwise bending and lead-lag bending, and two degrees of freedom in each of axial deformation and torsion. The new finite element is shown in Fig. 7.6.

The element mass matrix, damping matrix, stiffness matrix and load vector are given in Appendix 1. The natural frequencies are obtained by assembling the element mass and stiffness matrix, applying the cantilever boundary condition and solving the eigenvalue problem

$$\overline{\mathbf{K}}\phi = \omega^2 \overline{\mathbf{M}}\phi \tag{7.82}$$

where $\overline{\mathbf{K}}$ and $\overline{\mathbf{M}}$ are assembled stiffness and mass matrix, ω is the natural frequency and ϕ is the eigenvector.

7.5 Numerical Results

7.5.1 Analysis of Shape Functions

As a first step, we analyze the behavior of the new shape functions at different rotation speeds. For this comparison, the properties used for a rotating cantilever blade, similar to a typical helicopter blade, are given in Table 7.1. The other parameters, β_p, θ_0, e_g and e_t are taken as zero at this time. For the present analysis, four rotation speeds varying from a low to a very high value are considered. The lowest is $\Omega = 383$ rpm, which is typical of a helicopter rotor. The second, third and fourth rotation speeds are $\Omega = 1000$, 10,000 and 50,000 rpm, which are typical of turbine blades and high speed

Table 7.1 Properties of rotating blade

Parameter	Value
EI_y	66,837 Nm^2
EI_z	165,856 Nm^2
GJ	38,060 Nm^2
EA	12,162,177 N
K_{m_1}	0.06516 m
K_{m_2}	0.12201 m
m_0	6.46 kg/m
L	4.94 m

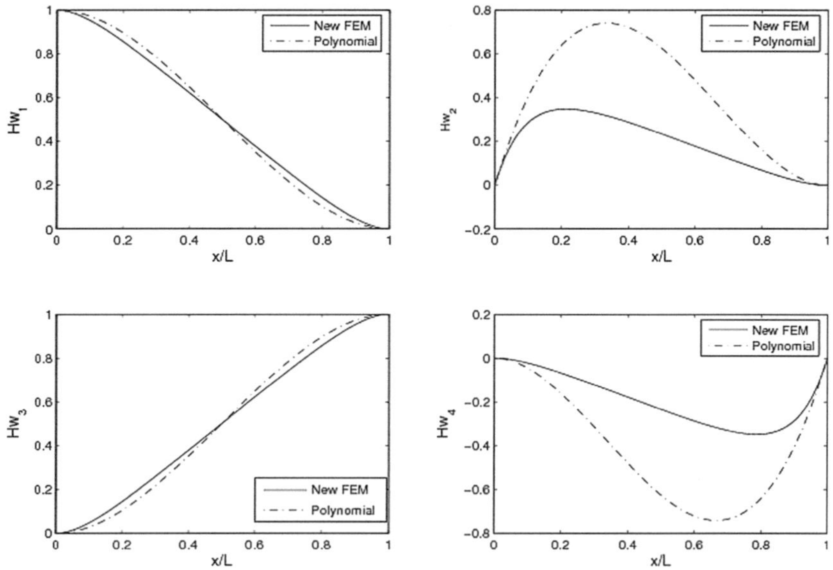

Fig. 7.7 Shape functions in flapwise bending at 383 rpm

rotating machinery. The objective here is to simply check the asymptotic behavior of the new shape functions, which has been theoretically observed in an earlier section.

The new shape functions, unlike the polynomial ones, change with the rotation speed, the material properties and the elemental length. The variation of the new shape functions and the polynomials for the cantilever blade at different rotation speeds is shown in Figs. 7.7, 7.8, 7.9, 7.10, 7.11, 7.12, 7.13, 7.14, 7.15, 7.16, 7.17 and 7.18.

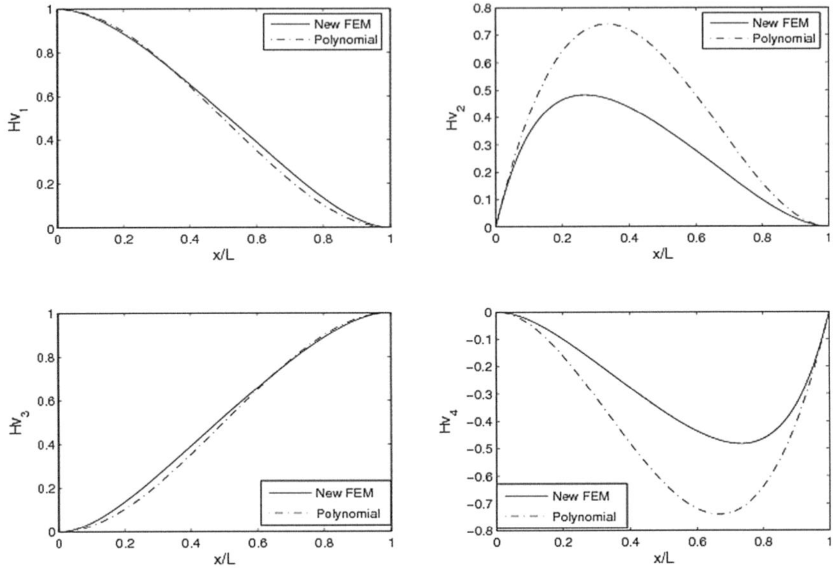

Fig. 7.8 Shape functions in lead-lag bending at 383 rpm

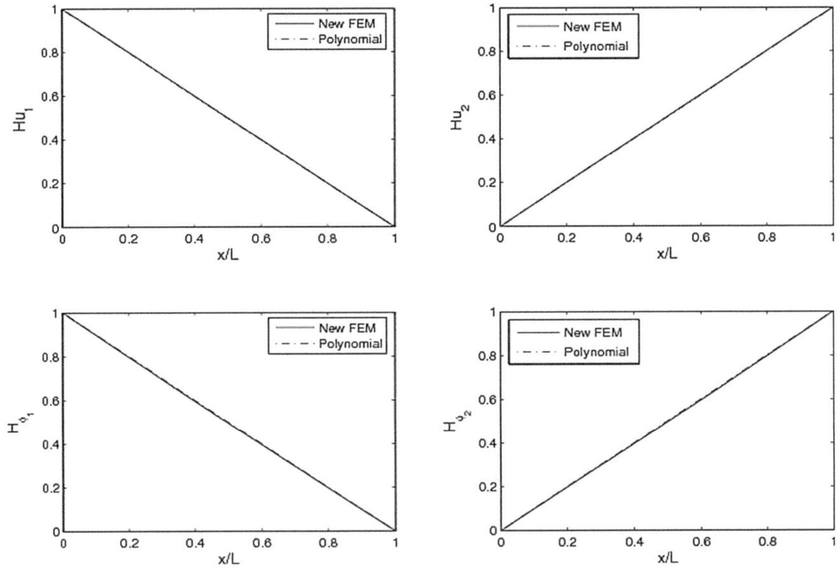

Fig. 7.9 Shape functions in axial deformation and torsion at 383 rpm

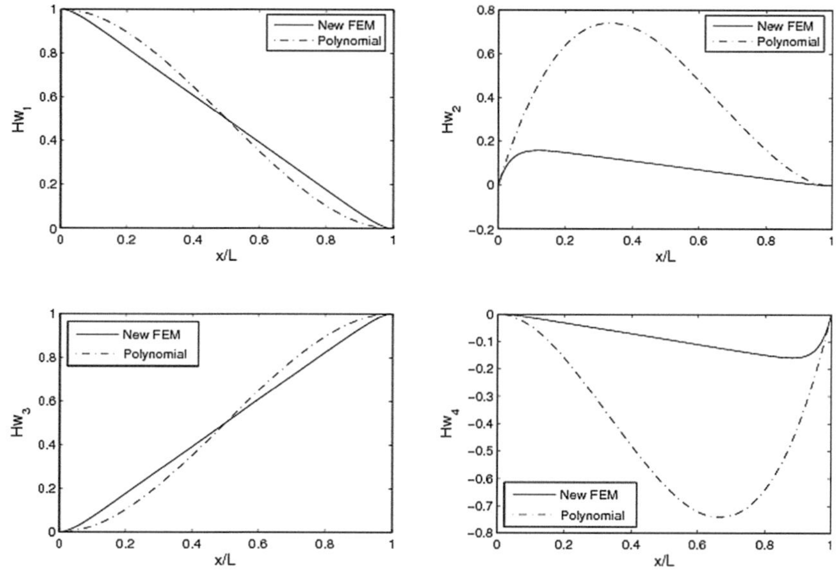

Fig. 7.10 Shape functions in flapwise bending at 1000 rpm

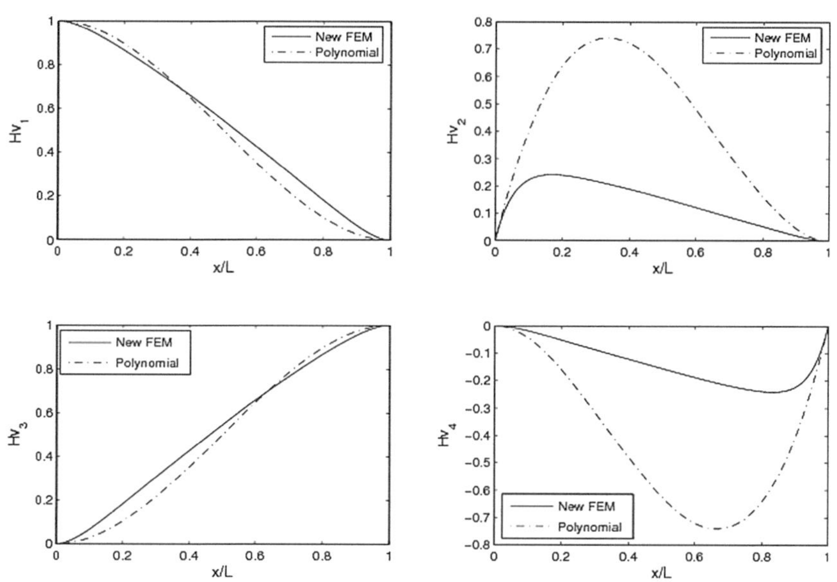

Fig. 7.11 Shape functions in lead-lag bending at 1000 rpm

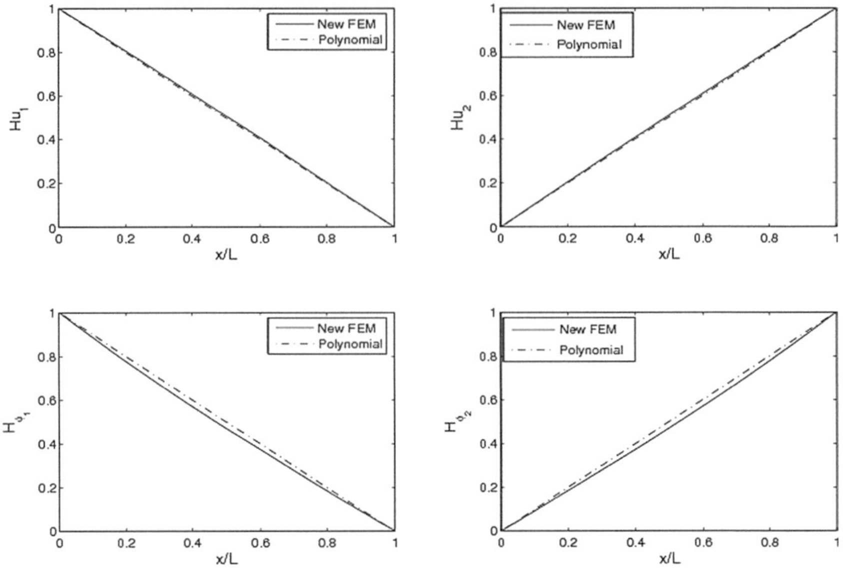

Fig. 7.12 Shape functions in axial deformation and torsion at 1000 rpm

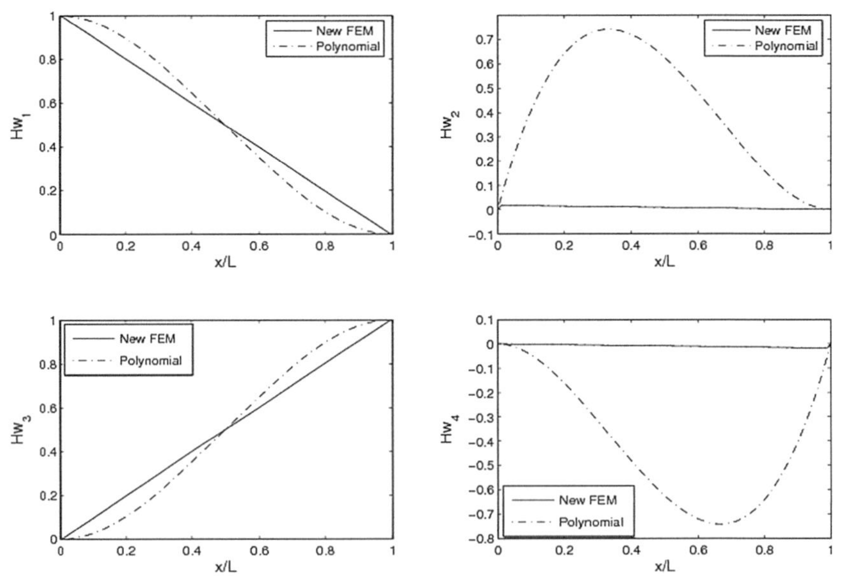

Fig. 7.13 Shape functions in flapwise bending at 10,000 rpm

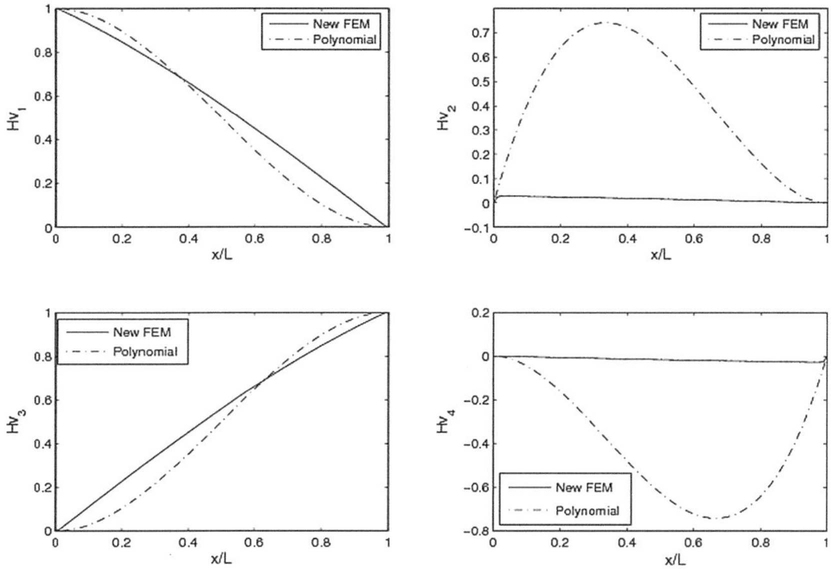

Fig. 7.14 Shape functions in lead-lag bending at 10,000 rpm

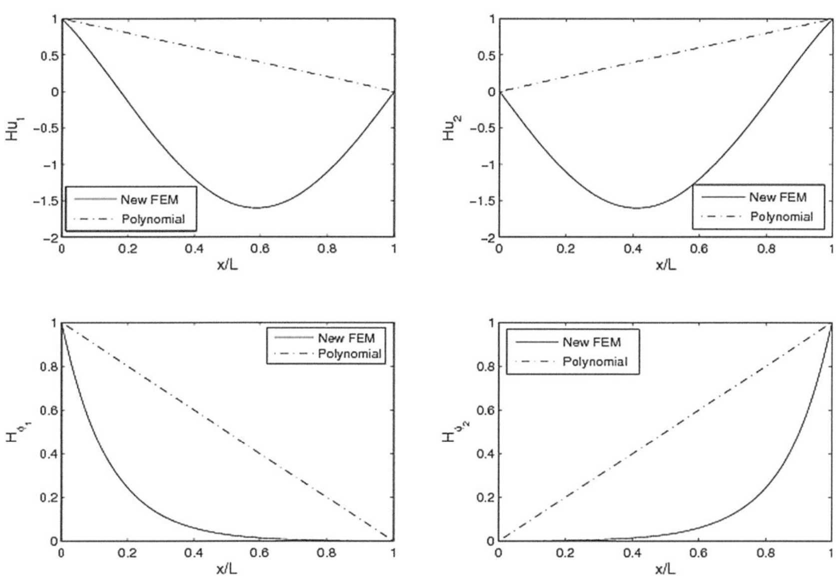

Fig. 7.15 Shape functions in axial deformation and torsion at 10,000 rpm

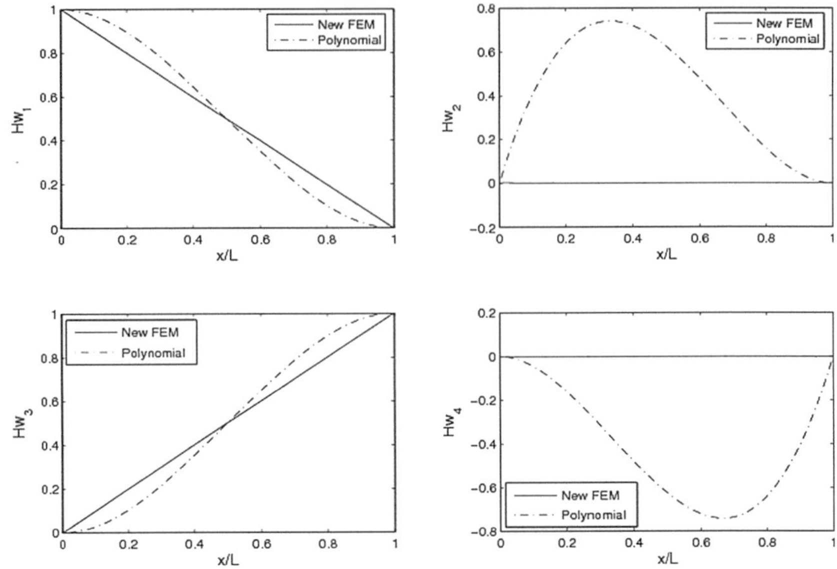

Fig. 7.16 Shape functions in flapwise bending at 50,000 rpm

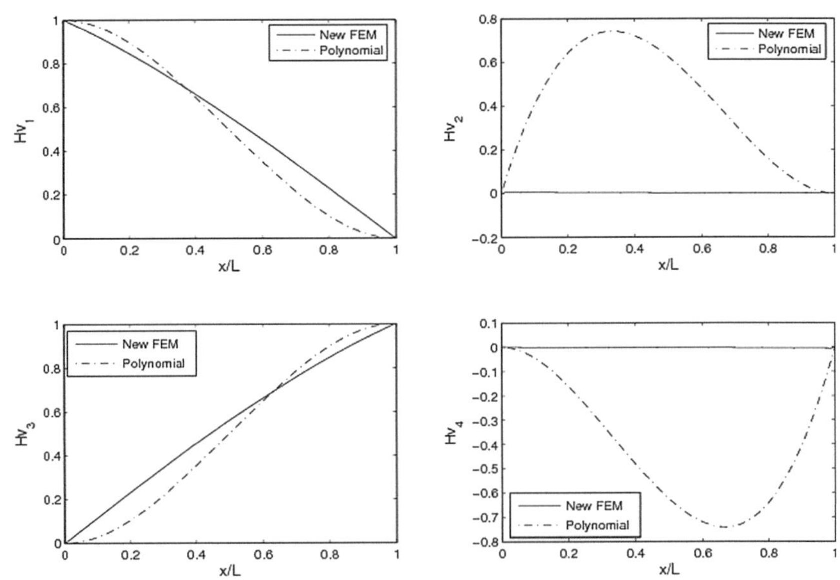

Fig. 7.17 Shape functions in lead-lag bending at 50,000 rpm

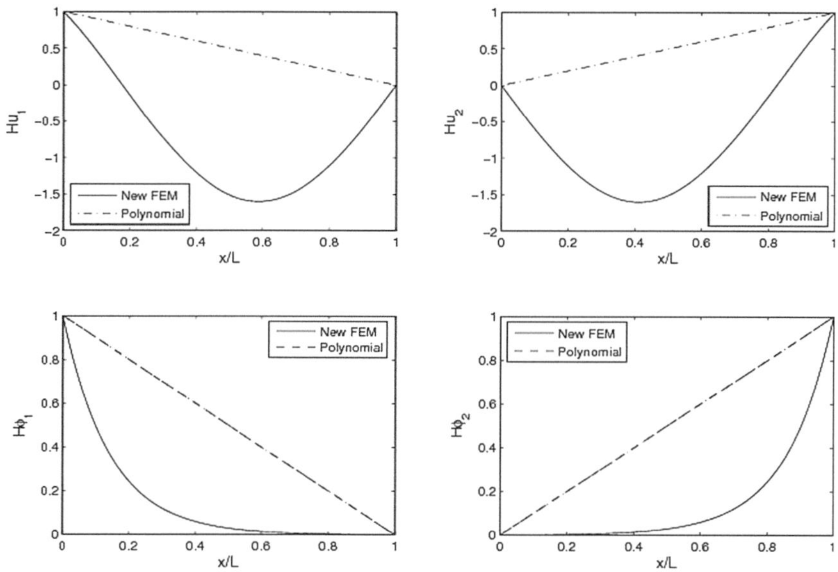

Fig. 7.18 Shape functions in axial deformation and torsion at 50,000 rpm

For comparison, Hermite cubic polynomials are used for the flap and the lag bending shape functions and linear polynomials are used for the axial and the torsion shape functions. At 383 and 1000 rpm, the new shape functions in flap and lead-lag mode vary significantly from the polynomials (Figs. 7.7, 7.8, 7.10, 7.11). But no significant variation is observed for the axial and the torsion shape functions at these rotation speeds (Figs. 7.9 and 7.12). At higher rotation speeds, say at 10,000 rpm (Figs. 7.13, 7.14 and 7.15) and 50,000 rpm (Figs. 7.16, 7.17 and 7.18), the centrifugal stiffening becomes significant and this is well captured by the new shape functions in all the modes. At 10,000 and 50,000 rpm, flapwise, lead-lag, axial and torsion shape functions are very much different from the polynomials. The flap and lag functions approach linear behavior at very high rotation speeds. At low rotation speeds, the primary changes are in the slope functions related to the slope for the bending degree of freedom. The effect of centrifugal stiffening on these shape functions is significant. This is the reason that these shape functions work well with less number of elements at higher rotation speeds.

As is clear from the shape function plots, they exhibit different shape function patterns at different rotation speeds. These variations are not exhibited by the conventional cubic polynomial or the linear shape functions. Therefore, the new shape functions can be used for a very wide span of the rotation speeds, upto very high values, without encountering any problems such as singularities or discontinuities. Their ability to capture the effects of the centrifugal stiffening and the blade stiffness through the displacement fields makes the new shape functions superior over the conventional shape functions for the finite element analysis of the rotating structures.

Table 7.2 Converged natural frequencies non-dimensionalized with Ω for new FEM and UMARC at 383 rpm

Mode	New FEM	UMARC
1st lag	0.7014	0.7014
1st flap	1.1258	1.1258
2nd flap	3.4216	3.4216
2nd lag	4.3000	4.3000
1st twist	4.4622	4.4622
3rd flap	7.6672	7.6672
1st axial	10.8321	10.8321
3rd lag	10.8996	10.8996

7.5.2 Validation Study

For the validation study, a uniform blade equivalent of the B0105 rotor is considered. The rotor blade material properties are given in Table 7.1 and are used for the analysis of the shape functions, except that the rotation speed is now fixed at $\Omega = 383$ rpm. A finite element code is implemented using the formulation derived in the previous section. The University of Maryland Advanced Rotorcraft Code (UMARC [32]) is chosen as a baseline for the validation of the new FEM code. The UMARC code employs a fifteen degrees of freedom model, four degrees of freedom in each of flapwise bending, lead-lag bending and axial deformation and three degrees of freedom in torsion, and is well validated [37, 38]. The new developed shape functions are exponential and trigonometric functions. Whereas, UMARC uses cubic polynomials for the flap and the lag bending, cubic polynomials for the axial deformation and quadratic polynomials for the torsion. The converged values of the frequencies for both the codes match exactly and are given in Table 7.2. Therefore, we note that the new FEM is validated with an existing analysis.

7.6 Convergence Study of New FEM Element and Polynomials

The validation with UMARC proves that the FE formulation is implemented correctly, but a comparison with UMARC cannot be used to accurately access the advantages of the new FEM element as the number of degrees of freedom are different for the two models. A convergence study is now undertaken for the new FEM element at different rotational speeds. The results are then compared with the same element but with polynomial shape functions. Both the models have twelve degrees of freedom. Hermite polynomials are used for flapwise and lead-lag bending. For axial deformation and torsion, linear polynomials are used. The chosen polynomials

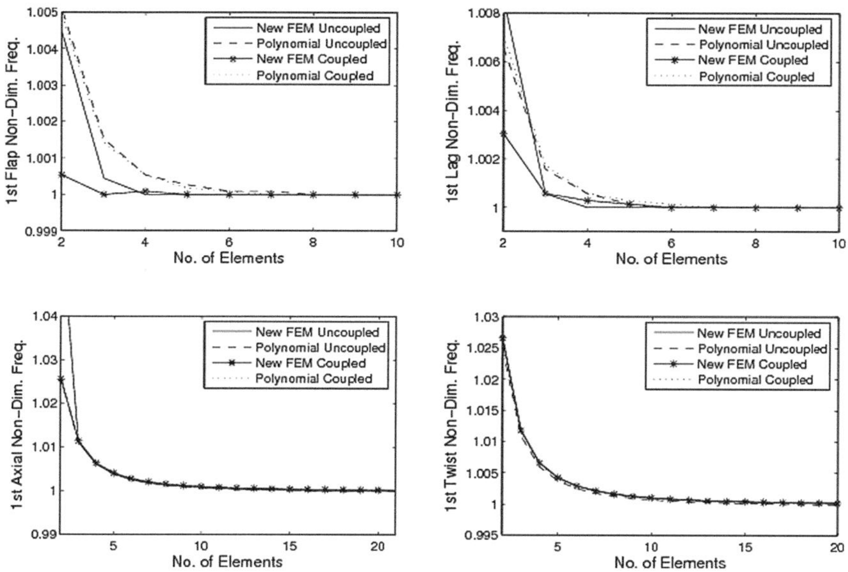

Fig. 7.19 Convergence study plots of the first natural frequencies at 383 rpm

correspond to the asymptotic nature of the new shape functions as the rotation speed tends to zero.

The B0105 rotor blade data presented in Table 7.1 is used for the convergence study for the new FEM element and the polynomial shape functions at 383 and 1000 rpm. The analysis is done for a coupled and an uncoupled problem. For the coupled case, coupling term θ_0 is taken as 10^0 and e_g/L is taken as 0.001. For the uncoupled case, these values are considered as zero. The flap, lag, axial and torsion natural frequencies are plotted at 383 rpm in Figs. 7.19 and 7.20, and at 1000 rpm in Figs. 7.21 and 7.22. The converged frequencies are shown in Table 7.3 and are used to normalize the results in these figures.

Figure 7.19 shows the fundamental natural frequencies for the four modes. The new FEM converges faster in all the cases. The polynomial shape functions result in a much slower convergence rate, especially for the flap mode which has a dominant effect of centrifugal stiffening. This effect for the flap mode is magnified in Fig. 7.21 where the results at 1000 rpm are shown. The convergence of the higher modes is somewhat similar for both the analysis as shown in Figs. 7.20 and 7.22. This is because the higher modes are more dependent on the structural stiffness as compared to the centrifugal stiffness and therefore the polynomial shape functions, which are exact for the non-rotating condition, work quite well. To mention about the superiority of the new FEM shape functions, at 383 rpm, the first twist natural frequency converges to four decimal places in eighteen elements for the new FEM, but it converges in fifty elements for the polynomials. The new shape functions therefore perform in a super-convergent manner as they accurately capture the centrifugal stiffening effect.

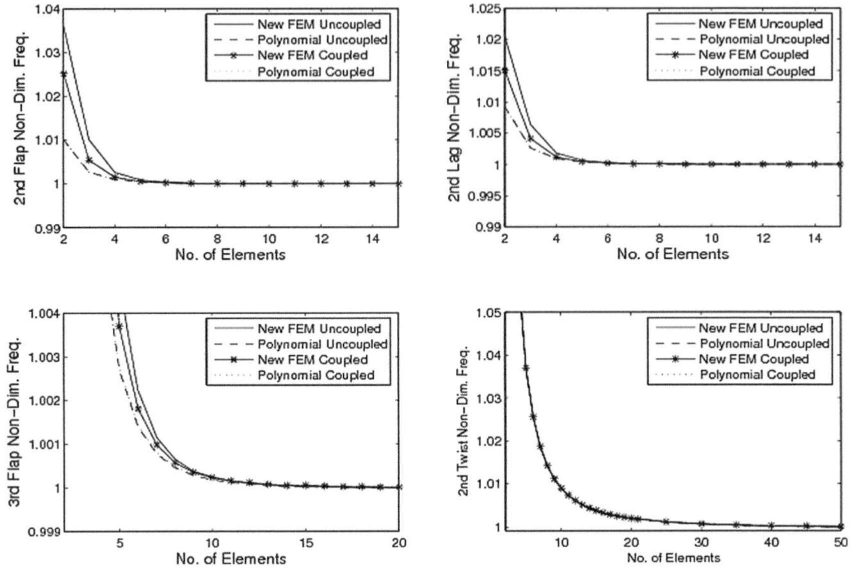

Fig. 7.20 Convergence study plots of the higher mode natural frequencies at 383 rpm

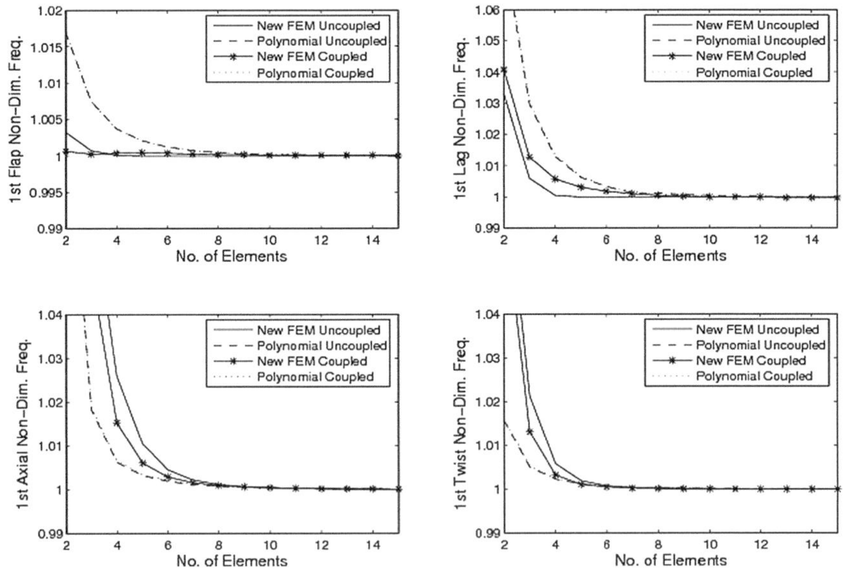

Fig. 7.21 Convergence study plots of the first natural frequencies at 1000 rpm

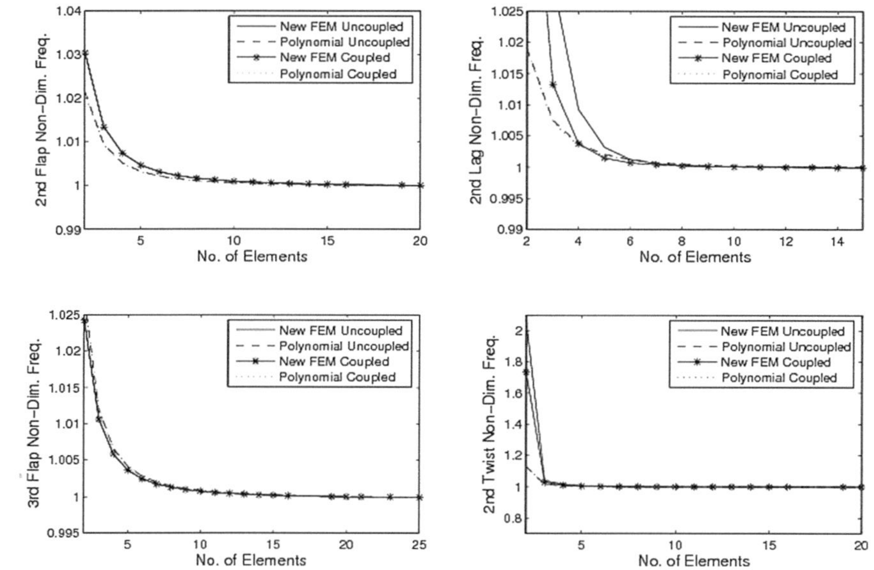

Fig. 7.22 Convergence study plots of the higher mode natural frequencies at 1000 rpm

The convergence study cannot be extended for the rotor blade at 10,000 and 50,000 rpm with the properties given in Table 7.1. As a numerical illustration, E and G are increased by 1000 times and the study is completed at these rotational speeds. The first frequencies in all the four modes at 10,000 and 50,000 rpm are plotted in Figs. 7.23 and 7.24 respectively for an uncoupled problem. Again it is noted that convergence is better for the new finite element for most cases. The converged values are given in Table 7.4 and are used to normalize the plotted results. The convergence of the torsion mode is better using polynomials for some cases, since the effect of centrifugal stiffening in this mode is typically much less than the structural stiffness. In contrast, for the flap mode, the effect of centrifugal force dominates and the blade becomes like a rotating string at high rotation speeds.

The shape functions derived in this paper provide a theoretically elegant alternative for the finite element analysis of rotating blades. The shape functions can also be used for the finite element analysis of rotating rods (using axial functions) and beams (using flap and/or lag functions). Moreover, it may be possible to use them as basis functions for other energy methods. While earlier works have shown the derivation of superconvergent elements for Timoshenko beam type problems where the exact solutions were obtained by only making a uniform property approximation within the element. The current work shows that the application of the concept is possible for rotating blades and other rotating structures. Non-rotating Timoshenko beams and blades are obvious candidates for extension of the current work.

Table 7.3 Converged natural frequencies non-dimensionalized with Ω for different modes at 383 and 1000 rpm

Mode	Frequency at 383 rpm		Frequency at 1000 rpm	
	Uncoupled	Coupled	Uncoupled	Coupled
1st lag	0.7014	0.6952	0.3840	0.3817
1st flap	1.1258	1.1296	1.0439	1.0446
2nd flap	3.4216	3.4163	1.8430	1.8357
2nd lag	4.3000	4.3040	2.6657	2.6446
1st twist	4.4622	4.4618	2.7062	2.7268
3rd flap	7.6672	7.6643	4.0456	4.0455
1st axial	10.8351	10.8330	4.7999	4.7916
2nd twist	13.2243	13.2314	5.6017	5.6047

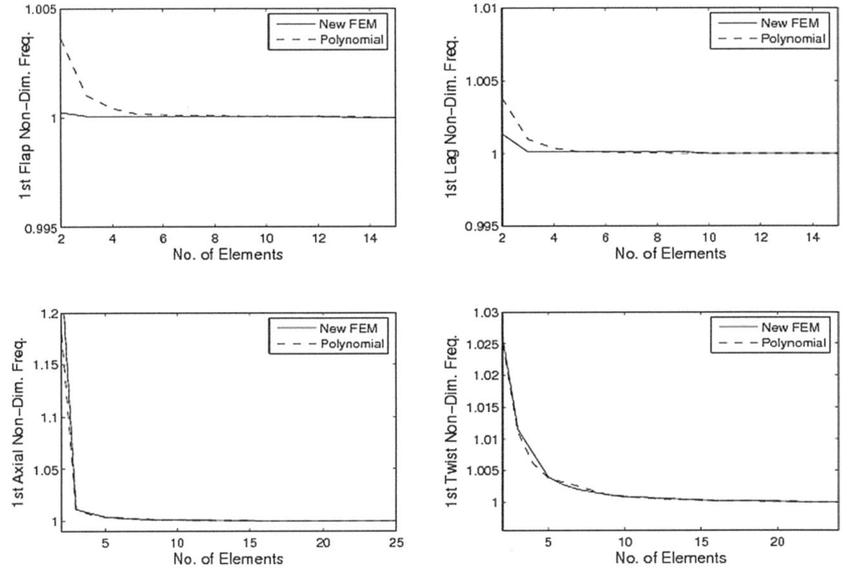

Fig. 7.23 Convergence study plots of the first natural frequencies at 10,000 rpm for uncoupled problem

7.7 Summary

In this chapter, new shape functions in axial deformation, flapwise bending, lead-lag bending and torsion are derived for a twelve degrees of freedom finite element model for a rotating blade. The exact solutions of the homogenous part of the governing static differential equations of a rotating blade are used to derive the shape functions. The new FEM model is validated using existing data for a rotor blade similar to a helicopter blade. The new shape functions are not only functions of the element

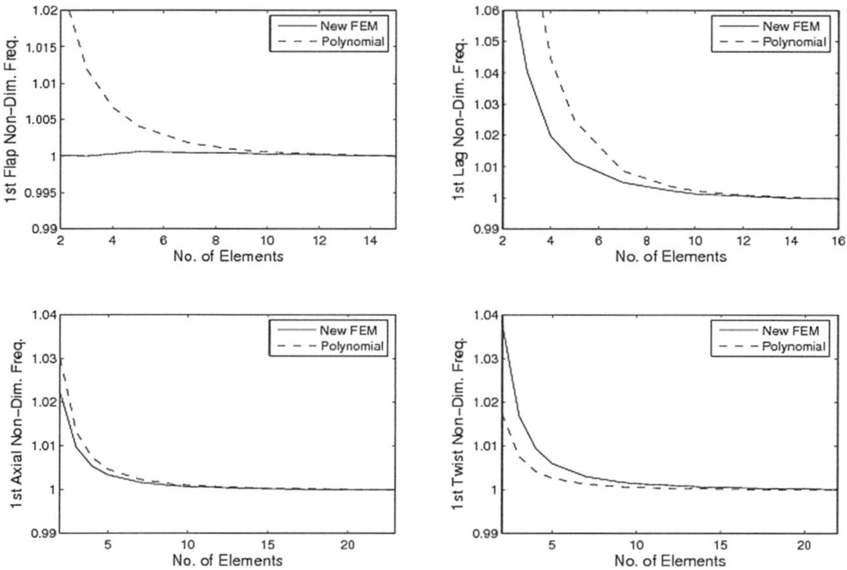

Fig. 7.24 Convergence study plots of the first mode natural frequencies at 50,000 rpm for uncoupled problem

Table 7.4 Converged natural frequencies non-dimensionalized with Ω for different modes at 10,000 and 50,000 rpm for uncoupled problem

Mode	Frequency at 10,000 rpm	Frequency at 50,000 rpm
1st lag	0.8093	0.2991
1st flap	1.1584	1.0274
1st twist	13.1394	1.3009
1st axial	5.3814	2.4384

length, but also of the rotation speed, the element location across the beam, the element mass, flexural and the torsional stiffnesses, the mass of outboard elements and the length of the beam. A convergence study is performed for a coupled and an uncoupled blade at different rotation speeds between the new finite element and a similar twelve degrees of freedom finite element model with polynomial shape functions. The new finite element shape functions capture the effect of centrifugal stiffening and show superior convergence for the first few modes at low as well as at high rotation speeds over the conventional polynomials. Thus, the new developed shape functions have a wide horizon of application as they change their behavior with the material properties and at the extremes of rotation speeds. They also provide basis functions for finite element analysis of rotating rods and beam structures, and for energy methods.

Appendix 1

Mass matrix is derived as,

$$[\mathbf{M}] = \begin{bmatrix} M_{uu} & 0 & 0 & 0 \\ 0 & M_{vv} & 0 & M_{v\phi} \\ 0 & 0 & M_{ww} & M_{w\phi} \\ 0 & M_{\phi v} & M_{\phi w} & M_{\phi\phi} \end{bmatrix} \tag{7.83}$$

where,

$$M_{uu} = \int_0^l m\mathbf{H}_u^T \mathbf{H}_u dx \tag{7.84}$$

$$M_{vv} = \int_0^l m\mathbf{H}_v^T \mathbf{H}_v dx \tag{7.85}$$

$$M_{v\phi} = -\int_0^l me_g \sin\theta_0 \mathbf{H}_v^T \mathbf{H}_\phi dx \tag{7.86}$$

$$M_{ww} = \int_0^l m\mathbf{H}_w^T \mathbf{H}_w dx \tag{7.87}$$

$$M_{w\phi} = \int_0^l me_g \cos\theta_0 \mathbf{H}_w^T \mathbf{H}_\phi dx \tag{7.88}$$

$$M_{\phi v} = -\int_0^l me_g \sin\theta_0 \mathbf{H}_\phi^T \mathbf{H}_v dx \tag{7.89}$$

$$M_{\phi w} = \int_0^l me_g \cos\theta_0 \mathbf{H}_\phi^T \mathbf{H}_w dx \tag{7.90}$$

$$M_{\phi\phi} = \int_0^l mK_m^2 \mathbf{H}_\phi^T \mathbf{H}_\phi dx \tag{7.91}$$

Stiffness matrix is derived as,

$$[\mathbf{K}] = \begin{bmatrix} K_{uu} & K_{uv} & K_{uw} & K_{u\phi} \\ K_{vu} & K_{vv} & K_{vw} & K_{v\phi} \\ K_{wu} & K_{wv} & K_{ww} & K_{w\phi} \\ K_{\phi u} & K_{\phi v} & K_{\phi w} & K_{\phi\phi} \end{bmatrix}$$

where,

$$K_{uu} = \int_0^l EA\mathbf{B}_u^T\mathbf{B}_u dx - \int_0^l m\mathbf{H}_u^T\mathbf{H}_u dx \tag{7.92}$$

$$K_{uv} = -\int_0^l EAe_A \cos\theta_0 \mathbf{B}_u^T\mathbf{N}_v dx \tag{7.93}$$

$$K_{uw} = -\int_0^l EAe_A \sin\theta_0 \mathbf{B}_u^T\mathbf{N}_w dx \tag{7.94}$$

$$K_{u\phi} = \int_0^l EAK_A^2\theta_0'\mathbf{B}_u^T\mathbf{B}_\phi dx \tag{7.95}$$

$$K_{vu} = -\int_0^l EAe_A \cos\theta_0 \mathbf{N}_v^T\mathbf{B}_u dx \tag{7.96}$$

$$K_{vv} = \int_0^l \left(EI_z\cos^2\theta_0 + EI_y\sin^2\theta_0\right)\mathbf{N}_v^T\mathbf{N}_v dx - \int_0^l m\mathbf{H}_v^T\mathbf{H}_v dx + \int_0^l \left[\int_x^l mxd\xi\right]\mathbf{B}_v^T\mathbf{B}_v dx \tag{7.97}$$

$$K_{vw} = \int_0^l \left(EI_z - EI_y\right)\sin\theta_0\cos\theta_0\mathbf{N}_v^T\mathbf{N}_w dx \tag{7.98}$$

$$K_{v\phi} = -\int_0^l EB_2\theta_0'\cos\theta_0\mathbf{N}_v^T\mathbf{B}_\phi dx + \int_0^l me_g\sin\theta_0\mathbf{H}_v^T\mathbf{H}_\phi dx - \int_0^l mxe_g\sin\theta_0\mathbf{B}_v^T\mathbf{H}_\phi dx \tag{7.99}$$

$$K_{wu} = -\int_0^l EAe_A\sin\theta_0\mathbf{N}_w^T\mathbf{B}_u dx \tag{7.100}$$

$$K_{wv} = \int_0^l \left(EI_z - EI_y\right)\sin\theta_0\cos\theta_0\mathbf{N}_w^T\mathbf{N}_v dx \tag{7.101}$$

$$K_{ww} = \int_0^l \left(EI_y\cos^2\theta_0 + EI_z\sin\theta_0\right)\mathbf{N}_w^T\mathbf{N}_w dx + \int_0^l\left[\int_x^l mxd\xi\right]\mathbf{B}_w^T\mathbf{B}_w dx \tag{7.102}$$

$$K_{w\phi} = -\int_0^l EB_2\theta_0'\sin\theta_0\mathbf{N}_w^T\mathbf{B}_\phi dx + \int_0^l mxe_g\cos\theta_0\mathbf{B}_w^T\mathbf{H}_\phi dx \tag{7.103}$$

$$K_{\phi u} = \int_0^l EAK_A^2\theta_0'\mathbf{B}_\phi^T\mathbf{B}_u dx \tag{7.104}$$

$$K_{v\phi} = -\int_0^l EB_2\theta_0'\cos\theta_0\mathbf{N}_v^T\mathbf{B}_\phi dx + \int_0^l me_g\sin\theta_0\mathbf{H}_v^T\mathbf{H}_\phi dx - \int_0^l mxe_g\sin\theta_0\mathbf{B}_v^T\mathbf{H}_\phi dx \tag{7.105}$$

$$K_{\phi w} = -\int_0^l EB_2\theta_0'\sin\theta_0\mathbf{B}_\phi^T\mathbf{N}_w dx + \int_0^l mxe_g\cos\theta_0\mathbf{H}_\phi^T\mathbf{B}_w dx \tag{7.106}$$

$$K_{\phi\phi} = \int_0^l \left(GJ + EB_1{\theta_0'}^2\right)\mathbf{B}_\phi^T\mathbf{B}_\phi dx - \int_0^l m\left(K_{m_1}^2 - K_{m_2}^2\right)cos2\theta_0\mathbf{H}_\phi^T\mathbf{H}_\phi dx \tag{7.107}$$

Damping matrix is derived as,

$$[\mathbf{C}] = \begin{bmatrix} 0 & C_{uv} & 0 & 0 \\ C_{vu} & C_{vv} & C_{vw} & 0 \\ 0 & C_{wv} & 0 & 0 \\ 0 & 0 & 0 & 0 \end{bmatrix} \tag{7.108}$$

where,

$$C_{uv} = -\int_0^l 2m\mathbf{H}_u^T\mathbf{H}_v dx \tag{7.109}$$

$$C_{vu} = \int_0^l 2m\mathbf{H}_v^T\mathbf{H}_u dx \tag{7.110}$$

$$C_{vv} = -\int_0^l 2me_g\cos\theta_0\mathbf{H}_v^T\mathbf{B}_v dx - \int_0^l 2me_g\cos\theta_0\mathbf{B}_v^T\mathbf{H}_v dx \tag{7.111}$$

$$C_{vw} = -\int_0^l 2m\beta_p\mathbf{H}_v^T\mathbf{H}_w dx - \int_0^l 2me_g\sin\theta_0\mathbf{H}_v^T\mathbf{B}_w dx \tag{7.112}$$

$$C_{wv} = -\int_0^l 2m\beta_p\mathbf{H}_w^T\mathbf{H}_v dx - \int_0^l 2me_g\sin\theta_0\mathbf{B}_w^T\mathbf{H}_v dx \tag{7.113}$$

Load vector is derived as,

$$\{\mathbf{R}\} = \begin{Bmatrix} -\int_0^l mx\mathbf{H}_u^T dx \\ -\int_0^l \left(me_g\cos\theta_0\mathbf{H}_v^T - me_g\sin\theta_0\ddot{\theta}_0\mathbf{H}_v^T + mxe_g\cos\theta_0\mathbf{B}_v^T\right)dx \\ \int_0^l \left(mx\beta_p\mathbf{H}_w^T + me_g\cos\theta_0\ddot{\theta}_0\mathbf{H}_w^T + mxe_g\sin\theta_0\mathbf{B}_w^T\right)dx \\ \int_0^l \left(\left(K_{m_2}^2 - K_{m_1}^2\right)\frac{sin2\theta_0}{2}m\mathbf{H}_\phi^T + mx\beta_p e_g\cos\theta_0\mathbf{H}_\phi^T\right)dx \end{Bmatrix} \tag{7.114}$$

References

1. Xiong JJ, Yu X (2007) Helicopter rotor-fuselage aeroelasticity modeling and solution using the partition-iteration method. J Sound Vib 302(4–5):821–840
2. Thakkar D, Ganguli R (2004) Dynamic response of rotating beams with piezoceramic actuation. J Sound Vib 270(4–5):729–753
3. Hwang HY, Kim C (2004) Damage detection in structures using a few frequency response measurements. J Sound Vib 270(1–2):1–14
4. Yoo HH, Kim JM, Chung J (2007) Equilibrium and modal analysis of rotating multibeam structures employing multiple reference frames. J Sound Vib 302(4–5):789–805
5. Das SK, Ray PC, Pohit G (2007) Free vibration analysis of a rotating beam with nonlinear spring and mass system. J Sound Vib 301(1–2):165–188
6. Choi SC, Park JS, Kim JH (2007) Vibration control of pretwisted rotating composite thin-walled beams with piezoelectric fiber composites. J Sound Vib 300(1–2):176–196
7. Rao AD (1993) A feasibility and assessment study for FT4000 humid air turbine (HAT), EPRI RP-325105

8. Thirupathi SR, Seshu P, Naganathan NG (1997) A finite element static analysis of smart turbine blades. Smart Mater Struct 6(5):607–615
9. Bernhard APF, Chopra I (2001) Analysis of a bending-torsion coupled actuator for a smart rotor with active blade tips. Smart Mater Struct 10(1):35–52
10. Leissa AW (1980) Vibrations of turbine engine blades by shell analysis. Shock Vib Dig 12(1):1–10
11. Ramamurti V, Sreenivasamurthy S (1980) Dynamic stress analysis of rotating twisted and tapered blades. J Strain Anal 15(3):117–126
12. Bogomolov SI, Lutsenko SS, Nazarenko SA (1982) Application of superparametric shell finite element to the calculation of turbine blade vibrations. Problemy Prochnosti 156(6):71–74
13. Bucco D, Mazumdar J (1983) Estimation of the fundamental frequencies of shallow shells by a finite element - isodeflection contour method. Comput Struct 17(3):441–447
14. Repetskii OV (1988) Numerical calculation of the natural blade vibrations of turbomachinery. Plenum Publishing Corporation, New York, pp 453–459
15. Jiang JJ, Hsiao CL, Shabana AA (1991) Calculation of non-linear vibration of rotating beams by using tetrahedral and solid finite elements. J Sound Vib 148(2):193–213
16. Qin F, Chen L, Li Y, Zhang X (2006) Fundamental frequencies of turbine blades with geometry mismatch in fir-tree attachments. J Turbomach 128:512–516
17. Wang G, Wereley NM (2004) Free vibration analysis of rotating blades with uniform tapers. AIAA J 42(12):2429–2437
18. Durocher LL, Kane J (1980) Preliminary design tools for pretwisted tapered beams or turbine blades. J Mech Des, Trans ASME 102(4):742–748
19. Ormiston RA, Hodges DH (1972) Linear flap-lag dynamics of hingeless helicopter rotor blades in hover. J Am Helicopter Soc 17:2–14
20. Qin QH, Mao CX (1996) Coupled torsional-flexural vibration of shaft systems in mechanical engineering-I, finite element model. Comput Struct 58(4):835–843
21. Ruhl RL, Booker JF (1972) A finite element model for distributed parameter turborotor systems. J Eng Ind, ASME 94:128–132
22. Thorkildson T (1974) Solution of a distributed mass and unbalanced rotor system using a consistent mass matrix approach, MSE Engineering Report, Arizona State University
23. Belo EM, Marques FD (1994) Analysis and vibration control of a helicopter rotor blade, Conference Publication No. 389, IEE, pp 1290–1295
24. Putter S, Manor H (1978) Natural frequencies of radial rotating beams. J Sound Vib 56:175–185
25. Ramamurti V, Kielb R (1984) Natural frequencies of twisted rotating plates. J Sound Vib 97:429–449
26. Yoo HH, Park JH, Park J (2001) Vibration analysis of rotating pre-twisted blades. Comput Struct 79:1811–1819
27. Banerjee JR (1999) Free vibration of centrifugally stiffened uniform and tapered beams using the dynamic stiffness method. J Sound Vib 233(5):857–875
28. Banerjee JR, Jackson DR (2006) Free vibration of rotating tapered beams using the dynamic stiffness method. J Sound Vib 298(4–5):1034–1054
29. Hashemi SM, Richard MJ (2001) Natural frequencies of rotating uniform beams with Coriolis effects. J Vib Acoust 123:444–455
30. Kuang JH, Hsu MH (2002) Eigensolutions of grouped turbo blades solved by the generalized differential quadrature method. J Eng Gas Turbines Power 124:1011–1017
31. Hodges DH, Dowell EH (1974) Nonlinear equations of motion for the elastic bending and torsion of twisted nonuniform rotor blades, NASA TN D-7818, Washington D.C
32. Bir G, Chopra I, Ganguli R et al (1994) University of Maryland advanced rotorcraft code (UMARC), Theory manual
33. Gopalakrishnan S (2000) A deep rod finite element for structural dynamics and wave propogation problems. Int J Numer Methods Eng 48(5):731–744
34. Chakraborty A, Gopalakrishnan S, Reddy JN (2003) A new beam finite element for the analysis of functionally graded materials. Int J Mech Sci 45(3):519–539

35. Mukherjee S, Reddy JN, Krishnamoorthy CS (2001) Convergence properties and derivative extraction of the superconvergent Timoshenko beam finite element. Comput Methods Appl Eng 190(26–27):3475–3500
36. Ghosh DP, Gopalakrishnan S (2007) A superconvergent finite element for composite beams with embedded magnetostrictive patches. Compos Struct 79(3):315–330
37. Bhat SR, Ganguli R (2004) Validation of comprehensive helicopter aeroelastic analysis with experimental data. Def Sci J 54(4):419–427
38. Ganguli R, Chopra I, Weller WH (1998) Comparison of calculated vibratory rotor hub loads with experimental data. J Am Helicopter Soc 43(4):312–318
39. Heck A (2003) Introduction to Maple, 3rd edn. Springer, New York

Chapter 8
Spectral Finite Element Method

Spectral finite element method (SFEM) is an efficient technique for solving problems where the frequency content of the input signal is very high. The spectral formulation requires that the assembled system of equations be solved in the frequency domain and utilizes the Fast Fourier Transform (FFT) to transform the time domain responses to the frequency domain and back. Typically, SFEM uses the exact solution of the governing differential equation in the frequency domain as the interpolating function for element formulation. As a result, the spectral element treats the distribution of mass and rotational inertia of the structural element exactly. In general, for simple problems, only one spectral element needs be placed between any two joints, substantially reducing the total number of degrees of freedom in the system. Spectral element method was originally proposed by [25] and popularized by [11]. Spectral element for elementary isotropic waveguides was formulated by [10, 12] and higher order waveguides by Gopalakrishnan et al. [14, 23]. Similarly, spectral element for elementary and higher order composite waveguides were formulated by [32, 33]. Spectral elements for inhomogeneous waveguides were formulated by Chakraborty and Gopalakrishnan [4, 5] and for carbon nanotubes (Chakraborty, Sivakumar and Gopalakrishnan [6]). These works show the versatility of SFEM in handling high frequency impact type short duration loads.

As mentioned before, most of the spectral finite element formulations use the exact solution to the governing differential equation as the element interpolating function. In more complicated problems such as a rotating beam or a beam of varying cross section or a combination of both, it is very difficult to find the exact solution, as the governing equation for such structures essentially give rise to variable coefficient differential equations. Here, we solve this problem using the weak form of the governing differential equation in which an assumed interpolating function is used for the transverse displacement. In this context, the method resembles the conventional finite element method. However in the spectral approach, assembly and solution of the equation has to be performed at each frequency. The solution is first obtained

R. Ganguli, *Finite Element Analysis of Rotating Beams*,
Foundations of Engineering Mechanics, DOI 10.1007/978-981-10-1902-9_8

for an unit impulse, which yields the transfer function as a direct by-product of the approach. The input is then convolved with the system transfer function to obtain the output in the frequency domain. The inverse FFT of the output will yield the time domain responses.

In the present chapter, an efficient spectral finite element method is developed and used to obtain the natural frequencies and response of rotating uniform and tapered beams, subjected to a short duration transverse impact load. The governing partial differential equation is first derived using the Hamilton's principle. Here, we use the frequency domain stationary potential energy principle [15] to formulate the dynamic stiffness matrix of a rotating beam element, where two different sets of interpolating functions for the spectral element formulation are used. In the first case, we use the exact solution to the governing equation of a rotating Euler–Bernoulli beam with a constant (maximum) centrifugal force, to generate the dynamic stiffness matrix. This element was discussed in [36] and we call this element as SFER. In the second case, we use the exact solution of the non-rotating Euler–Bernoulli beam governing equation in the frequency domain, to obtain the element dynamic stiffness matrix. We call this element SFEN. These elements are used to perform both free vibration and wave propagation studies.

8.1 Governing Differential Equation

A rotating blade can be represented as a rotating beam having vibration displacements perpendicular to the plane of rotation (flapping motion). Considering the elementary theory of beams, the axial and the transverse displacement fields are represented as

$$u(x, y, z, t) = u_0(x, t) - z\frac{\partial w(x, t)}{\partial x} \tag{8.1}$$

$$w(x, y, z, t) = w(x, t) \tag{8.2}$$

where $u(x, y, t)$ and $w(x, y, t)$ are the axial and the transverse displacements respectively and $u_0(x, t)$ is the mid-plane axial displacement.

The governing equation can be derived from the Hamilton's principle where the action integral $\int_{t_1}^{t_2} L dt$ assumes a stationary value i.e. $\delta \int_{t_1}^{t_2} L dt = 0$, where L is the Lagrangian given as

$$L = (KE - PE) \tag{8.3}$$

where KE is the kinetic energy and PE is the potential energy. On substituting the kinetic and potential energy expressions for the rotating beam in Eq. (8.2) and using the Hamilton's principle, we get the governing differential equation for transverse displacement $w(x, t)$ as

$$\frac{\partial^2}{\partial x^2}\left(EI(x)\frac{\partial^2 w}{\partial x^2}\right) - \frac{\partial}{\partial x}\left(T(x)\frac{\partial w}{\partial x}\right) + \rho A(x)\frac{\partial^2 w}{\partial t^2} = 0 \tag{8.4}$$

where E, $I(x)$, $T(x)$, ρ and $A(x)$ are the Young's modulus, moment of inertia, axial force due to centrifugal stiffening, mass density and area of cross-section of the beam respectively.

The axial force due to centrifugal stiffening, $T(x)$ is given as

$$T(x) = \int_x^L \rho A(x)\Omega^2 x dx \tag{8.5}$$

where Ω is the rotation speed of the beam.

8.2 Spectral Finite Element Formulation

The spectral element here is formulated following the procedure adopted by [15]. For constructing the weak form of the governing differential equation (Eq. 8.4), we first write the total energy of a non-conservative rotating system in transverse motion, in the frequency domain as

$$\hat{\Pi} = e^{i\omega_n t}\sum_{n=1}^{N}\left(\frac{1}{2}E\int_0^L I(x)\left(\frac{d^2\hat{w}_n}{x^2}\right)^2 dx + \frac{1}{2}\int_0^L T(x)\left(\frac{d\hat{w}_n}{dx}\right)^2 dx \right.$$
$$\left. + \frac{1}{2}\rho\int_0^L -A(x)\omega_n^2\hat{w}_n^2 dx + \frac{1}{2}\eta\int_0^L i\omega_n\hat{w}_n^2 dx - [\hat{F}]^T\{\hat{W}\}\right) \tag{8.6}$$

where $[\hat{F}]^T, \{\hat{W}\}$ represents the externally applied nodal force and nodal displacement vectors in frequency domain, ω_n the circular frequency of the nth sampling point, and $\hat{w}_n$ is the spatially dependent Fourier coefficient and η is the damping force per unit length, per unit velocity. According to principle of minimum potential energy in the frequency domain, we have

$$\delta\hat{\Pi} = 0$$

i.e.
$$E\int_0^L I(x)\frac{d^2\hat{w}_n}{dx^2}\frac{d^2}{dx^2}(\delta\hat{w}_n)dx + \int_0^L T(x)\frac{d\hat{w}_n}{dx}\frac{d}{dx}(\delta\hat{w}_n)dx$$
$$+\rho\int_0^L -A(x)\omega_n^2\hat{w}_n\delta\hat{w}_n dx + \eta\int_0^L i\omega_n\hat{w}_n\delta\hat{w}_n dx - \delta\{\hat{W}\}[\hat{F}]^T = 0 \tag{8.7}$$

To obtain the dynamic stiffness matrix in frequency domain, we need to substitute an interpolating function for the transverse displacement ($\hat{w}_n$) in the above Eq. (8.8). As mentioned earlier, we considered two choices for the interpolating function and hence two types of elements viz: the SFER and SFEN are formulated.

8.2.1 Interpolating Function for SFER

To obtain the interpolating function for SFER, we assume the beam to be uniform and replace $T(x)$ by the maximum centrifugal force, T_{max} [36]

$$T_{max} = \int_0^L \rho A(x)\Omega^2 x dx = \frac{\rho A\Omega^2 L^2}{2} \tag{8.8}$$

This allows us to represent Eq. (8.4) with damping, as a constant coefficient PDE, which can be written as

$$EI\frac{\partial^4 w}{\partial x^4} - T_{max}\frac{\partial^2 w}{\partial x^2} + \rho A\frac{\partial^2 w}{\partial t^2} + \eta\frac{\partial w}{\partial t} = 0 \tag{8.9}$$

where η is the damping force per unit length, per unit velocity. The above Eq. (8.9) is analogous to the "stiff string" equation where the tension is constant along the length [30]. The exact solution of this equation, in frequency domain is given as

$$\hat{w}_n(x) = C_{1n}e^{-ik_{1n}x} + C_{2n}e^{-ik_{4n}x} + C_{3n}e^{ik_{2n}(L-x)} + C_{4n}e^{ik_{3n}(L-x)} \tag{8.10}$$

where k_{pn} is the wave number, obtained from the relation.

$$k_{pn} = \pm\sqrt{\frac{-\left(\frac{\rho A\Omega^2 L^2}{2EI}\right) \pm \sqrt{\left(\frac{\rho A\Omega^2 L^2}{2EI}\right)^2 - 4\left(\frac{i\omega_n\eta}{EI} - \frac{\rho A\omega_n^2}{EI}\right)}}{2}} \tag{8.11}$$

Here, p represents the modes of wave propagation. The above Eq. (8.11) is the interpolating function for SFER in the frequency domain.

8.2.2 Interpolating Function for SFEN

Similar to SFER, we can obtain the interpolating function for SFEN, by assuming the beam to be uniform and non-rotating. This can be obtained by substituting $T_{max} = 0$ in Eq. (8.9), which then reduces to the form

$$EI\frac{\partial^4 w}{\partial x^4} + \rho A\frac{\partial^2 w}{\partial t^2} + \eta\frac{\partial w}{\partial t} = 0 \tag{8.12}$$

The exact solution of the above Eq. (8.12) can be obtained in the similar form as Eq. (8.13) and is given by

$$\hat{w}_n(x) = C_{1n}e^{-ik_{1n}x} + C_{2n}e^{-ik_{4n}x} + C_{3n}e^{ik_{2n}(L-x)} + C_{4n}e^{ik_{3n}(L-x)} \tag{8.13}$$

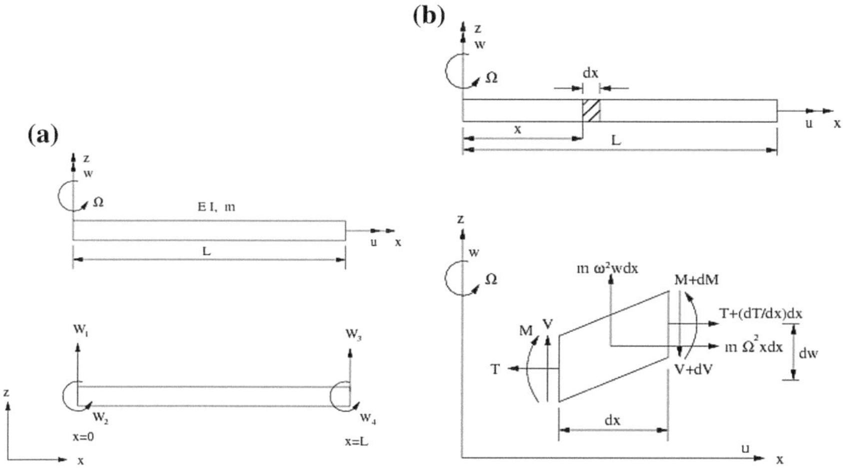

Fig. 8.1 **a** Spectral finite element for a beam under rotation, **b** Forces acting on an incremental length dx of the rotating beam

But here, the wavenumber is different from the previous case and is given by

$$k_{pn} = \pm\sqrt{\pm\sqrt{\left(\frac{\rho A\omega_n^2}{EI}\right) - \left(\frac{i\omega_n\eta}{EI}\right)}} \tag{8.14}$$

where p represents the mode of wave propagation. The Eq. (8.13) is the frequency domain interpolating function for SFEN.

8.2.3 Dynamic Stiffness Matrix in Frequency Domain

In finite element terms, Eqs. (8.13) and (8.14) represents the shape function and obviously it is dependent on the frequency ω_n. The shape function is now forced to satisfy the geometric boundary conditions which results in establishing a relation between the unknown frequency dependent coefficients C_{pn}, of Eqs. (8.12) and (8.14) and the nodal degrees of freedom. The detailed analysis is given in [36] and hence it is not discussed here. The relation can be written as

$$[C] = [A]^{-1}\{\hat{W}\} \tag{8.15}$$

We can represent Eq. (8.13) in matrix form as

$$\hat{w}_n(x) = \left[e^{-ik_{1n}x}\ e^{-ik_{4n}x}\ e^{ik_{2n}(L-x)}\ e^{ik_{3n}(L-x)}\right]\{C\} \tag{8.16}$$

or

$$\hat{w}_n(x) = [N]\{C\} \tag{8.17}$$

Substituting Eq. (8.15) in (8.17) we get the interpolating functions for the nth sampling frequency in the form

$$\hat{w}_n(x) = [N][G]\{\hat{W}\} \tag{8.18}$$

Here, $G = [A]^{-1}$ is independent of x. The interpolating functions obtained from the above (Eq. (8.18)), on substituting in Eq. (8.8), we get the corresponding dynamic stiffness matrix for SFER or SFEN and is given as

$$\begin{aligned}
&E\int_0^L I(x)[G]^T\left[\frac{\partial^2 N}{\partial x^2}\right]^T\left[\frac{\partial^2 N}{\partial x^2}\right][G]\{\hat{W}\}dx \\
&+\int_0^L T(x)[G]^T\left[\frac{\partial N}{\partial x}\right]^T\left[\frac{\partial N}{\partial x}\right][G]\{\hat{W}\}dx \\
&-\omega_n^2\rho\int_0^L A(x)[G]^T[N]^T[N][G]\{\hat{W}\}dx \\
&+i\omega_n\eta\int_0^L [G]^T[N]^T[N][G]\{\hat{W}\}dx - \{\hat{F}\} = 0
\end{aligned} \tag{8.19}$$

The above expression (Eq. 8.19) on integration and simplification gives the frequency dependent dynamic stiffness relation and can be symbolically written as

$$[\hat{K}]\{\hat{W}\} = \{\hat{F}\} \tag{8.20}$$

where $[\hat{K}]$, $\{\hat{W}\}$ and $\{\hat{F}\}$ are the dynamic stiffness matrix, nodal degrees of freedom, input nodal forces, in frequency domain respectively. The dynamic stiffness matrix obtained is complex and symmetric. Similar to an eigenvalue problem, the vanishing determinant of the dynamic stiffness matrix indicates resonances. If the assumed amplitude field ($\hat{w}_n$) is the exact solution for the governing differential equation, the present method will yield the exact dynamic stiffness matrix. Since it is impossible to find the closed form exact solution for this problem, the accuracy of the FE model can be improved by increasing the number of finite elements. Thus, for a rotating beam problem, unlike the usual spectral finite element approach, the length of the element is a deciding factor in the accuracy of the solution. For a uniform beam, if n is the total number of finite elements each of length l_e and j is the element under consideration, we can write $T(x)$ and T_{max} in discrete form as

$$T(x)_j = \frac{\rho A\Omega^2}{2}\left[(nl_e)^2 - ((j-1)l_e)^2\right] - \frac{\rho A\Omega^2}{2}\left[2(j-1)l_e x + x^2\right] \tag{8.21}$$

$$T_{(max)j} = \frac{\rho A\Omega^2}{2}\left[(nl_e)^2 - ((j-1)l_e)^2\right] \tag{8.22}$$

The dynamic global structural stiffness matrix is formed from the element stiffness matrices, assembled in a way analogous to that followed in CFEM. The major difference is that the structural matrix has to be assembled and evaluated for each frequency component up to the Nyquist frequency.

8.3 Free Vibration Results

In this section, results of free vibration analysis are presented for both uniform and tapered rotating beams. The results are obtained for different λ, varying from 0 to 12, where λ is a non-dimensional rotation parameter defined as, $\lambda^2 = \left(\frac{\Omega^2}{\frac{EI_0}{\rho A_0 L^4}}\right)$. The results of the present analysis are compared with [37]. Both cantilever and hinged boundary conditions are considered in the present study, as they correspond to hingeless and articulated rotor blades.

8.3.1 Uniform Beam

In this section, we validate the spectral finite element model using the free vibration results of uniform rotating beams. To compare the spectral element models, SFER and SFEN, a preliminary convergence study (Table 8.1) at the non-dimensional rotation speed $\lambda = 12$, is performed. The study reveals that SFER has better convergence than SFEN, although the difference in the predicted frequencies are small and hence, we will use SFER for further study in this section. The resonances can be determined by plotting the behavior of the determinant of the dynamic stiffness matrix. A typical plot for uniform beam is given in Fig. 8.2. Here, $Log(1/det(\hat{K}))$ of structural dynamic stiffness matrix is plotted against ω, for different rotation speeds. The results are summarized in Tables 8.2, 8.3, 8.4 and 8.5 and they are compared with the values of

Table 8.1 Convergence study on the natural frequencies of uniform beam for $\lambda = 12$, using the spectral elements SFER (rotating beam basis functions assuming constant(maximum) centrifugal force) and SFEN (non-rotating beam basis functions)

Modes	SFER	SFEN	SFER	SFEN	Exact
	$n = 8$		$n = 16$		
1	13.1731	13.1731	13.1731	13.1731	13.1702
2	37.6017	37.6113	37.6017	37.6017	37.6031
3	79.6136	79.6328	79.6136	79.6136	79.6145
4	140.5318	140.5702	140.5318	140.5318	140.5342
5	220.5385	220.6056	220.5385	220.5385	220.536

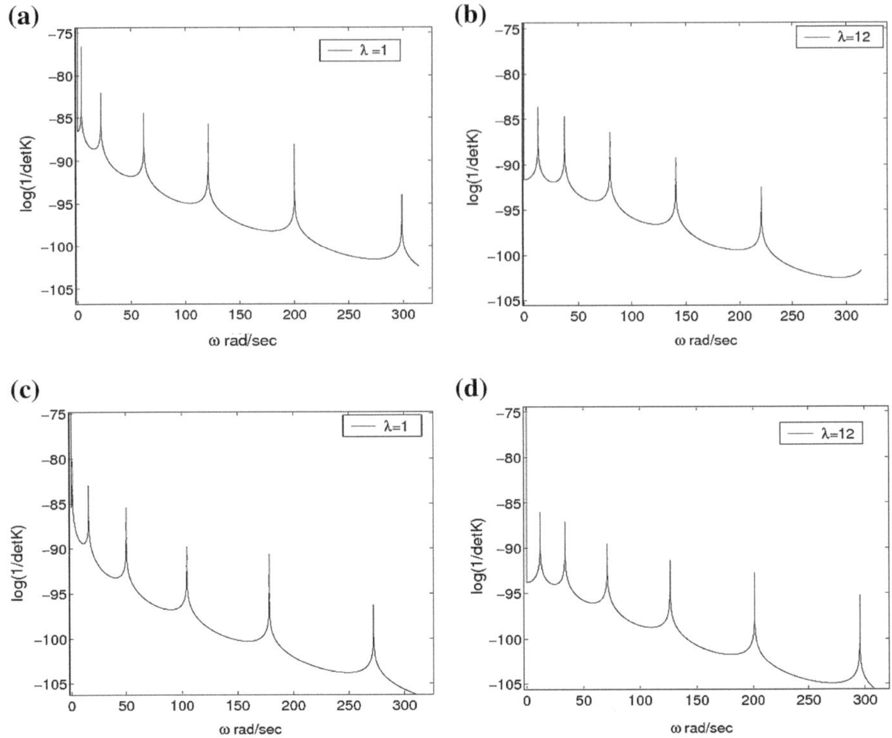

Fig. 8.2 **a** and **b** Resonant peaks of uniform cantilever beam, **c** and **d** Resonant peaks of uniform hinged beam

Table 8.2 Validation of natural frequencies of rotating uniform cantilever beam for different non-dimensional rotation speeds, λ: Modes 1 to 3

λ	First mode			Second mode			Third mode		
	Present (SFER)	Wang	Wright	Present (SFER)	Wang	Wright	Present (SFER)	Wang	Wright
0	3.5186	3.5160	3.5160	22.0318	22.0345	22.0345	61.6948	61.6972	61.6972
1	3.6816	N/A	3.6817	22.1852	N/A	22.1810	61.8386	N/A	61.8418
2	4.1417	N/A	4.1373	22.6166	N/A	22.6149	62.2796	N/A	62.2732
3	4.7937	N/A	4.7973	23.3165	N/A	23.3203	62.9891	N/A	62.9850
4	5.5799	N/A	5.5850	24.2752	N/A	24.2734	63.9670	N/A	63.9668
5	6.4523	N/A	6.4495	25.4449	N/A	25.4461	65.2038	N/A	65.2050
6	7.3631	N/A	7.3604	26.8063	N/A	26.8091	66.6898	N/A	66.6840
7	8.3027	N/A	8.2996	28.3403	N/A	28.3341	68.3868	N/A	68.3860
8	9.2614	N/A	9.2568	29.9989	N/A	29.9954	70.2947	N/A	70.2930
9	10.2201	N/A	10.2257	31.7726	N/A	31.7705	72.3847	N/A	72.3867
10	11.1981	N/A	11.2023	33.6421	N/A	33.6404	74.6473	N/A	74.6493
11	12.1856	N/A	12.1843	35.5883	N/A	35.5890	77.0634	N/A	77.0638
12	13.1731	13.1702	13.1702	37.6017	37.6031	37.6031	79.6136	79.6145	79.6145

Table 8.3 Validation of natural frequencies of rotating uniform cantilever beam for different non-dimensional rotation speeds, λ: Modes 4 and 5

λ	Fourth mode			Fifth mode		
	Present (SFER)	Wang	Wright	Present (SFER)	Wang	Wright
0	120.8969	120.902	120.902	199.8585	199.860	199.860
1	121.0503	N/A	121.051	200.0119	N/A	200.012
2	121.4913	N/A	121.497	200.4721	N/A	200.467
3	122.2391	N/A	122.236	210.2199	N/A	201.223
4	123.265	N/A	123.261	202.2745	N/A	202.277
5	124.5688	N/A	124.566	203.6168	N/A	203.622
6	126.1412	N/A	126.140	205.2562	N/A	205.253
7	127.9723	N/A	127.972	207.1641	N/A	207.161
8	130.043	N/A	130.049	209.3405	N/A	209.338
9	132.3634	N/A	132.358	211.7756	N/A	211.775
10	134.8848	N/A	134.884	214.4601	N/A	214.461
11	137.6173	N/A	137.6173	217.3843	N/A	217.385
12	140.5318	140.534	140.5342	220.5381	220.536	220.536

Table 8.4 Validation of natural frequencies of uniform rotating hinged beam for different non-dimensional rotation speeds, λ: Modes 1 to 3

λ	First mode			Second mode			Third mode		
	Present (SFER)	Wang	Wright	Present (SFER)	Wang	Wright	Present (SFER)	Wang	Wright
0	0	0	0	15.4165	15.4182	15.4182	49.9694	49.9649	49.9649
1	0.9971	N/A	1	15.6274	N/A	15.6242	50.142	N/A	50.1437
2	1.9942	N/A	2	16.2218	N/A	16.2261	50.6789	N/A	50.6760
3	3.0009	N/A	3	17.1806	N/A	17.1807	51.5485	N/A	51.5498
4	3.9979	N/A	4	18.4269	N/A	18.4313	52.7498	N/A	52.7463
5	5.0046	N/A	5	19.9226	N/A	19.9197	54.2454	N/A	54.2419
6	6.0017	N/A	6	21.5908	N/A	21.5944	56.0095	N/A	56.0099
7	6.9988	N/A	7	23.4124	N/A	23.4133	58.0228	N/A	58.0223
8	7.9959	N/A	8	25.349	N/A	25.3436	60.2471	N/A	60.2513
9	9.0026	N/A	9	27.3624	N/A	27.3601	62.6727	N/A	62.6705
10	9.9997	N/A	10	29.4428	N/A	29.4439	65.2517	N/A	65.2554
11	11.0063	N/A	11	31.5808	N/A	31.5809	67.9841	N/A	67.9842
12	12.0034	12	12	33.7668	33.7603	33.7603	70.8412	70.8373	70.8373

Table 8.5 Validation of natural frequencies of uniform rotating hinged beam for different non-dimensional rotation speeds, λ: Modes 4 and 5

λ	Fourth mode			Fifth mode		
	Present (SFER)	Wang	Wright	Present (SFER)	Wang	Wright
0	104.2532	104.248	104.248	178.2677	178.270	178.270
1	104.4257	N/A	104.420	178.4403	N/A	178.440
2	104.9339	N/A	104.936	178.9485	N/A	178.949
3	105.7872	N/A	105.789	179.7921	N/A	179.794
4	106.976	N/A	106.971	180.9714	N/A	180.970
5	108.4716	N/A	108.469	182.467	N/A	182.469
6	110.2740	N/A	110.270	184.2886	N/A	184.283
7	112.3545	N/A	112.356	186.3978	N/A	186.401
8	114.7034	N/A	114.709	188.8139	N/A	188.812
9	117.3112	N/A	117.313	191.5079	N/A	191.504
10	120.149	N/A	120.146	194.4608	N/A	194.462
11	123.1978	N/A	123.192	197.6726	N/A	197.6730
12	126.4384	126.431	126.431	201.1241	201.122	201.122

[37, 38]. It can be seen that the correlation of frequency predictions are excellent. The tabulated values up to the fifth mode are obtained from a model which used only 8 elements or 18 degrees of freedom. For the same beam, [37] have used 50 terms in the Frobenius series, where as the number of terms used by [38] is not mentioned in their paper.

8.3.2 *Tapered Beam 1-Linear Mass and Cubic Flexural Stiffness Variation*

Polynomial variations of both mass and flexural stiffness are typically used to model taper in rotating beams. Here we consider the tapered beam used by [37] for the free vibration analysis. The mass distribution is assumed linear and is given as

$$m(x) = \rho A_0 \left[1 + \alpha \left(\frac{x}{L}\right)\right] \tag{8.23}$$

where $\alpha = -0.5$, which is a taper parameter for mass distribution, A_0 is the area of the initial cross-section and ρ is the mass density. The flexural stiffness variation is given by

$$EI(x) = EI_0 \left[1 + \beta \left(\frac{x}{L}\right)\right]^3 \tag{8.24}$$

where $\beta = -0.5$, which is a taper parameter for stiffness distribution, I_0 is the moment of inertia of the initial cross-section and E is the Young's modulus. As before, the convergence of the two different models (SFEN and SFER) is first performed and the results are shown in Table 8.6. The table shows that the convergence of both the models are equally good and the model SFER is used for further study, in this subsection. Here, both the elements have got equal convergence because, the bending stiffness variation is comparable or more than the centrifugal stiffness variation. Note that the basis functions assume a beam with uniform properties. The present approach requires 11 elements or 24 degrees of freedom to achieve matching results for the first three modal frequencies using SFER. In contrast, [37] used 80 terms in the Frobenius series to predict the first three modes. The number of terms required may go up further, for accurately predicting the higher modes. The results are tabulated in Table 8.7 and it matches with the literature values.

Table 8.6 Convergence study on the natural frequencies of tapered beam 1 for $\lambda = 12$, using the spectral elements SFER and SFEN

Modes	SFER	SFEN	SFER	SFEN	Exact
	$n = 11$		$n = 13$		
1	13.4703	13.4703	13.4703	13.4703	13.4711
2	34.0879	34.0879	34.0879	34.0879	34.0877
3	65.5297	65.5297	65.5297	65.5297	65.5237

Table 8.7 Validation of natural frequencies of rotating cantilever tapered beam 1 for different non-dimensional rotation speeds, λ: Modes 1 to 3

λ	First mode			Second mode			Third mode		
	Present (SFER)	Wang	Wright	Present (SFER)	Wang	Wright	Present (SFER)	Wang	Wright
0	3.8254	3.8238	3.8238	18.3167	18.3173	18.3173	47.2658	47.2648	47.2648
1	3.9883	3.9866	3.9866	18.4749	18.4740	18.4740	47.4192	47.4143	47.4143
2	4.439	4.4368	4.4368	18.9351	18.9366	18.9366	47.8746	47.8716	47.8716
3	5.0909	5.0927	5.0927	19.6829	19.6839	19.6839	48.6224	48.6190	48.6190
4	5.8711	5.8788	5.8788	20.6848	20.6852	20.6852	49.6483	49.6456	49.6456
5	6.7447	6.7434	6.7345	21.9072	21.9053	21.9053	50.9377	50.9338	50.9338
6	7.6555	7.6551	7.6551	23.3069	23.3093	23.3093	52.4669	52.4633	52.4633
7	8.5951	8.5956	8.5956	24.8649	24.8647	24.8647	54.2166	54.2124	54.2124
8	9.5538	9.5540	9.5540	26.5427	26.5437	26.5437	56.1629	56.1595	56.1595
9	10.5269	10.5239	10.5239	28.3211	28.3227	28.3227	58.2865	58.2833	58.2833
10	11.5048	11.5015	11.5015	30.1811	30.1827	30.1827	60.5683	60.5639	60.5639
11	12.4828	12.4845	12.4845	32.1081	32.1085	32.1085	62.9891	62.9829	62.9829
12	13.4703	13.4711	13.4711	34.0879	34.0877	34.0877	65.5297	65.5237	65.5237

Table 8.8 Convergence study on the natural frequencies of tapered beam 2 for λ = 12, using the spectral elements SFER and SFEN

Modes	SFER	SFEN	SFER	SFEN	Exact
	$n = 20$		$n = 22$		
1	14.0295	14.0295	14.0295	14.0295	14.0313
2	35.9079	35.9079	35.9079	35.9079	35.9064
3	72.8579	72.8579	72.8579	72.8579	72.8565
4	126.40	126.40	126.40	126.40	126.401
5	196.8864	196.8864	196.8864	196.8864	196.880

8.3.3 Tapered Beam 2-Linear Mass and Flexural Stiffness Variation

In this section we use a tapered beam with linear variation in both mass and stiffness, which is represented as

$$m(x) = \rho A_0 \left[1 + \alpha \left(\frac{x}{L}\right)\right] \tag{8.25}$$

where $\alpha = -0.8$, which is a taper parameter for mass distribution, A_0 is the area of the initial cross-section.

$$EI(x) = EI_0 \left[1 + \beta \left(\frac{x}{L}\right)\right] \tag{8.26}$$

where $\beta = -0.95$, which is a taper parameter for stiffness distribution, I_0 is the moment of inertia of the initial cross-section. Similar to tapered beam 1, it is found that both the spectral elements are equally good in convergence (Table 8.8) and as before, SFER is used for further analysis. Analytical predictions up to the fifth mode (Tables 8.9, 8.11, 8.10 and 8.12) are obtained for cantilever and hinged boundary conditions using SFER. The spectral model had 20 elements with a total of 42 degrees of freedom. For the same beam, [37] had used 350 terms in the Frobenius series for the analysis.

The frequency predictions obtained from our work matches very well with the literature values. From these results, it is obvious that the present formulation (both spectral element models) require lesser computational effort compared to other approaches. Wang and Wereley [37] used 50, 80 and 350 terms in the Frobenius series where as the present formulation used 8, 11 and 20 elements for the uniform beam, tapered beam 1 and tapered beam 2 respectively. The basic difference between the proposed method and the method by [37] lies in the choice of interpolating function. Wang and Wereley [37] used the series solution of the variable coefficient governing differential equation, obtained using the Frobenius method, as the interpolating function. A single element was used to get the modal frequencies.

Table 8.9 Validation of natural frequencies of rotating cantilever tapered beam 2 for different non-dimensional rotation speeds, λ: Modes 1 to 3

λ	First mode			Second mode			Third mode		
	Present (SFER)	Wang	Wright	Present (SFER)	Wang	Wright	Present (SFER)	Wang	Wright
0	5.2731	5.2738	5.2738	24.0068	24.0041	24.0041	59.9772	59.9708	59.9701
1	5.3945	5.3903	5.3903	24.1091	24.1069	24.1069	60.0681	60.0703	60.0696
2	5.7269	5.7249	5.7249	24.4159	24.4129	24.4130	60.3685	60.3676	60.3669
3	6.2382	6.2402	6.2402	24.9144	24.9148	24.9149	60.8607	60.8598	60.8590
4	6.8901	6.8928	6.8928	25.6047	25.6013	25.6013	61.5382	61.5420	61.5412
5	7.6443	7.6443	7.6443	26.4612	26.4580	26.4581	62.4075	62.4078	62.4069
6	8.4625	8.4653	8.4653	27.471	27.4692	27.4693	63.4493	63.4494	63.4483
7	9.3317	9.3347	9.3347	28.6215	28.6184	28.6185	64.6573	64.6579	64.6566
8	10.2393	10.2379	10.2379	29.8871	29.8893	29.8894	66.0251	66.0238	66.0222
9	11.1597	11.1651	11.1650	31.2676	31.2667	31.2669	67.5335	67.5370	67.5351
10	12.0157	12.1092	12.0192	32.7377	32.7367	32.7369	69.1825	69.1875	69.1851
11	13.0644	13.0657	13.0657	34.2845	34.2868	34.2871	70.9594	70.9653	70.9622
12	14.0295	14.0313	14.0313	35.9079	35.9060	35.9064	72.8577	72.8604	72.8565

Table 8.10 Validation of natural frequencies of rotating hinged tapered beam 2 for different non-dimensional rotation speeds, λ: Modes 1 to 3

λ	First mode			Second mode			Third mode		
	Present (SFER)	Wang	Wright	Present (SFER)	Wang	Wright	Present (SFER)	Wang	Wright
0	0	0	0	16.7332	16.7328	16.7328	48.4674	48.4696	48.4691
1	0.9971	1	1	16.8738	16.8711	16.8711	48.5889	48.5876	48.5872
2	2.0006	2	2	17.2828	17.2793	17.2794	48.9404	48.9399	49.9395
3	3.004	3	3	17.9348	17.9389	17.9388	49.522	49.5214	49.5210
4	4.0011	4	4	18.7232	18.8233	18.8233	50.3274	50.3239	50.3234
5	4.9982	5	5	19.9034	19.9022	19.9022	51.3372	51.3366	51.3360
6	6.0017	6	6	21.1434	21.1456	21.1456	52.5427	52.5471	52.5463
7	6.9988	7	7	22.524	22.5259	22.5260	53.9386	53.9415	54.9405
8	8.0023	8	8	24.0196	24.0194	24.0195	55.5045	55.5054	55.5042
9	8.9994	9	9	25.6047	25.6059	25.6060	57.2239	57.2245	57.2230
10	9.9964	10	10	27.2665	27.2690	27.2692	59.0838	59.0847	59.0828
11	10.9935	11	11	28.9922	28.9953	28.9956	61.0716	61.0727	61.0700
12	12.0034	12	12	33.7755	33.7741	33.7745	63.1744	63.1758	63.1722

Table 8.11 Validation of natural frequencies of rotating cantilever tapered beam 2 for different non-dimensional rotation speeds, λ: Modes 4 and 5

λ	Fourth mode			Fifth mode		
	Present (SFER)	Wang	Wright	Present (SFER)	Wang	Wright
0	112.9138	112.892	112.909	183.0295	183.473	183.024
1	113.0097	112.992	113.009	183.1317	183.576	183.124
2	113.310	113.290	113.307	183.4258	183.887	183.424
3	113.8022	113.784	113.803	183.9243	184.404	183.923
4	114.4925	114.472	114.492	184.621	185.127	184.619
5	115.3745	115.351	115.372	185.5094	186.055	185.509
6	116.4419	116.414	116.439	186.596	187.186	186.591
7	117.6883	117.658	117.685	187.8679	188.520	187.862
8	119.0608	119.075	119.107	189.3188	190.055	189.316
9	120.6987	120.659	120.696	190.955	191.793	190.950
10	122.450	122.403	122.446	192.7639	193.734	192.759
11	124.3547	124.298	124.350	194.7389	195.882	194.737
12	126.400	126.336	126.401	196.8864	198.243	196.880

Table 8.12 Validation of natural frequencies of rotating hinged tapered beam 2 for different non-dimensional rotation speeds, λ: Modes 4 and 5

λ	Fourth mode			Fifth mode		
	Present (SFER)	Wang	Wright	Present (SFER)	Wang	Wright
0	97.1713	97.1599	97.1704	162.9996	163.277	163.002
1	97.2799	97.2717	97.2823	163.1133	163.388	163.111
2	97.6187	97.6062	97.6172	163.4457	163.723	163.438
3	98.1748	98.1611	98.1727	163.9825	164.281	163.983
4	98.9418	98.9324	98.9450	164.7432	165.059	164.741
5	99.9261	99.9147	99.9287	165.7147	166.055	165.771
6	101.1149	101.102	101.117	166.8907	167.268	166.889
7	102.5083	102.485	102.504	168.2713	168.695	168.269
8	104.0806	104.058	104.079	169.850	170.333	169.847
9	105.8319	105.809	105.835	171.6205	172.180	171.618
10	107.7621	107.730	107.762	173.5763	174.236	173.573
11	109.8522	109.810	109.850	175.7111	176.500	175.708
12	112.0956	112.040	112.090	178.0185	178.978	178.105

However, since the Frobenius series have infinite terms, the accuracy of the solution depends on the number of terms used. The number of terms required in the power series increases depending on the modes, taper and the rotation speed. Again, the governing equation is of fourth order and four terms need to be separately expanded using the power series. Furthermore, the exact solution should be made available for each sampling frequency/iteration, up to the Nyquist frequency. Also, if our interest is in higher modes, which is quite common in structural health monitoring applications, the Nyquist frequency selected should be sufficiently high. Unlike the series solution of [37], here, we exploit the advantages of the variational principle and FE procedure to formulate an approximate SFEM that has convergence comparable to [37]. Next we will see how these elements perform for wave propagation studies, where the fast convergence of higher modes is very critical for accurate response estimation.

8.4 Wave Propagation Study

8.4.1 Convergence Study

A short duration unit impact pulse with high frequency content, as shown in Fig. 8.3a is applied at the beam tip for getting the impact response. Figure 8.3b shows the FFT of the pulse or the input amplitude spectrum. The general procedure for obtaining the wave propagation response is as follows. First, the input loading is fed into a FFT program to obtain the force spectrum, which is read in and stored. The dynamic stiffness matrix is generated for all the elements and assembled and solved for unit impulse, which gives the system transfer function. Note that all these operations of element stiffness matrix generation, assembly and solution have to be performed for

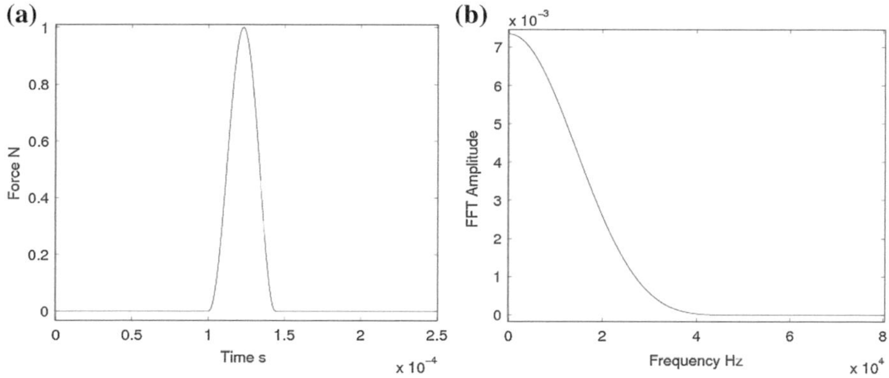

Fig. 8.3 **a** Input force pulse **b** FFT of the input force pulse

Table 8.13 Convergence study on wave propagation response, using spectral elements SFER and SFEN; R: SFER, N: SFEN

Beam	$\Omega = 1000$		$\Omega = 3000$		$\Omega = 6000$		$\Omega = 8000$		$\Omega = 10000$	
	R	N	R	N	R	N	R	N	R	N
Uniform, $L0.6$ m n=	4	7	13	24	18	32	20	34	21	37
Taper 2, $L0.6$ m n=	25	25	25	25	25	27	25	30	25	35
Taper 2, $L1.2$ m n=	45	45	49	58	58	74	61	84	65	95

Table 8.14 Beam Properties of uniform beam

Length, m	0.6
Cross sectional area, m^2	240×10^{-6}
Moment of Inertia, m^4	2000×10^{-12}
E,GPa	200
Mass density, kg/m^3	7840

each sampled frequency. This is not a major issue since the system size involved in SFEM is very small. The obtained transfer function is then convolved with the input force spectrum to obtain the required output in the frequency domain. Inverse FFT of the output will give the time domain response.

As done in the earlier section, the two elements, SFER and SFEN are used for studying the convergence of the wave propagation response. Properties of the uniform beam considered in this section are given in Table 8.14. The taper parameters of the tapered beam 2 is same as mentioned earlier with E, ρ, I_0 and A_0 values equal to that of uniform beam. The elements are compared for the convergence on response peaks and the time of reflection Fig. 8.4. The number of elements required for convergence for uniform beam and tapered beam 2 with lengths of 0.6 m, at different rotation speeds, are given in Table 8.13. Also, a case of tapered beam 2 with a length of 1.2 m is presented, to study the effect of the length of the beam on the convergence for a given taper. The study reveals that the SFER gives faster convergence than SFEN especially at higher speeds. The two spectral elements are equally good when the bending stiffness and the centrifugal stiffness distributions are comparable, as in the case of short beams. But in situations like high rotation speeds, long beams or long beams rotating at high speeds, SFER model is found to be superior to the SFEN model. In general, we can say that the SFER is a better choice, when the centrifugal stiffness variation dominates the flexural stiffness variation.

8.4.2 Numerical Results

Numerical results for wave propagation response of rotating beams are now obtained using SFER. The uniform beam and tapered beam 2, each of length 0.6 m are taken to illustrate the wave propagation behavior and wave propagation response of the beams. Properties of the uniform beam and tapered beam considered here is the same as that given in the previous section. The wave propagation response analysis of the beams for different rotation speeds bring out the difference in wave propagation behavior of the beam qualitatively.

From Figs. 8.5 and 8.6, it is obvious that, as rotation speed increases, the spectrum relation becomes non-dispersive. The gradual evolvement of this behavior is illustrated in these figures and is due to the relatively dominant contribution of the centrifugal term in the expression for wavenumber (Eq. (8.12)). Also, the peak of the velocity response marginally dips at high rotation speed Ω, due to the effect of centrifugal stiffening which has been introduced into the stiffness matrix through the energy approach. Again, for the tapered beam 2, for a given rotation speed, the

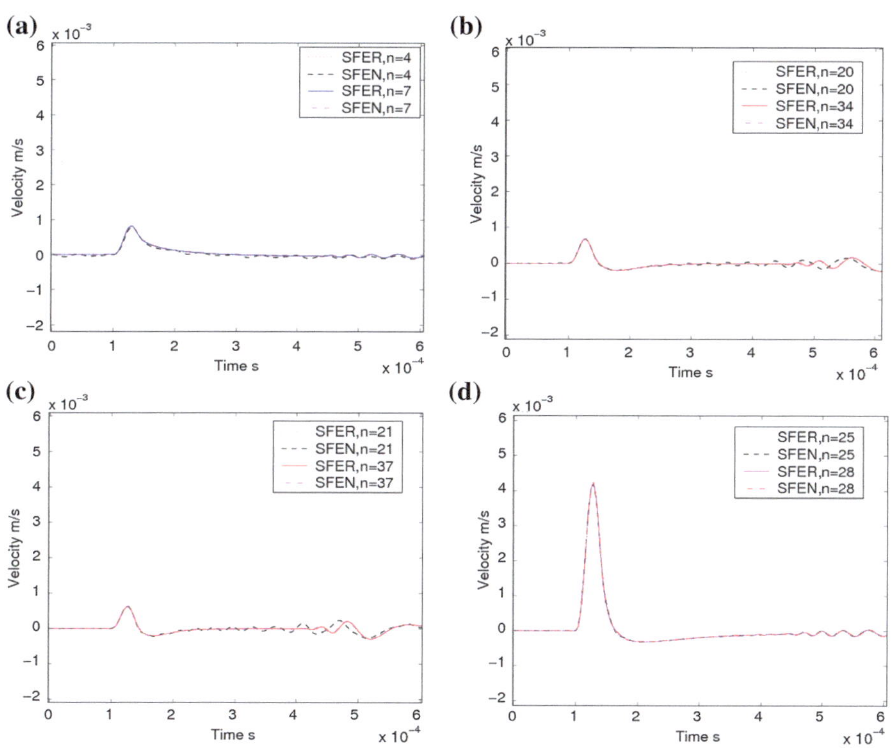

Fig. 8.4 Impact response convergence of uniform beam, $L0.6$ m **a** $\Omega = 1000$ rad/s **b** $\Omega = 8000$ rad/s **c** $\Omega = 10000$ rad/s; Tapered beam 2, $L0.6$ m **d** $\Omega = 1000$ rad/s **e** $\Omega = 8000$ rad/s **f** $\Omega = 10000$ rad/s; Tapered beam 2, $L1.2$ m **g** $\Omega = 1000$ rad/s **h** $\Omega = 8000$ rad/s **i** $\Omega = 10000$ rad/s

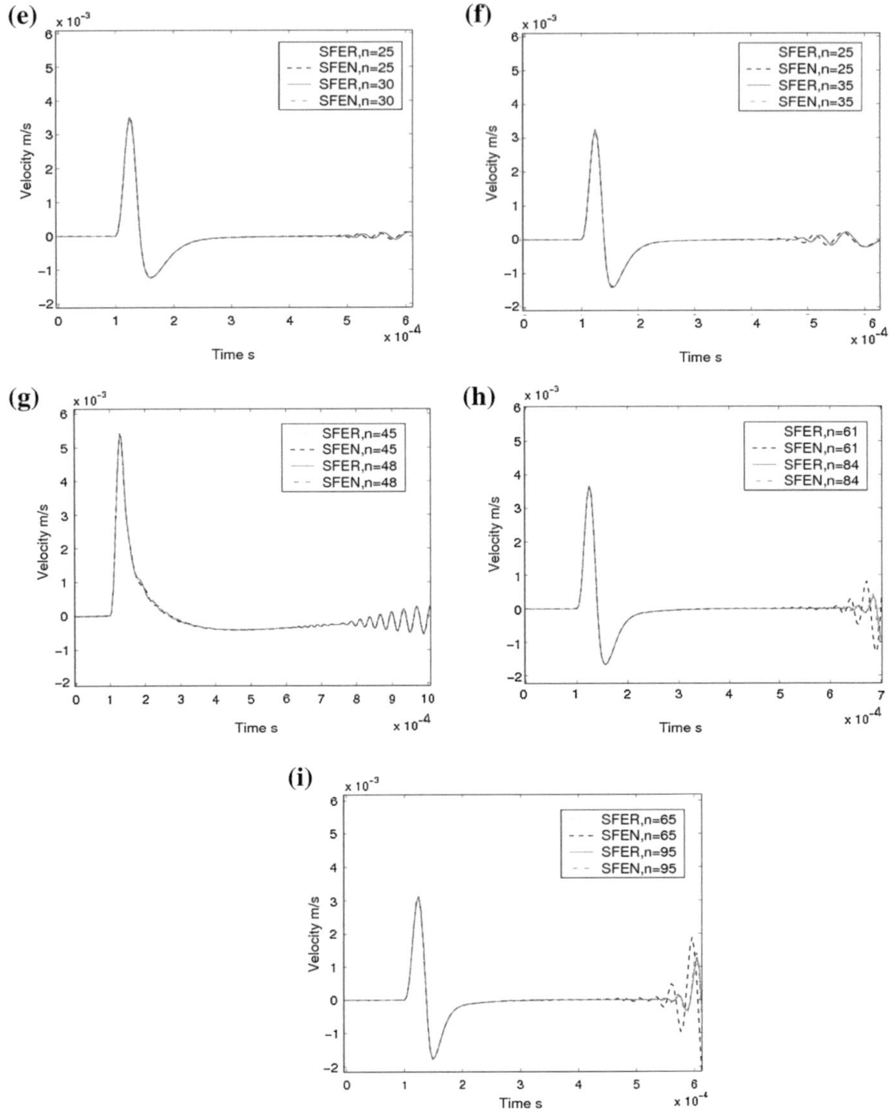

Fig. 8.4 (continued)

velocity peak is higher than that of the uniform beam due to the reduced bending and centrifugal stiffness of the tapered beam. However the overall behavior of uniform and tapered beams are similar.

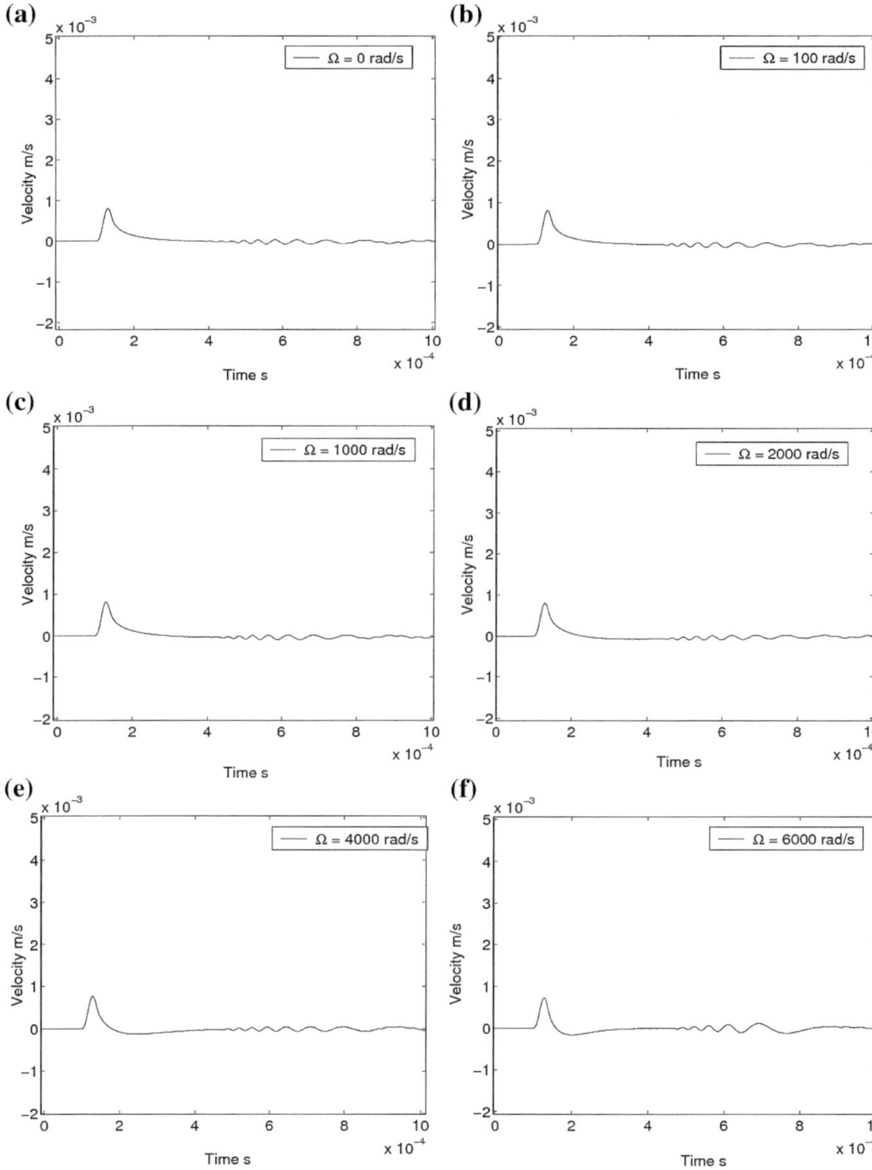

Fig. 8.5 Impulse response of uniform beam for various rotation speeds

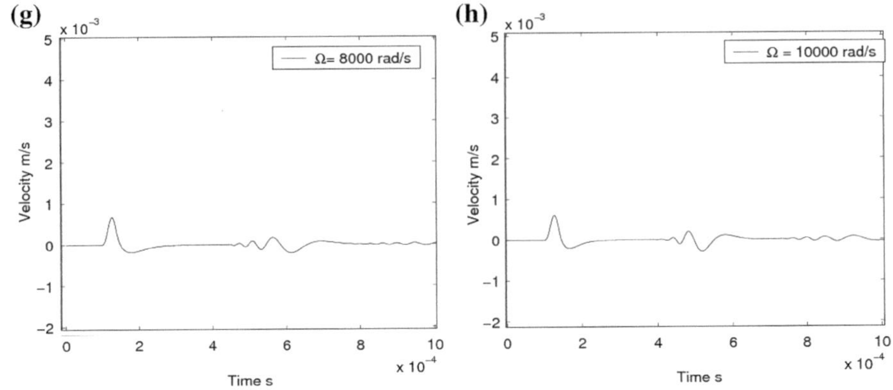

Fig. 8.5 (continued)

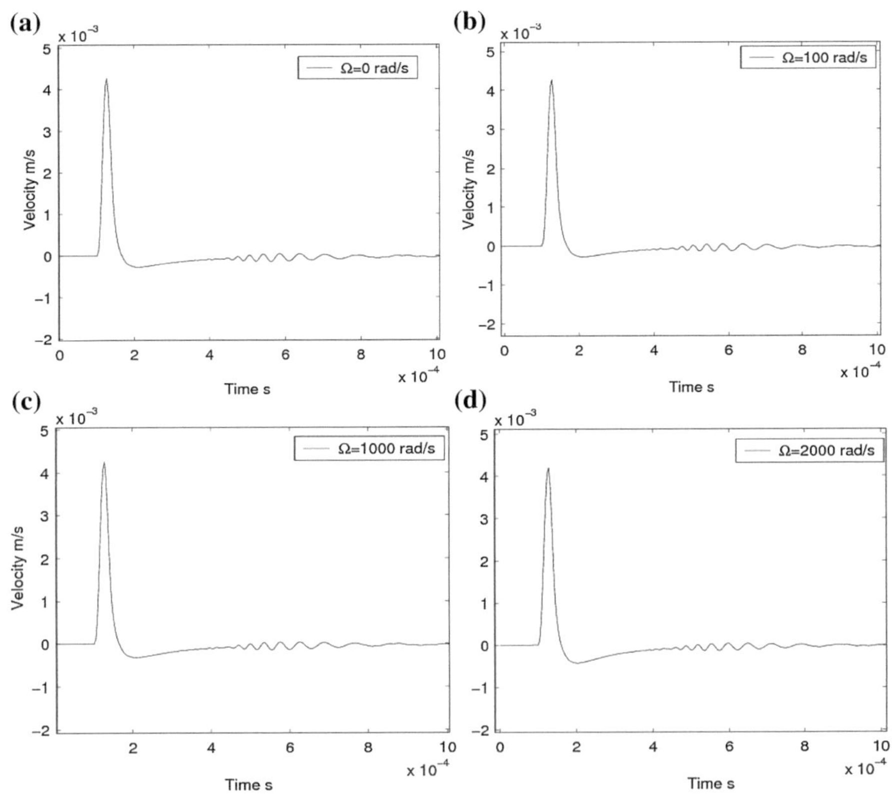

Fig. 8.6 Impulse response of tapered beam 2 for various rotation speeds

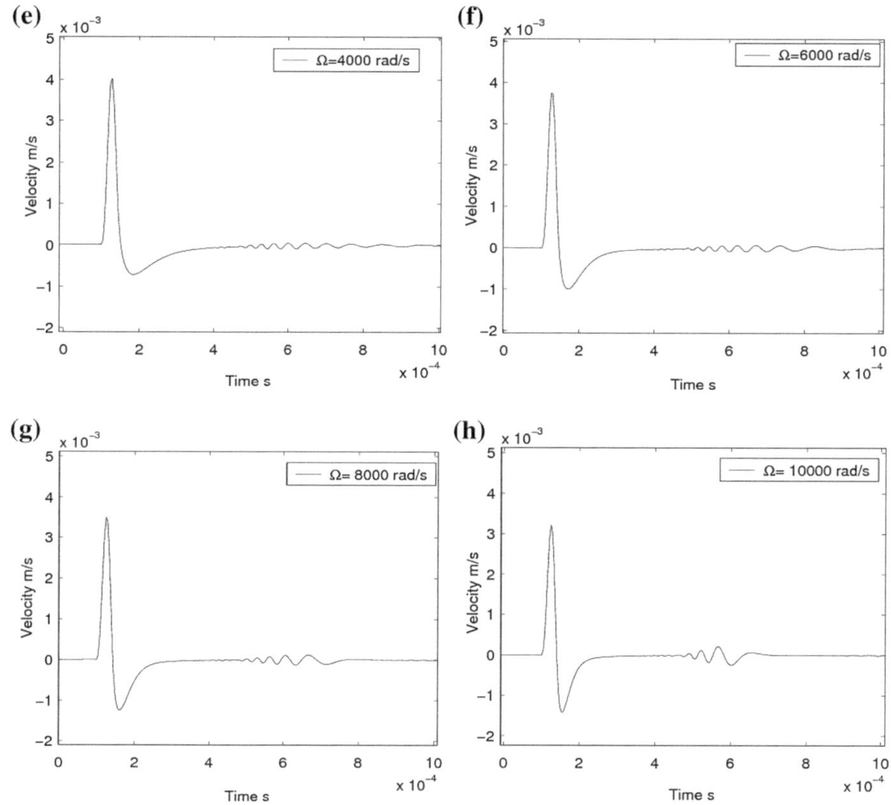

Fig. 8.6 (continued)

8.5 Summary

The rotating beam vibration analysis problem is formulated using the spectral finite element method. The governing differential equation for a rotating beam is arrived at using Hamilton's principle. The weak form of the governing equation, obtained through the principle of minimum stationary potential in frequency domain is utilized to formulate the dynamic stiffness matrix. The natural frequencies are obtained from the determinant plot. Here, we used two interpolating functions in the frequency domain for the transverse displacement field. These two functions are used to formulate two different spectral elements (SFER and SFEN). SFER includes the effect of centrifugal stiffening in an approximate sense whereas SFEN ignores it. The convergence study on frequency predictions shows that both the elements converge in a similar fashion for tapered beams. However, for uniform beams, SFER converges better, since only centrifugal stiffness variation is present along the beam length. The natural frequency predictions obtained for the uniform and tapered, cantilever and hinged beams are compared with the literature values and are found to agree very well.

The beam is then subjected to an impact load and the wave propagation response is analyzed for different rotation speeds. The study reveals that at high rotation speeds, the beam behaves non-dispersively. Also, the velocity response amplitude decreases with the rotation speed. A convergence study is performed to assess the quality of the two formulated spectral elements. It is seen that, the SFER converges faster than SFEN, especially for higher rotation speeds, long beams or long beams rotating at high speeds. It can be concluded that the SFER is a better choice for problems where the centrifugal stiffness variation dominates the bending stiffness variation, because it captures the physics of centrifugal effects better.

The present approach provides a computationally efficient and simple way of obtaining the natural frequencies and response of a rotating beam, unlike the other methods suggested in the literature. Even for very high rotation speeds, the number of elements required is quite low. The method becomes especially useful when the higher modes are required such as for impact response analysis and structural health monitoring studies.

References

1. Banerjee JR (2000) Free vibration of centrifugally stiffened uniform and tapered beams using the dynamic stiffness method. J Sound Vib 233(5):857–875
2. Boyce WE (1956) Effect of hub radius on the vibration of uniform bar. J Appl Mech 23(2):287–290
3. Cesnik CES, Shin S (2001) On the modeling of integrally actuated helicopter blades. Int J Solids Str 38(10–13):1765–1789
4. Chakraborty A, Gopalakrishnan S (2003) A spectrally formulated finite element for wave propagation analysis in functionally graded beams. Int J Solid Str 40:2421–2448
5. Chakraborty A, Gopalakrishnan S (2005) A spectral finite element for axial-flexural-shear coupled wave propagation analysis in lengthwise graded beam. Comput Mech 36:1–12
6. Chakraborty A, Sivakumar MS, Gopalakrishnan S (2006) Spectral element based model for wave propogation analysis in multi-wall carbon nanotubes. Int J Solid Str 43(2):279–294
7. Cheng J, Xu H (2006) Inner mass impact damper for attenuating structure vibration. Int J Solid Str 43(17):5355–5369
8. Chung J, Yoo HH (2002) Dynamic analysis of a rotating cantilever beam by using the finite element method. J Sound Vib 249(1):147–164
9. Datta PK, Ganguli R (1990) Vibration characteristics of rotating gas turbine blades with localized damage including the effects of shear deformation and rotary inertia. Comput Str 36(6):1129–1133
10. Doyle JF (1988) A spectrally formulated finite element for longitudinal wave propagation. Int J Anal Exp Modal Anal 5:99–107
11. Doyle JF (1989) Wave Propagation in Structures: an FFT based spectral analysis methodology. Springer, New York
12. Doyle JF, Farris TN (1990) A spectrally formulated element for flexural wave propagation in beams. Int J Anal Exp Modal Anal 3:1–5
13. Fox CHJ, Burdess JS (1979) The natural frequencies of a thin rotating cantilever with offset root. J Sound Vib 65:151–158

14. Gopalakrishnan S, Martin M, Doyle JF (1992) A matrix methodology for spectral analysis of wave propagation in multiple connected Timoshenko beams. J Sound Vib vol 158(l). pp 11–24
15. Gopalakrishnan S, Doyle JF (1994) Wave propagation in connected waveguides of varying cross-section. J Sound Vib 175(3):347–363
16. Gudla PK, Ganguli R (2006) Discontinuous Galerkin finite element in time for solving periodic differential equations. Comput Method Appl Mech Eng 196(1–3):682–696
17. Hodges DH, Rutkowski MJ (1981) Free-vibration analysis of rotating beams by a variable-order finite element method. AIAA J 19(11):1459–1466
18. Hodges DH (1979) Vibration response of non-uniform rotating beams with discontinuities. J Am Helicopter Soc 24(5):43–50
19. Hunter WF (1970) Integrating matrix method for determining the natural vibration characteristics of propeller blades. NASA TN D-6064
20. Khulief YA (1989) Vibration frequencies of a rotating tapered beam with end mass. J Sound Vib 134(1):87–97
21. Lawther R (2007) Assessing how changes to a structure can create gaps in the natural frequency spectrum. Int J Solid Str 44(2):614–635
22. Lee U, Oh H (2005) Dynamics of an axially moving viscoelastic beam subject to axial tension. Int J Solid Str 42(8):2381–2398
23. Martin M, Gopalakrishanan S, Doyle JF (1994) Wave propagation in multiply connected deep wave guides. J Sound Vib 174(4):521–538
24. Naguleswaran N (1994) Lateral vibration of a centrifugally tensioned uniform Euler-Bernoulli beam. J Sound Vib 176(5):613–624
25. Narayanan GV, Beskos DE (1982) Numerical operational methods for time-dependant linear problems. Int J Numer Method Eng 18:1829–1854
26. Pawar PM, Ganguli R (2003) Genetic fuzzy system for damage detection in beams and helicopter rotor blades. Comput Methods Appl Mech Eng 192:2031–2057
27. Pawar PM, Ganguli R (2007) On the effect of progressive damage on composite helicopter rotor system behavior. Compos Str 78(3):410–423
28. Pnueli D (1972) Natural bending frequency comparable to rotational frequency in rotating cantilever beam. J Appl Mech 39(2):604–606
29. Putter S, Manor H (1978) Natural frequencies of radial rotating beams. J Sound Vib 56:175–185
30. Rossing TD, Fletcher NH (1995) Principles of vibration and sound. Springer, New York
31. Roy N, Ganguli R (2005) Helicopter rotor blade frequency evolution with damage growth and signal processing. J Sound Vib 283(3–5):821–851
32. Roy Mahapatra D, Gopalakrishnan S, Sankar TS (2000) Spectral-element-based solutions for wave propagation analysis of multiply connected unsymmetric laminated composite beams. J Sound Vib 237(5):819–836
33. Roy Mahapatra D, Gopalakrishnan S (2003) A spectral finite element model for analysis of axial-flexural-shear coupled wave propagation in laminated composite beams. Compos Str 59:67–88
34. Stephen NG, Zhang Y (2006) Eigen analysis and continuum modeling of pre-twisted repetitive beam-like structures. Int J Solid Str 43(13):3832–3855
35. Thakkar D, Ganguli R (2004) Dynamic response of rotating beams with piezoceramic actuation. J Sound Vib 270(4–5):729–753
36. Vinod KG, Gopalakrishnan S, Ganguli R (2006) Wave propagation characteristics of rotating uniform Euler-Bernoulli beams. Computer Modeling in Engineering and Sciences. Accepted
37. Wang G, Wereley NM (2004) Free vibration analysis of rotating blades with uniform tapers. AIAA J 42(12):2429–2437
38. Wright AD, Smith CE, Thresher RW, Wang JLC (1982) Vibration modes of centrifugally stiffened beams. J Appl Mech 49(2):197–202

Chapter 9
Violin String Shape Functions

The previous chapters focused on Euler–Bernoulli beams which are good models for long and slender blades. Gas and steam turbine blades are short and rigid and can be modeled as rotating Timoshenko beams [1]. Rotating Euler–Bernoulli beams only consider centrifugal force in addition to inertial and elastic forces for vibration analysis. The secondary effects such as shear deformation and rotary inertia have a small effect on lower modes but have considerable effect on higher modes [1]. Hence for accurate prediction of higher modes, Timoshenko beam models are employed.

Vibration analysis of non-rotating Timoshenko beams has been addressed by a some investigators using the finite element method. Many of these studies experienced difficulty in incorporating all the boundary conditions. Kapur [1] used a finite element model approach for finding the frequencies and mode shapes for free vibrations of a uniform or nonuniform Timoshenko beam for various boundary conditions. Thomas and Abbas [2] analyzed a uniform Timoshenko beam by taking total deflection, total slope, bending slope and derivative of bending slope as internal degrees of freedom. The frequency equation for Timoshenko beams was developed by van Rensburg and van der Merve [3] and properties of natural frequencies and mode shapes such as double eigenvalues, significance of dimensionless parameters and estimates of large and small eigenvalues were studied for various boundary conditions. Bokanian [4] developed exact frequency equations for axially loaded beams. Friedman and Kosmatka [16] developed a two node finite element for a Timoshenko beam which satisfied the static homogeneous governing differential equations. Kosmatka [5] also developed a two node finite element for an axially loaded Timoshenko beam. Although some of the methods presented in cited references take into account axial loads, they are restricted to non-rotating tapered beams.

R. Ganguli, *Finite Element Analysis of Rotating Beams*,
Foundations of Engineering Mechanics, DOI 10.1007/978-981-10-1902-9_9

The dynamics of rotating Timoshenko beams differs from that of non-rotating Timoshenko beams since the centrifugal forces effect the computed frequencies and mode shapes in addition to shear deformation and rotary inertia. Also, the partial differential equation for rotating beams has variable coefficients compared to that of non-rotating beam which has constant coefficients. Timoshenko beams are better models for many rotating beam structures. Yokohama [6] studied free vibration characteristics of rotating Timoshenko beams addressing the hub radius and setting angle effects. Lin et al. [7] studied the instability and vibration of a rotating Timoshenko beam with precone. Banerjee and Sobey [8] gave energy expressions for rotating tapered Timoshenko beams. Banerjee [9] extended the dynamic element formulation of a rotating Euler–Bernoulli beam to account for secondary effects. Du et al. [10] solved the partial differential equation of a rotating Timoshenko beam using the power series method. Kaya [11] extended their works for rotating uniform and tapered Timoshenko beams using DTM. Rao and Gupta [12] used finite element approach to study effects of twist, offset, speed of rotation and variation of breadth and taper ratios on natural frequencies and mode shapes. Bazoune and Khulief [13] used finite element for vibration analysis of double tapered Timoshenko beams for various boundary conditions. Bambill et al. [14] used differential quadrature method for vibration analysis of rotating Timoshenko beams. In this method the derivatives are approximated using the weighted linear summation of functional values at all sample points in overall domain. Using this approximation the differential equation is transformed into a set of algebraic equations. The number of equations depends on number of sample points taken in the domain.

In this chapter, we seek to find better shape functions for rotating Timoshenko beams using the static part of governing differential equation of a violin string. Maezawa et al. [15] have shown that Timoshenko beam theory gives a better estimate of the inharmonicity effects in violin strings compared to Euler–Bernoulli beam theory. Inharmonicity in violin strings occurs because flexural stiffness and other effects which are present in real strings cause the frequency of the nth mode to deviate from the exact nth multiple of the fundamental natural frequency. Interestingly, we found that the violin string equations can be obtained by assuming *constant tension* in the rotating beam equation. Moreover, the violin string static homogenous differential equations have an *exact solution*. As the violin string is the closest approximation to rotating Timoshenko beams for which an exact solution exists, shape functions derived from them can exhibit better convergence compared to cubic elements for predicting the frequencies.

9.1 Timoshenko Rotating Beam and Violin String

The governing equation for free vibrations of a rotating Timoshenko beam is given by [5]

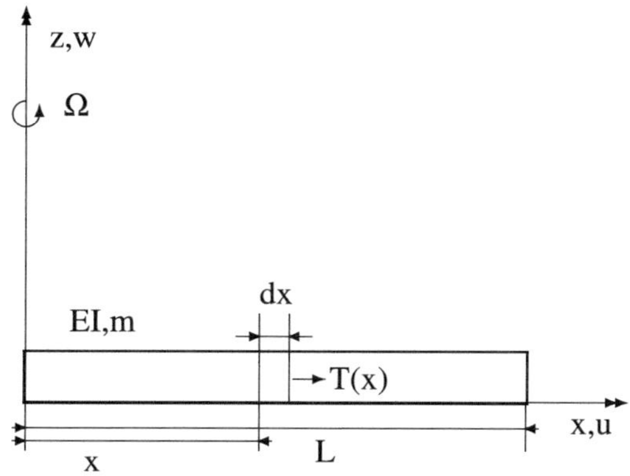

Fig. 9.1 Rotating beam

$$\rho A(x)\frac{\partial^2 w(x,t)}{\partial t^2} - \frac{\partial}{\partial x}\left(T(x)\frac{\partial w(x,t)}{\partial x}\right) - \frac{\partial}{\partial x}\left(GA(x)k\left(\frac{\partial w(x,t)}{\partial x} - \theta(x,t)\right)\right) = 0 \tag{9.1}$$

$$\rho I(x)\frac{\partial^2 \theta(x,t)}{\partial t^2} - GA(x)k\left(\frac{\partial w(x,t)}{\partial x} - \theta(x,t)\right) - \frac{\partial}{\partial x}\left(EI(x)\frac{\partial \theta(x,t)}{\partial x}\right) = 0 \tag{9.2}$$

where $T(x)$ is axial force due to centrifugal stiffening and is given by

$$T(x) = \int_x^L \rho A(x)\Omega^2\, dx$$

Here L is the length of beam, Ω is the rotational speed, $\theta(x,t)$ is angle of rotation of cross section, $w(x,t)$ is the vertical displacement of beam, ρ is the density, E and G are elastic constants, k is the shear coefficient, $A(x)$ is area of cross section, $I(x)$ is moment of inertia of cross section as shown in Fig. 9.1. The non-dimensional constants are $\nu = \frac{AL^2}{I}$, $\beta = \frac{GAkL^2}{EI}$ and $\gamma = \frac{\beta}{\nu} = \frac{Gk}{E}$, $\kappa^2 = \frac{\rho A\Omega^2 L^4}{EI}$. For beams with constant tension, Eqs. (9.1) and (9.2) reduce to

$$\begin{aligned} &\rho A(x)\frac{\partial^2 w(x,t)}{\partial t^2} - T\frac{\partial^2 w(x,t)}{\partial x^2} - \frac{\partial}{\partial x}\left(GA(x)k\left(\frac{\partial w(x,t)}{\partial x} - \theta(x,t)\right)\right) = 0 \\ &\rho I(x)\frac{\partial^2 \theta(x,t)}{\partial t^2} - GA(x)k\left(\frac{\partial w(x,t)}{\partial x} - \theta(x,t)\right) - \frac{\partial}{\partial x}\left(EI(x)\frac{\partial \theta(x,t)}{\partial x}\right) = 0 \end{aligned} \tag{9.3}$$

where T is the constant axial tension. The derivation of this equation is given by Banerejee [9] and is also given in Appendix C for completeness. We call these the violin string equations in an analogy to the typical Euler–Bernoulli stiff string equations. The stiff strings are studied in the acoustic analysis of piano strings. Piano strings have flexural stiffness but are typically quite long. In fact, the most prized pianos are those with the largest strings. Violins are much smaller than pianos. Timoshenko theory gives a better representation of inharmonicity of violin strings compared to Euler–Bernoulli beam theory [15]. The static part of Eqs. (9.4) and (9.5) is obtained by ignoring the inertial term which leads to

$$T\frac{d^2w}{dx^2} + \frac{d}{dx}\left(GA(x)k\left(\frac{dw}{dx} - \theta\right)\right) = 0 \tag{9.4}$$

$$GA(x)k\left(\frac{dw}{dx} - \theta\right) + \frac{d}{dx}\left(EI(x)\frac{d\theta}{dx}\right) = 0 \tag{9.5}$$

The classical approach in finite element analysis is to use polynomial shape functions which are obtained using the governing equation of non-rotating Timoshenko beam given by

$$\frac{d}{dx}\left(GA(x)k(\frac{dw}{dx} - \theta)\right) = 0 \tag{9.6}$$

$$GA(x)k\left(\frac{dw}{dx} - \theta\right) + \frac{d}{dx}\left(EI(x)\frac{d\theta}{dx}\right) = 0 \tag{9.7}$$

Putting Eq. (9.7) in Eq. (9.6) yields,

$$\frac{d^2}{dx^2}\left(EI(x)\frac{d\theta}{dx}\right) = 0 \tag{9.8}$$

For a uniform beam,

$$\frac{d^3\theta}{dx^3} = 0 \Longrightarrow \theta(x) = a_0 + a_1x + a_2x^2 \tag{9.9}$$

Also, for a uniform beam, Eqs. (9.8) and (9.9) give

$$\frac{d^2w}{dx^2} = \frac{d\theta}{dx} \tag{9.10}$$

$$\frac{dw}{dx} - \theta + \frac{EI}{GAk}\frac{d^2\theta}{dx^2} = 0 \tag{9.11}$$

From Eqs. (9.12) and (9.13), we see that the typical cubic shape functions used for the displacement field $w(x)$ satisfies the non-rotating Timoshenko beam equation. It therefore works quite well for a non-rotating beam. Friedman and Kosmatka [16] also ensured that the $w(x)$ and $\theta(x)$ displacement fields satisfied both Eqs. (9.8) and (9.9) for a uniform beam. In the case of axially loaded beams, Kosmatka used the same shape functions and did not include the axial force. This yields

$$w(x) = a_3 + \left(a_0 - \frac{2EI}{GAk}a_2\right)x + \frac{a_1}{2}x^2 + \frac{a_2}{3}x^3 \tag{9.12}$$

as the displacement field. We see that four constants a_0, a_1, a_2 and a_3 are present in $w(x)$ and $\theta(x)$ which can be calculated from four nodal values for an element.

We propose to use Eqs. (9.6) and (9.7) to derive new shape functions which are suitable for rotating Timoshenko beams. The presence of $T(x)$ in Eq. (9.1) presents a difficulty in obtaining the analytical solution. While the assumption of $T(x) = T = constant$ in Eq. (9.6) may appear too simplistic for the whole beam, it is reasonable if the beam is divided into several finite elements and *T is assumed constant within each element*. Thus, the rotating beam is assumed as a sequence of violin strings for the calculation of shape functions.

Using the finite element method, the beam is divided into N number of elements of equal length. The tension $T(x)$ is assumed as constant (T_i) within the element i, which reduces the rotating beam equation to the violin string equation. For an ith element along the beam, the relationship between global and local coordinates is given by

$$x = x_i + \bar{x} \tag{9.13}$$

where $x_i = (i-1)l$ is the distance from the root to the left edge of element i, l is length of the element and $\bar{x}$ is the local element coordinate. This constant tension is taken as the maximum centrifugal force an element experiences which is given by

$$T(x) = T_i = \int_{x_i}^{L} \rho A(x)\Omega^2 x\, dx = \sum_{j=i}^{j=N} \int_{x_j}^{x_{j+1}} m_j(\bar{x})\Omega^2 \bar{x}\, d\bar{x} \tag{9.14}$$

The tension in the inboard elements is high and decreases for the outboard elements. Assuming $EI(x) = EI_i$ and $\rho A(x) = \rho A_i$ within an element as constant, Eqs. (9.6) and (9.7) give

$$T_i \frac{d^2 w}{d\bar{x}^2} + GAk\left(\frac{d^2 w}{d\bar{x}^2} - \frac{d\theta}{d\bar{x}}\right) = 0 \tag{9.15}$$

$$GAk\left(\frac{dw}{d\bar{x}} - \theta\right) + EI\frac{d^2\theta}{d\bar{x}^2} = 0 \tag{9.16}$$

From Eqs. (9.15) and (9.16), we get

$$\frac{d^3\theta}{d\bar{x}^3} = \frac{T_i}{EI}\frac{d^2 w}{d\bar{x}^2} \tag{9.17}$$

Note that this equation relates $\theta(x)$ and $w(x)$ is quite different from Eq. (9.9). Also, if $T_i = 0$, Eq. (9.17) yields Eq. (9.9) for the non-rotating beam. Using Eq. (9.17) in Eqs. (9.15) and (9.16) to remove the coupling between w and θ yields,

$$\frac{d^4w}{d\bar{x}^4} - \frac{T_i}{EI_i(1+\frac{T_i}{GA_ik})}\frac{d^2w}{d\bar{x}^2} = 0 \tag{9.18}$$

$$\frac{d^3\theta}{d\bar{x}^3} - \frac{T_i}{EI_i(1+\frac{T_i}{GA_ik})}\frac{d\theta}{d\bar{x}} = 0 \tag{9.19}$$

The solutions of which are

$$\theta(\bar{x}) = b_0 + b_1e^{-d\bar{x}} + b_2e^{d\bar{x}} \tag{9.20}$$

where $d = \sqrt{\frac{T_i}{EI_i(1+\frac{T_i}{GA_ik})}} = \sqrt{\frac{C^2}{(1+C^2m_3)}}$ and $C = \sqrt{\frac{T_i}{EI_i}}, m_3 = \frac{EI_i}{GA_ik}$. Using Eqs. (9.17), (9.18) and (9.20), the displacement field $w(x)$ is given by

$$\begin{aligned} w(\bar{x}) = b_0x + b_1\frac{e^{-dx}}{(1+C^2m_3)}\left(\left(\frac{-1}{d}\right) - \alpha x(1+C^2m_3)\right) \\ +b_2\frac{e^{dx}}{(1+C^2m_3)}\left(\left(\frac{1}{d}\right) - \alpha x(1+C^2m_3)\right) + a_0\frac{1}{(1+C^2m_3)} \end{aligned} \tag{9.21}$$

where $\alpha = m_3d^2 + \frac{1}{(1+C^2m_3)} - 1$.

Equations (9.21) and (9.20) satisfy Eqs. (9.15) and (9.16) and Eqs. (9.18) and (9.19) and therefore satisfy the static homogeneous violin string equations within the element.

9.2 Violin String Shape Functions

A two node, four degree freedom finite element is taken as shown in Fig. 9.2. Applying finite element boundary conditions at nodes for the element of length l and eliminating constants give interpolation fields for displacements and rotations. The conditions are given by $w(0) = w_1, \theta(0) = \theta_1, w(l) = w_2, \theta(l) = \theta_2$. Substituting for boundary conditions in Eqs. (9.20) and (9.21) and writing in matrix form yields

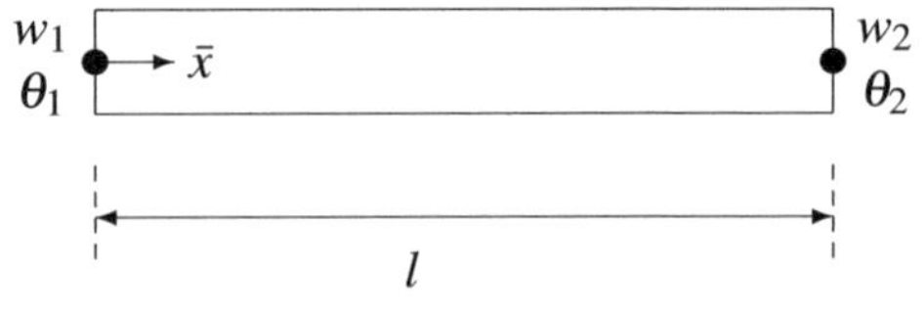

Fig. 9.2 Beam element

$$\begin{bmatrix} w_1 \\ \theta_1 \\ w_2 \\ \theta_2 \end{bmatrix} = \begin{bmatrix} 0 & \frac{-1}{d(1+C^2m_3)} & \frac{1}{d(1+C^2m_3)} & \frac{1}{(1+C^2m_3)} \\ 1 & 1 & 1 & 0 \\ l & \frac{e^{-dl}(\frac{-1}{d}-\alpha l(1+C^2m_3))}{(1+C^2m_3)} & \frac{e^{dl}(\frac{1}{d}-\alpha l(1+C^2m_3))}{(1+C^2m_3)} & \frac{1}{(1+C^2m_3)} \\ 1 & e^{-dl} & e^{dl} & 0 \end{bmatrix} \begin{bmatrix} b_0 \\ b_1 \\ b_2 \\ a_0 \end{bmatrix} \tag{9.22}$$

Eliminating the unknown constants gives interpolation fields for displacement and rotation field in terms of nodal degrees of freedom as

$$w = \begin{bmatrix} x & \frac{e^{-dx}(\frac{-1}{d}-\alpha x(1+C^2m_3))}{(1+C^2m_3)} & \frac{e^{dx}(\frac{1}{d}-\alpha x(1+C^2m_3))}{(1+C^2m_3)} & \frac{1}{(1+C^2m_3)} \end{bmatrix} c_1^{-1} \begin{bmatrix} w_1 \\ \theta_1 \\ w_2 \\ \theta_2 \end{bmatrix} \tag{9.23}$$

$$\theta = \begin{bmatrix} 1 & e^{-dx} & e^{dx} & 0 \end{bmatrix} c_1^{-1} \begin{bmatrix} w_1 \\ \theta_1 \\ w_2 \\ \theta_2 \end{bmatrix} \tag{9.24}$$

where

$$c_1 = \begin{bmatrix} 0 & \frac{-1}{d(1+C^2m_3)} & \frac{1}{d(1+C^2m_3)} & \frac{1}{(1+C^2m_3)} \\ 1 & 1 & 1 & 0 \\ l & \frac{e^{-dl}(\frac{-1}{d}-\alpha l(1+C^2m_3))}{(1+C^2m_3)} & \frac{e^{dl}(\frac{1}{d}-\alpha l(1+C^2m_3))}{(1+C^2m_3)} & \frac{1}{(1+C^2m_3)} \\ 1 & e^{-dl} & e^{dl} & 0 \end{bmatrix}$$

Equations (9.23) and (9.24) simplify to

$$w(x) = H_{w1}w_1 + H_{w2}\theta_1 + H_{w3}w_2 + H_{w4}\theta_2 \tag{9.25}$$

$$\theta(x) = H_{\theta 1}w_1 + H_{\theta 2}\theta_1 + H_{\theta 3}w_2 + H_{\theta 4}\theta_2 \tag{9.26}$$

where
$H_{w1} = \frac{N_{w1}}{D}$, $H_{w2} = \frac{N_{w2}}{dD}$, $H_{w3} = \frac{N_{w3}}{D}$, $H_{w4} = \frac{N_{w4}}{dD}$ and
$H_{\theta 1} = \frac{N_{\theta 1}}{D}$, $H_{\theta 2} = \frac{N_{\theta 2}}{dD}$, $H_{\theta 3} = \frac{N_{\theta 3}}{D}$, $H_{\theta 4} = \frac{N_{\theta 4}}{dD}$

Here

$$\begin{aligned} N_{w1} &= 2 + e^{dl}\alpha ld + e^{-dx} - e^{-dl} - e^{-dl}\alpha ld - dxe^{dl}C^2m_3 \\ &+e^{d(-l+x)}\alpha xdC^2m_3 - e^{-d(-l+x)} + e^{dx} - e^{-dl}\alpha ldC^2m_3 + \\ &e^{-dx}\alpha xd + e^{dl}\alpha ldC^2m_3 - dxe^{dl} + dxe^{-dl} - e^{-d(-l+x)}\alpha xdC^2m_3 \\ &+dxe^{-dl}C^2m_3 + e^{-dx}\alpha xdC^2m_3 - e^{dx}\alpha xdC^2m_3 - e^{dl} - e^{d(-l+x)} + \\ &e^{dl}ld + e^{dl}ldC^2m_3 - e^{-dl}ld - e^{dx}\alpha xd - e^{-dl}ldC^2m_3 + e^{d(-l+x)}\alpha xd \\ &-e^{-d(-l+x)}\alpha xd \end{aligned}$$

$$N_{w2} = (1 + C^2m_3)(-e^{d(-l+x)}\alpha ldC^2m_3 - e^{d(-l+x)}ld + e^{dl}\alpha ld$$
$$-e^{-dx} + 2dxC^2m_3 + e^{-dl} + e^{-dl}\alpha ld - e^{d(-l+x)}\alpha ld+$$
$$2e^{d(-l+x)}\alpha^2xd^2lC^2m_3 - e^{-d(-l+x)}ld - e^{-d(-l+x)}ldC^2m_3$$
$$-2e^{-d(-l+x)}\alpha xd^2lC^2m_3 - dxe^{dl}C^2m_3 + e^{d(-l+x)}\alpha xdC^2m_3-$$
$$e^{-d(-l+x)}\alpha xd^2C^4m_3^2l - e^{d(-l+x)}ldC^2m_3 + 2e^{d(-l+x)}\alpha xd^2lC^2m_3$$
$$-e^{-d(-l+x)}\alpha^2xd^2l + e^{-d(-l+x)} + e^{dx} + e^{-dl}\alpha ldC^2m_3 - e^{-dx}\alpha xd+$$
$$e^{dl}\alpha ldC^2m_3 - dxe^{dl} - dxe^{-dl} + e^{-d(-l+x)}\alpha xdC^2m_3$$
$$-e^{-d(-l+x)}\alpha^2xd^2C^4m_3^2l + e^{d(-l+x)}\alpha xd^2l - e^{-d(-l+x)}\alpha ldC^2m_3+$$
$$e^{d(-l+x)}\alpha xd^2C^4m_3^2l - e^{-d(-l+x)}\alpha ld - dxe^{-dl}C^2m_3$$
$$-e^{-dx}\alpha xdC^2m_3 - e^{dx}\alpha xdC^2m_3 - e^{dl} - e^{d(-l+x)} + 2dx + e^{dl}ld$$
$$+e^{dl}ldC^2m_3 + e^{-dl}ld - 2e^{-d(-l+x)}\alpha^2xd^2lC^2m_3 - e^{-d(-l+x)}\alpha xd^2l$$
$$-e^{dx}\alpha xd + e^{-dl}ldC^2m_3 + e^{d(-l+x)}\alpha^2xd^2C^4m_3^2l + e^{d(-l+x)}\alpha^2xd^2l$$
$$+e^{d(-l+x)}\alpha xd + e^{-d(-l+x)}\alpha xd)$$

$$N_{w3} = dxe^{dl} + dxe^{dl}C^2m_3 - dxe^{-dl} - dxe^{-dl}C^2m_3 - e^{-dx} + e^{-dx+dl}$$
$$-e^{-dx}\alpha xd + e^{-dx+dl}\alpha xd - e^{-dx}\alpha xdC^2m_3 + e^{-dx+dl}\alpha xdC^2m_3$$
$$+e^{dx-dl} - e^{dx} - e^{dx-dl}\alpha xd + e^{dx}\alpha xd - e^{dx-dl}\alpha xdC^2m_3+$$
$$e^{dx}\alpha xdC^2m_3 - e^{dl} + 2 - e^{-dl}$$

$$N_{w4} = (1 + C^2m_3)(-e^{dx}\alpha xd^2l - e^{dl}\alpha ld + e^{-dx} - 2dl - 2ldC^2m_3$$
$$+2dxC^2m_3 + e^{dx-dl}\alpha ld - e^{-dl} - e^{-dl}\alpha ld + e^{dx-dl}\alpha ldC^2m_3$$
$$+e^{-dx+dl}\alpha ldC^2m_3 - dxe^{dl}C^2m_3 - e^{-dx+dl} - 2e^{dx-dl}\alpha^2xd^2lC^2m_3$$
$$-e^{dx} + e^{-dx+dl}\alpha^2xd^2l - e^{-dl}\alpha ldC^2m_3 + e^{dx-dl} - e^{dx-dl}\alpha xd + e^{-dx}\alpha xd$$
$$+e^{-dx+dl}\alpha^2xd^2C^4m_3^2l - e^{dl}\alpha ldC^2m_3 - dxe^{dl} - dxe^{-dl} + e^{dx}ld + e^{-dx}ld$$
$$-e^{dx-dl}\alpha xdC^2m_3 - e^{-dx+dl}\alpha xd - 2\alpha xd^2e^{-dl}lC^2m_3 - \alpha xd^2C^4m_3^2e^{-dl}l$$
$$+e^{dx}ldC^2m_3 - 2e^{dx}\alpha xd^2lC^2m_3 - e^{dx}\alpha xd^2C^4m_3^2l + e^{-dx}ldC^2m_3$$
$$+e^{-dx}\alpha xd^2l + e^{-dx}\alpha xd^2C^4m_3^2l + 2e^{-dx}\alpha xd^2lC^2m_3 - e^{-dx+dl}\alpha xdC^3m_3$$
$$-dxe^{-dl}C^2m_3 - dxe^{-dl}C^2m_3 + e^{-dx}\alpha xdC^2m_3 + e^{dx}\alpha xdC^2m_3 + e^{dl}$$
$$+2e^{-dx+dl}\alpha^2xd^2lC^2m_3 - e^{dx-dl}\alpha^2xd^2l + 2dx - e^{dx-dl}\alpha^2xd^2C^4m_3^2l$$
$$+\alpha xd^2C^4m_3^2e^{dl}l + \alpha xd^2e^{dl}l + 2\alpha xd^2e^{dl}lC^2m_3 - \alpha xd^2e^{-dl}l$$
$$+e^{-dx+dl}\alpha ld + e^{dx}\alpha xd)$$

are displacement shape functions. Similarly

$$N_{\theta 1} = -(-e^{dl} + e^{-dl})(1 + C^2m_3) - e^{-dx}(-1 + e^{dl})(1 + C^2m_3$$
$$+e^{dx}(e^{-dl} - 1)(1 + C^2m_3))$$

$$N_{\theta 2} = -(-e^{dl} + 2e^{-dl}e^{dl} - e^{-dl}) - e^{-dx}(e^{dl}dl + e^{dl}ldC^2m_3 + 1$$
$$-e^{dl} + e^{dl}\alpha ld + e^{dl}\alpha ldC^2m_3) + e^{dx}(e^{-dl}dl + e^{-dl}ldC^2m_3 - 1$$
$$+e^{-dl} + e^{-dl}\alpha ld + e^{-dl}\alpha ldC^2m_3)$$

$$N_{\theta 3} = (-e^{dl} + e^{-dl})(1 + C^2m_3) + e^{-dx}(-1 + e^{dl})(1 + C^2m_3)$$
$$-e^{dx}(e^{-dl} - 1)(1 + C^2m_3)$$

$$N_{\theta 4} = (-2 + e^{dl} - e^{dl}\alpha ld - e^{dl}\alpha ldC^2m_3 + e^{-dl} + e^{-dl}\alpha ld$$
$$+e^{-dl}\alpha ldC^2m_3) + e^{-dx}(1 - e^{dl} + e^{dl}\alpha ld + e^{dl}\alpha ldC^2m_3$$
$$+dl + ldC^2m_3) - e^{dx}(-1 + e^{-dl} + e^{-dl}\alpha ld + e^{-dl}\alpha ldC^2m_3$$
$$+dl + ldC^2m_3)$$

are rotation shape functions. Here

$$D = -2e^{dl} + 4 - 2e^{-dl} + e^{dl}ld + e^{dl}ldC^2m_3 - e^{-dl}ld - e^{-dl}ldC^2m_3$$
$$+e^{dl}\alpha ld + e^{dl}\alpha ldC^2m_3 - e^{-dl}\alpha ld - e^{-dl}\alpha ldC^2m_3$$

These *violin string shape functions* are functions of non-dimensional rotational speed, mass of outboard elements, beam length, element mass, flexural stiffness, shear stiffness, shear coefficient and location of element. As the beam becomes slender $\beta = \frac{EI}{GAkL^2}$ approaches 0; the shape functions tend to those derived for an Euler–Bernoulli stiff string. The variations of violin string shape functions along the elements is shown in Figs. 9.3, 9.4, 9.5 and 9.6 for $N = 1$ for different slenderness ratios at rotational speeds of $\kappa = 12$ and $\kappa = 100$. We can see the significant effect of slenderness ratio and rotation speed on shape functions.

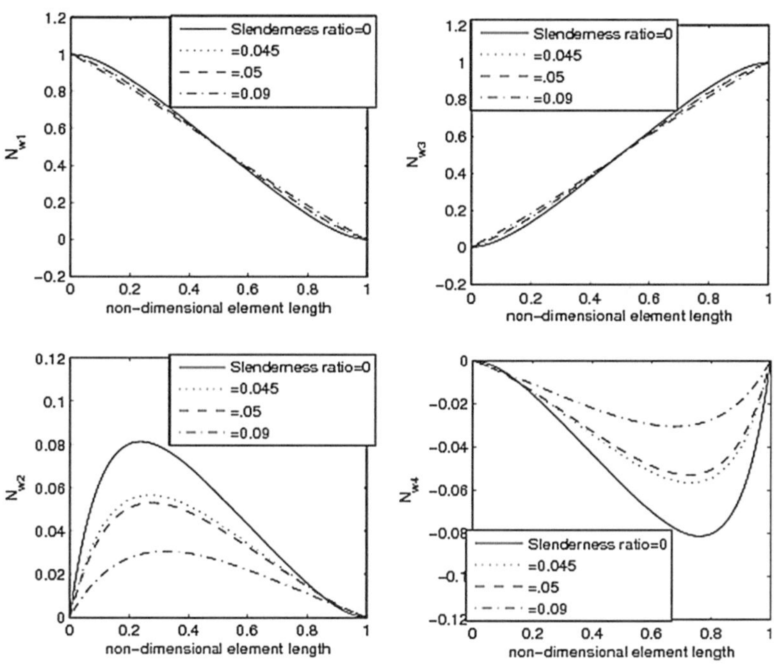

Fig. 9.3 Displacement shape functions at different slenderness ratios for $\kappa = 12$ (N = 1)

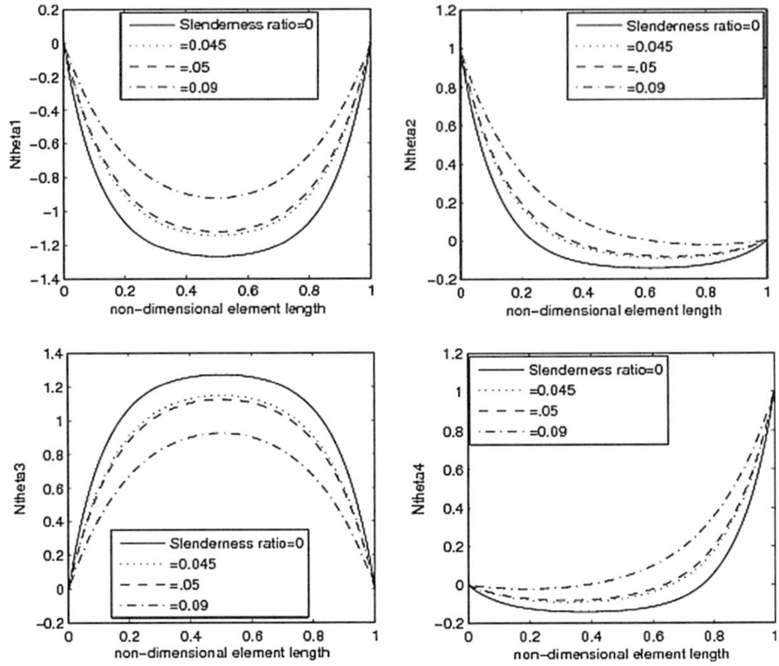

Fig. 9.4 Rotation shape functions at different slenderness ratios for $\kappa = 12$ (N = 1)

The mass and stiffness matrices can be obtained using the energy expressions. The kinetic energy($K.E$) for a Timoshenko beam is given by

$$K.E = \int_0^L \frac{1}{2}\rho\left[\dot{w}(x,t)^2 + \dot{u}(x,t)^2\right]dx \tag{9.27}$$

The strain energy (U) expression is given by

$$U = \frac{1}{2}EI\int_0^L (\frac{d\theta}{dx})^2 dx + \frac{1}{2}GAk\int_0^L (\frac{dw}{dx} - \theta)^2 dx + \frac{1}{2}T\int_0^L (\frac{dw}{dx})^2 dx \tag{9.28}$$

Here $x = x_i + \bar{x}$ and $x_i = (i-1)l$, where l is the length of element. The mass and stiffness matrices (M_i and K_i) for a beam element can be obtained from the energy expressions in Eqs. (9.27) and (9.28) for a uniform beam. The calculations for these matrices involve evaluating the following integrals:

$$M_i = \int_0^l \rho A_i(\bar{x})(N_w)^T(N_w)\,d\bar{x} + \int_0^l \rho I_i(\bar{x})(N_\theta)^T(N_\theta)\,d\bar{x} \tag{9.29}$$

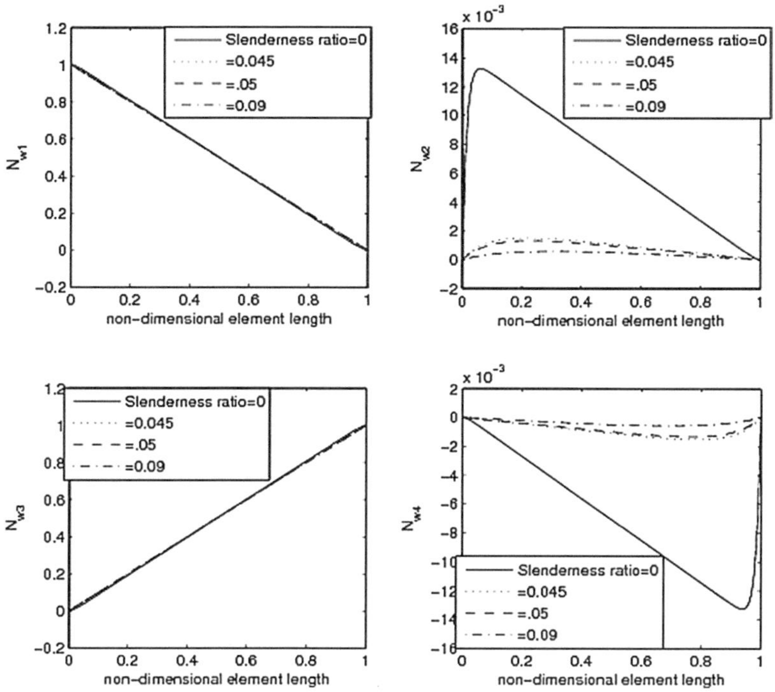

Fig. 9.5 Displacement shape functions at different slenderness ratios for $\kappa = 100$ (N = 1)

$$K_i = \int_0^l EI_i(\bar{x}) \left(\frac{dN_\theta}{d\bar{x}}\right)^T \left(\frac{dN_\theta}{d\bar{x}}\right) d\bar{x} + \int_0^l GA_i(\bar{x})k \left(\frac{dN_w}{d\bar{x}} - N_\theta\right)^T \left(\frac{dN_w}{d\bar{x}} - N_\theta\right) d\bar{x} + \int_0^l T_i(\bar{x}) \left(\frac{dN_w}{d\bar{x}}\right)^T \left(\frac{dN_w}{d\bar{x}}\right) d\bar{x} \tag{9.30}$$

where centrifugal tension $T_i(\bar{x})$ acting on an element is given by

$$T_i(\bar{x}) = \sum_{j=1}^{N} \int_{x_j}^{x_{j+1}} m_j(\bar{x})\Omega^2 \bar{x} d\bar{x} - \int_{x_i}^{x_i+\bar{x}} m_i(\bar{x})\Omega^2 \bar{x} d\bar{x} \tag{9.31}$$

The *assumption of constant tension is only made during formulation of the shape functions*. But no such assumptions are made while deriving the stiffness matrix of the element. The element mass and stiffness matrices are assembled to get global stiffness **K** and mass matrix **M**. The resulting eigenvalue problem is solved for natural frequencies and mode shapes.

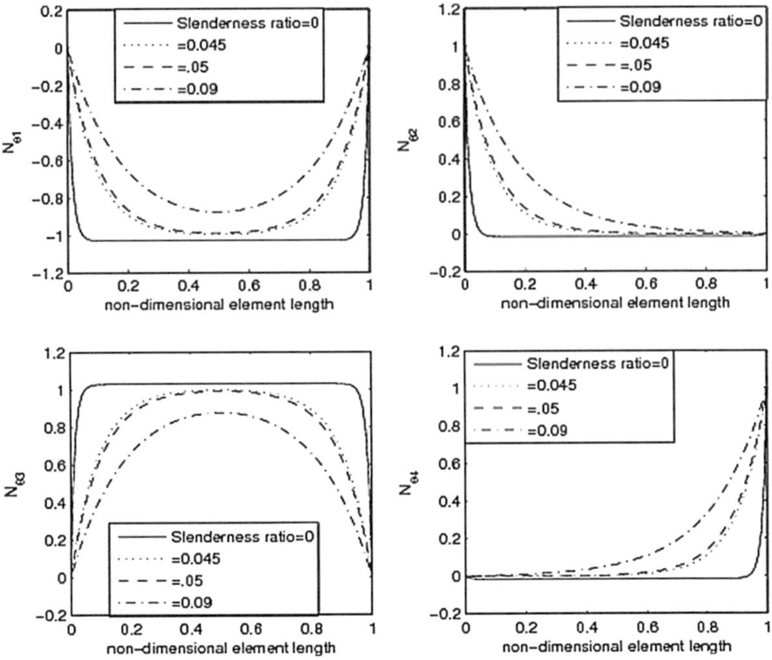

Fig. 9.6 Rotation shape functions at different slenderness ratios for $\kappa = 100$ (N = 1)

$$\mathbf{K}\phi = \omega^2 \mathbf{M}\phi \tag{9.32}$$

The non dimensional frequencies are obtained as $\lambda = \frac{\omega L^2}{\sqrt{\frac{EI}{\rho A}}}$.

9.3 Results and Discussion

The properties of the Timoshenko beam used in the analysis are: slenderness ratio $\frac{1}{\nu} = 0.09, 0.05, 0.045$; $\gamma = 0.25$

9.3.1 Uniform Beam

Tables 9.1, 9.2 and 9.3 shows the comparison of non-dimensional frequencies obtained for Timoshenko beam with various slenderness ratio of 0.045, 0.05 and 0.09 and for rotational speeds of $\kappa = 4$, 12 and 100, respectively. Convergence of first five modes is achieved using 80 finite elements. The results in parenthesis are from Du et al. [10]. A convergence study is done to compare the present element with the

Table 9.1 Non-dimensional frequencies of first five modes for uniform beam at $\kappa = 4$ for various slenderness ratios, $\frac{1}{\nu}$

Mode	Slenderness ratio, $\frac{1}{\nu} = 0.09$	0.05	0.045
1	5.24943(5.249)	5.46293(5.463)	5.48459
2	17.25698	20.99124	21.48647
3	35.34860	47.42615	49.41892
4	52.89319	77.95112	82.38964
5	67.59790	110.44737	117.92903

Table 9.2 Non-dimensional frequencies of first five modes for uniform beam at $\kappa = 12$ for various slenderness ratios, $\frac{1}{\nu}$

Mode	Slenderness ratio, $\frac{1}{\nu} = 0.09$	0.05	0.045
1	12.48715(12.487)	12.82721(12.827)	12.87493
2	30.69199	34.36649	34.82098
3	53.50637	64.87082	66.55376
4	66.58940	1.0007833e + 02	1.0386890e + 02
5	83.31214	1.3705581e + 02	1.4366688e + 02

Table 9.3 Non-dimensional frequencies of first five modes for uniform beam at $\kappa = 100$ for various slenderness ratios, $\frac{1}{\nu}$

Mode	Slenderness ratio, $\frac{1}{\nu} = 0.09$	0.05	0.045
1	63.06658	99.72437	1.0009367e + 02
2	80.27038	1.8810892e + 02	2.2690680e + 02
3	1.0068832e + 02	2.1491474e + 02	3.4020408e + 02
4	1.0653987e + 02	2.4634100e + 02	3.4711095e + 02
5	1.3670214e + 02	2.6293160e + 02	4.0269847e + 02

polynomial non-rotating Timoshenko beam finite element. From Figs. 9.7 and 9.8, it is seen that at low rotational speeds, the violin string shape functions lead to better convergence than the polynomial for the fundamental mode over various slenderness ratios. The convergence behavior for the fundamental mode does not change much as the slenderness ratio decreases. But for other modes, the conventional cubic element shows equivalent to superior convergence compared to the present element.

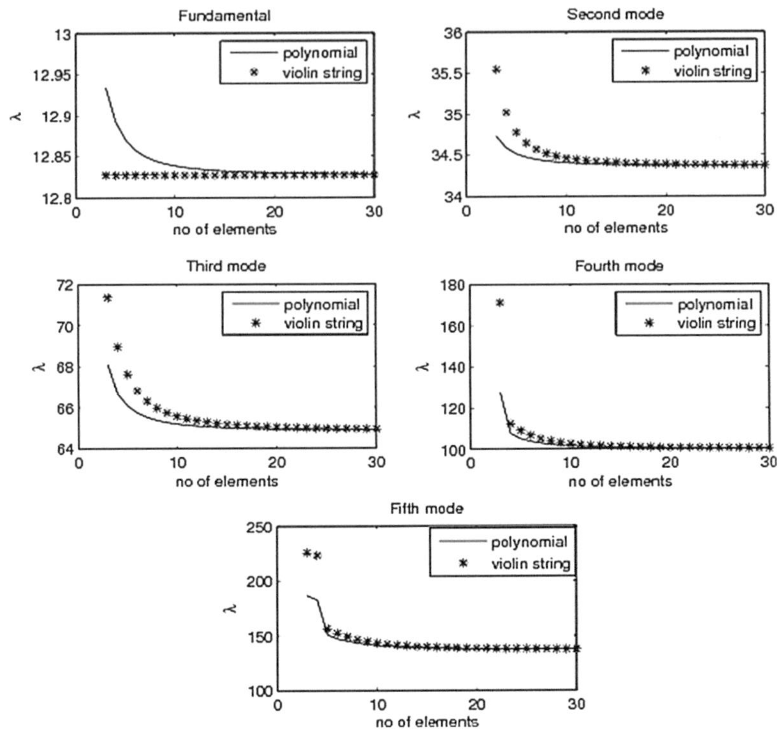

Fig. 9.7 Convergence of first five modes for slenderness ratio $\frac{1}{\nu} = 0.05$ at $\kappa = 12$

Figures 9.9 and 9.10 show that at higher rotational speed, the fundamental mode is better captured using the present element compared to the polynomial element. The other modes which are affected due to flexural bending also converge faster for the present element. The present element performs better at higher slenderness ratios as Timoshenko effects are included in the shape functions along with centrifugal effects. Overall, the violin string basis functions show superior convergence at high rotational speeds when centrifugal effects dominate. They are therefore very useful for turbo machinery applications where high rotation speed is the norm.

9.3.2 Tapered Beam

Tapered beam properties used in analysis are given by taper ratio $c_h = 0.5$, slenderness ratios of $\frac{1}{\nu} = 0.09, 0.08, 0.05, 0.045$ and $\gamma = 0.3269$. Tables 9.4, 9.5 and 9.6 shows the values of non-dimensional frequency obtained at rotational speeds of

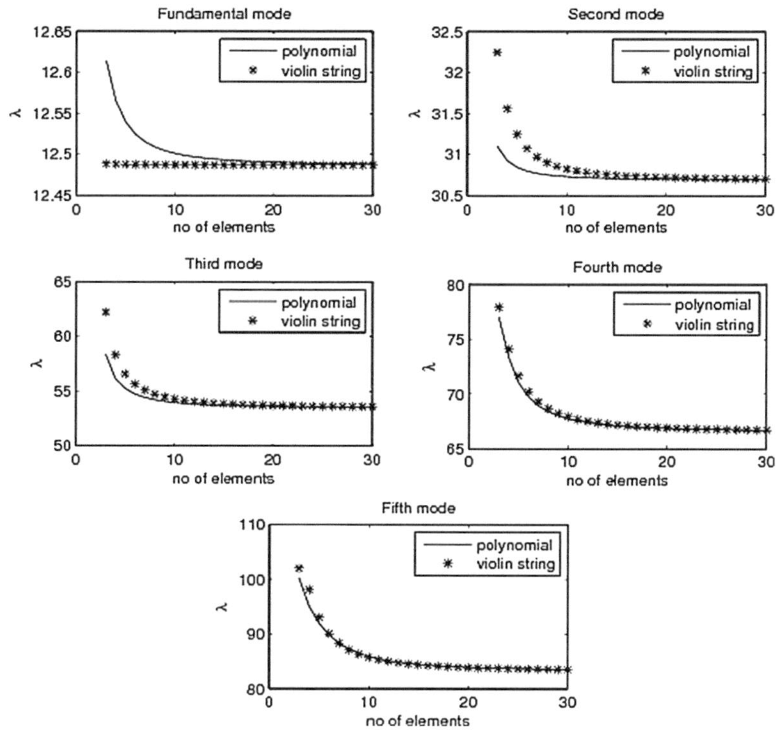

Fig. 9.8 Convergence of first five modes for slenderness ratio $\frac{1}{\nu} = 0.09$ at $\kappa = 12$

$\kappa = 5$, 10 and 100, respectively. A convergence study similar to the uniform case is performed for rotational speeds of $\kappa = 5$ and 100 at slenderness ratios of 0.05 and 0.09. From Figs. 9.11 and 9.12 it can be seen that at low rotational speeds, the violin string shape function is slightly better in capturing the fundamental mode. The convergence of other modes, affected due to flexural bending is similar to polynomial for various slenderness ratios considered.

Figure 9.13 shows that at high rotational speeds, all modes except the second show better convergence with the new shape functions. Note that tapering stiffens the structure, lowering the Timoshenko effects. Figure 9.14 gives even better results as Timoshenko effects important at higher slenderness ratio are well captured by the new basis functions.

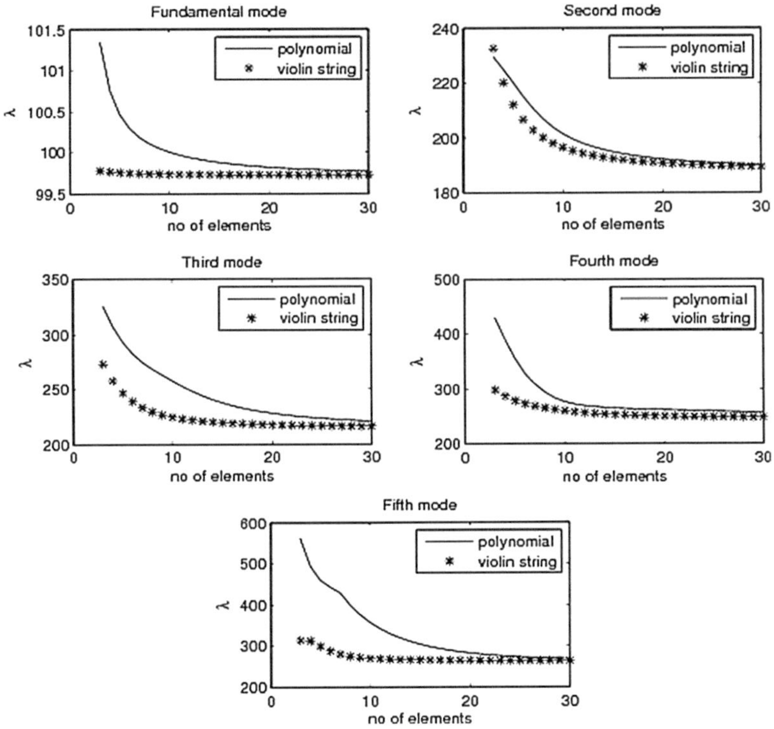

Fig. 9.9 Convergence of first five modes for slenderness ratio of $\frac{1}{\nu} = 0.05$ at $\kappa = 100$

Table 9.4 Non-dimensional frequencies of first five modes for uniform beam at $\kappa = 5$ for various slenderness ratios, $\frac{1}{\nu}$

Mode	$\frac{1}{\nu} = 0.09$	$\frac{1}{\nu} = 0.08$	$\frac{1}{\nu} = 0.05$	$\frac{1}{\nu} = 0.045$
1	6.44226	6.49643	6.63771	6.65679
2	18.24326	18.78224	20.40638	20.65571
3	35.52302	37.26258	43.39974	44.49768
4	55.27350	58.78815	72.45273	75.19202
5	76.35562	81.99292	1.0532239e + 02	1.1038658e + 02

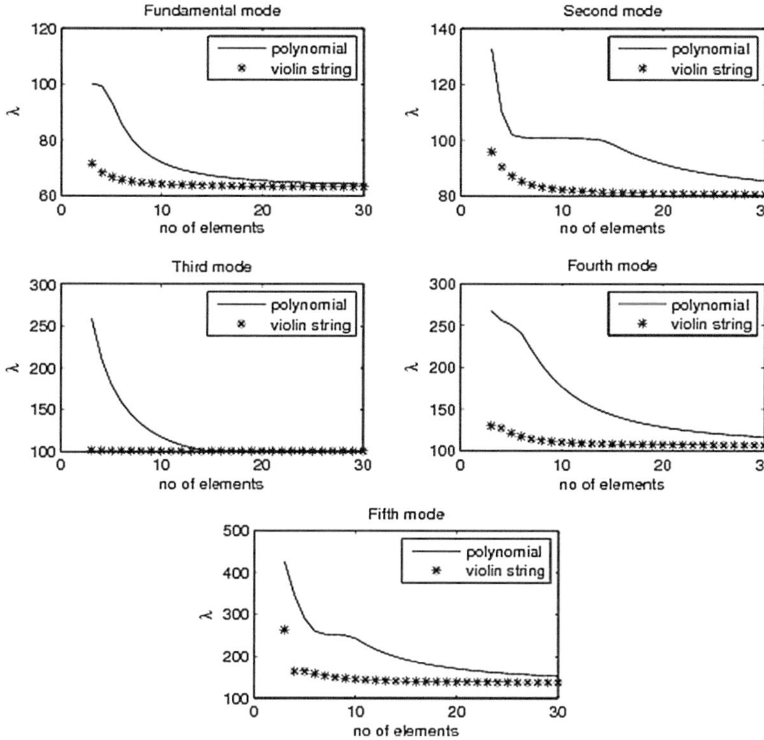

Fig. 9.10 Convergence of first five modes for slenderness ratio of $\frac{1}{\nu} = 0.09$ at $\kappa = 100$

Table 9.5 Non-dimensional frequencies of first five modes for uniform beam at $\kappa = 10$ for various slenderness ratios, $\frac{1}{\nu}$

Mode	$\frac{1}{\nu} = 0.09$	$\frac{1}{\nu} = 0.08$	$\frac{1}{\nu} = 0.05$	$\frac{1}{\nu} = 0.045$
1	10.96516	11.04290	11.27927	11.31593
2	26.54922	27.04032	28.59061	28.84280
3	46.49090	48.00319	53.41685	54.41845
4	68.92987	72.12647	84.23079	86.70439
5	92.19335	97.84211	1.1888447e + 02	1.234884e + 02

Table 9.6 Non-dimensional frequencies of first five modes for uniform beam at $\kappa = 100$ for various slenderness ratios, $\frac{1}{\nu}$

Mode	$\frac{1}{\nu} = 0.09$	$\frac{1}{\nu} = 0.08$	$\frac{1}{\nu} = 0.05$	$\frac{1}{\nu} = 0.045$
1	93.77694	98.85174	98.85174	1.0001682e + 02
2	1.0361440e + 02	1.2171681e + 02	1.2171681e + 02	2.2523472e + 02
3	1.1854724e + 02	1.4459206e + 02	1.4459206e + 02	2.8583622e + 02
4	1.3624639e + 02	1.6395177e + 02	1.6395177e + 02	3.1729580e + 02
5	1.5891659e + 02	1.8722420e + 02	1.8722420e + 02	3.50223769e + 02

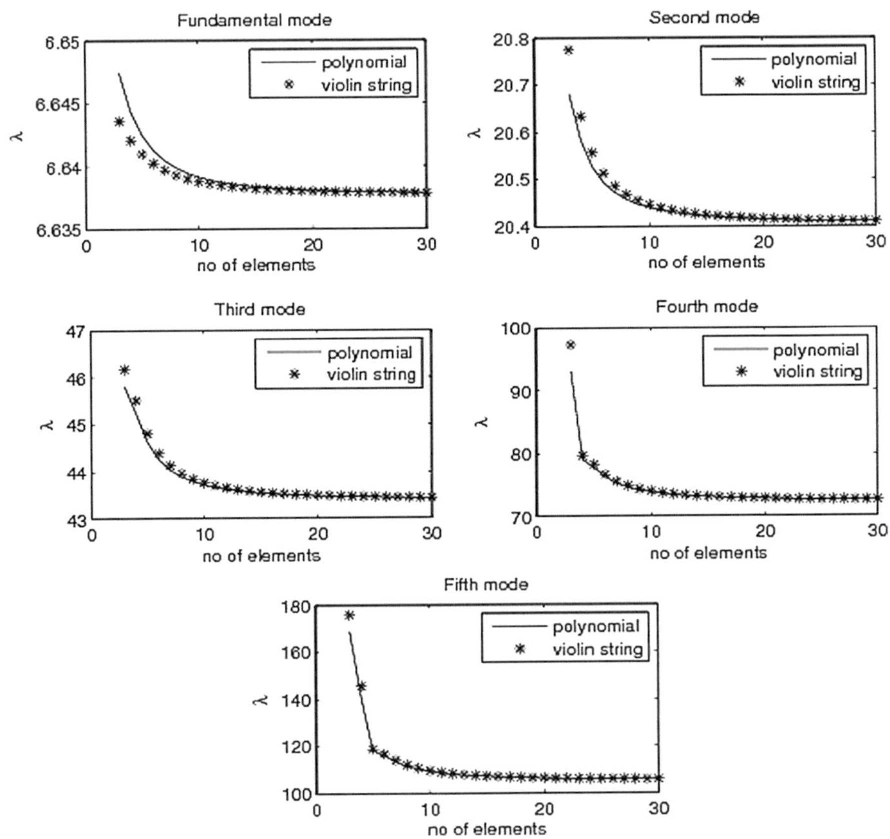

Fig. 9.11 Convergence of first five modes for slenderness ratio of $\frac{1}{\nu} = 0.05$ at $\kappa = 5$

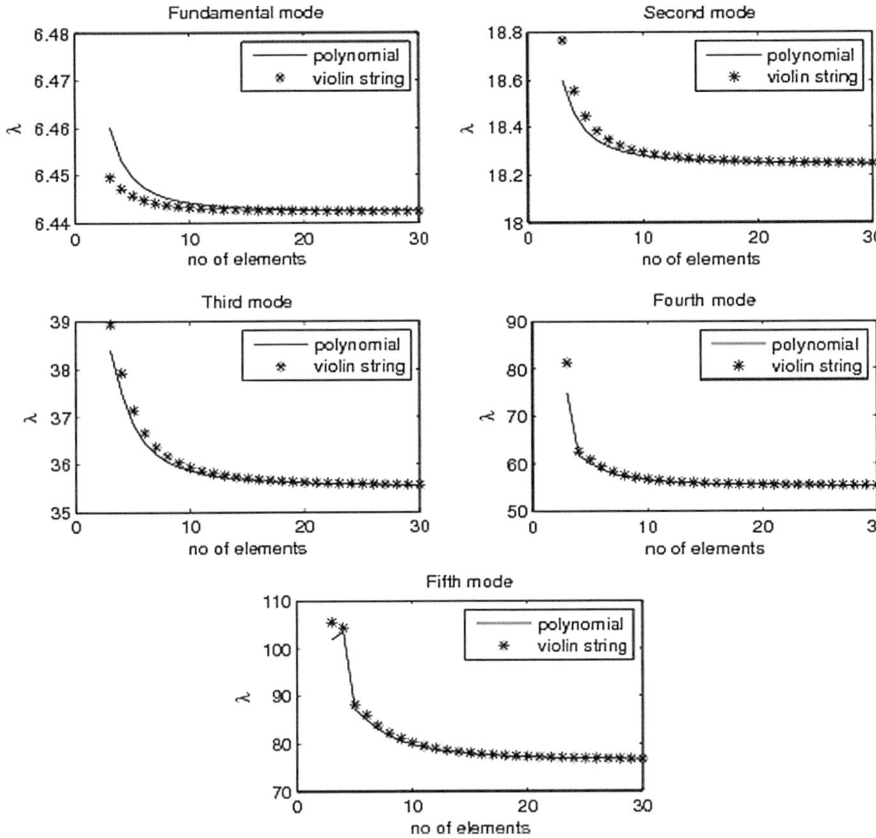

Fig. 9.12 Convergence of first five modes for slenderness ratio of $\frac{1}{\nu} = 0.09$ at $\kappa = 5$

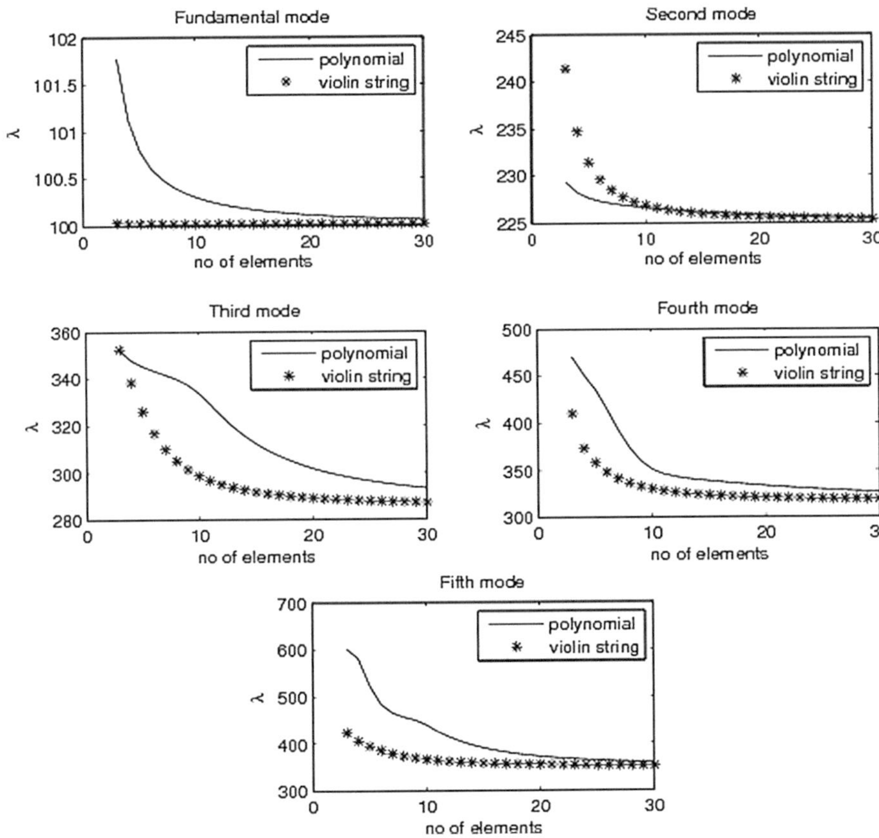

Fig. 9.13 Convergence of first five modes for slenderness ratio of $\frac{1}{\nu} = 0.05$ at $\kappa = 100$

9.4 Summary

Violin string shape functions are obtained using the governing differential equation of Timoshenko beam with constant tension. These new basis functions show superior convergence of the fundamental mode compared to that of conventional polynomial Timoshenko shape functions at low and high rotational speeds. However, at high speeds, the violin string basis functions show a better convergence rate compared to the cubic for various slenderness ratios for uniform and tapered beams. The violin string basis functions derived in this chapter are therefore computationally efficient for vibration analysis of high speed rotating beams. Timoshenko beams are typically used to model short and stiff beams such as those used in turbo machinery, which work

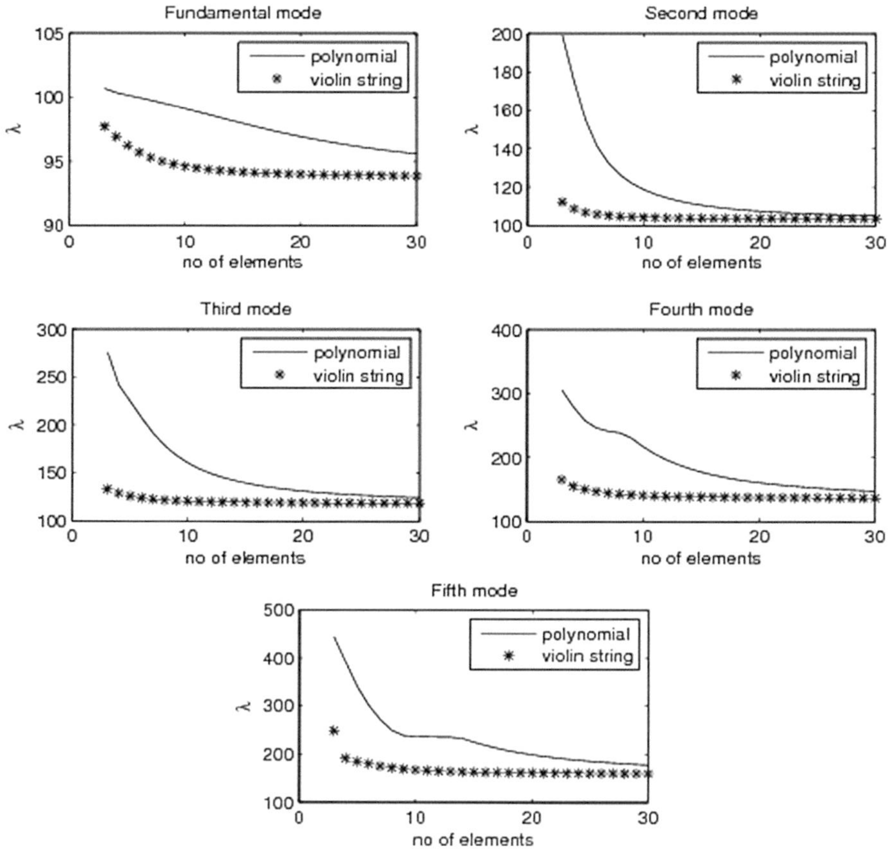

Fig. 9.14 Convergence of first five modes for slenderness ratio of $\frac{1}{\nu} = 0.09$ at $\kappa = 100$

at high rotation speeds. The new basis functions are also useful for the finite element analysis of short and stiff strings used in violins and other musical instruments as Timoshenko beam theory is a better model for predicting the inharmonicity of musical instruments.

References

1. Kapur K (1966) Vibrations of Timoshenko beam using finite element approach. J Acoust Soc Am 40(83):1058–1063
2. Thomas J, Abbas BAH (1975) Finite element model for dynamic analysis of Timoshenko beam. J Sound Vib 41(3):291–299
3. van Rensburg NFJ, van der Merve AJ (2006) Natural frequencies and modes of a Timoshenko beam. Wave Motion 44(1):58–69
4. Bokaian A (1990) Natural frequencies of beams under tensile loads. J Sound Vib 142(3):481–498

5. Kosmatka JB (1995) An Improved two-node finite element for stability and natural frequencies of axially loaded Timoshenko beams. Comput Struct 57(1):141–149
6. Yokoyama T (1988) Free vibration characteristics of rotating Timoshenko beams. Int J Mech Sci 30(10):734–755
7. Lin SM, Lin YS, Lee SY (2009) Instability and vibration of a rotating Timoshenko beam with precone. Comput Model Eng Sci 51(2):114–121
8. Banerjee JR, Sobey AJ (2002) Energy expression for rotating tapered Timoshenko beams. J Sound Vib 254(4):818–822
9. Banerjee JR (2001) Dynamics stiffness formulation and free vibration analysis of centrifugally stiffened Timoshenko beams. J Sound Vib 247(1):97–115
10. Du H, Lim MK, Liew KM (1994) A power series solution for vibration of a rotating Timoshenko Beam. J Sound Vib 175(4):505–523
11. Kaya MO (2006) Free vibration analysis of a rotating Timoshenko beam by differential transform method. Aircr Eng Aerosp Tech 78(3):194–203
12. Rao SS, Gupta RS (2001) Finite element vibration analysis of rotating Timoshenko beams. J Sound Vib 242(1):103–124
13. Bazoune A, Khulief YAF (1992) A finite beam element for vibration analysis of rotating tapered Timoshenko beams. J Sound Vib 156(1):141–164
14. Bambill DV, Felix DH, Rossi RE (2010) Vibration analysis of rotating Timoshenko beams by means of a differential quadrature method. Struct Eng Mech 24(2):231–245
15. Maezawa S, Temma K (1988) Vibration of bowed string: application of Timoshenkko's beam theory and transmission of sound energy through the bridge. J Soc Mech Eng Int J Ser 3, 31(1):30–38
16. Friedman Z, Kosmatka JB (1993) An improved two-node Timoshenko beam finite element. Comput Struct 47(3):473–481

Appendix A
Stiffness Matrix

Strain energy of an element is given by,

$$V = \frac{1}{2}\int_{0}^{l} EI\left(\frac{\partial^2 w}{\partial x^2}\right) dx + \frac{1}{2}\int_{0}^{l} T(x)\left(\frac{\partial w}{\partial x}\right)^2 dx \tag{A.1}$$

$$w' = \sum H_i' q_i \quad and \quad w'' = \sum H_i'' q_i$$

$$V = \frac{1}{2}\int_{0}^{l}\left(EI\sum_{i=1}^{4}\{H_i'' q_i\}\sum_{i=1}^{4}\{H_j'' q_j\}\right)dx + \frac{1}{2}\int_{0}^{l}\left(T(x)\sum_{i=1}^{4}\{H_i' q_i\}\sum_{i=1}^{4}\{H_j' q_j\}\right)dx$$

$$V = \frac{1}{2}q^T K q$$

$$V = \frac{1}{2}\sum_{i=1}^{4}\sum_{j=1}^{4} K_{ij} q_i q_j$$

$$K_{ij} = \int_{0}^{l} EIH_i'' H_j'' dx + \int_{0}^{l} T(x) H_i' H_j' dx \tag{A.2}$$

where H_1, H_2, H_3 *and* H_4 are known as shape functions (Figs. A.1 and A.2).

$$H_1 = 2\left(\frac{x}{l}\right)^3 - 3\left(\frac{x}{l}\right)^2 + 1$$

$$H_2 = \left[\left(\frac{x}{l}\right)^3 - 2\left(\frac{x}{l}\right)^2 + \frac{x}{l}\right] l$$

$$H_3 = -2\left(\frac{x}{l}\right)^3 + 3\left(\frac{x}{l}\right)^2$$

R. Ganguli, *Finite Element Analysis of Rotating Beams*,
Foundations of Engineering Mechanics, DOI 10.1007/978-981-10-1902-9

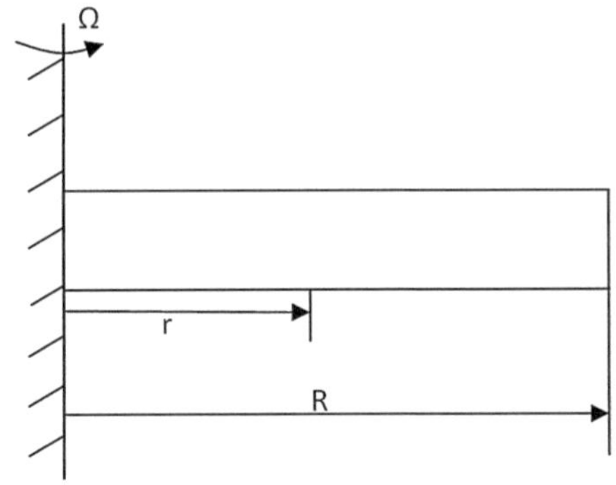

Fig. A.1 Cantilever Beam

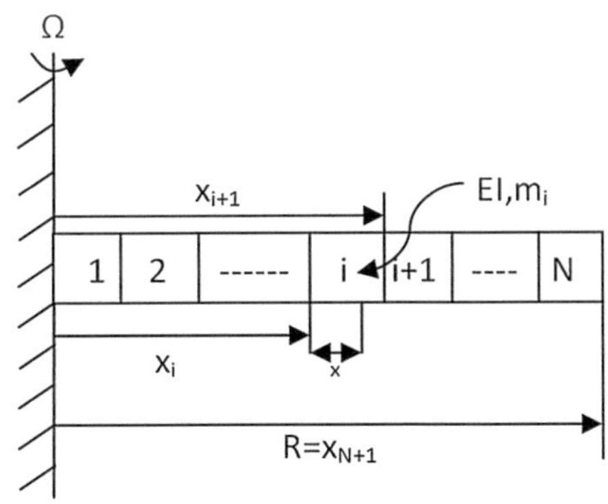

Fig. A.2 Elements in Cantilever Beam

$$H_4 = \left[\left(\frac{x}{l}\right)^3 - \left(\frac{x}{l}\right)^2\right] l$$

$$T(r) = \int_r^R mr\Omega^2 dr$$

$$= \frac{m\Omega^2}{2}\left(R^2 - r^2\right)$$

$$T(x) = m\Omega^2 \frac{\left(R^2 - r^2\right)}{2} \tag{A.3}$$

$$T = \int_{x_i+x}^{l} m\Omega^2 \rho d\rho \tag{A.4}$$

$$T = \int_{x_i}^{l} m\Omega^2 \rho d\rho - \int_{x_i}^{x_i+x} m\Omega^2 \rho d\rho$$

$$T = \sum_{j=i}^{N} \int_{x_j}^{x_{j+1}} m_j \Omega^2 \rho d\rho - \int_{x_i}^{x_i+x} m_i \Omega^2 \rho d\rho$$

$$T = \sum_{j=i}^{N} \left\{ m_j \Omega^2 \left(\frac{x_{j+1}^2 - x_j^2}{2} \right) \right\} - \left\{ m_i \Omega^2 \left(\frac{(x_i + x)^2 - x_i{}^2}{2} \right) \right\}$$

$$T = \frac{\Omega^2 A_i}{2} - \frac{m_i \Omega^2}{2} \left[x^2 + 2xx_i \right] \tag{A.5}$$

where

$$A_i = \sum_{j=i}^{N} \left\{ m_j \left(x_{j+1}^2 - x_j^2 \right) \right\}$$

Substitute, T in main equation we get,

$$K_{ij} = \int_0^l EIH_i'' H_j'' dx + \frac{\Omega^2 A_i}{2} \int_0^l H_i' H_j' dx - \frac{m_i \Omega^2}{2} \int_0^l \left(x^2 + 2xx_i \right) H_i' H_j' dx \tag{A.6}$$

Lets consider for simplicity,

$$K_{(ij)1} = \int_0^l EIH_i'' H_j'' dx$$

$$K_{(ij)2} = \frac{\Omega^2 A_i}{2} \int_0^l H_i' H_j' dx$$

$$K_{(ij)3} = -\frac{m_i \Omega^2}{2} \int_0^l \left(x^2 + 2xx_i \right) H_i' H_j' dx$$

$$K_{ij} = K_{(ij)1} + K_{(ij)2} + K_{(ij)3}$$

Now we obtain each term of the stiffness matrix,

1. i = 1, j = 1

$$H_1 = 2\left(\frac{x}{l}\right)^3 - 3\left(\frac{x}{l}\right)^2 + 1$$

$$H_1' = \left\{6\left(\frac{x}{l}\right)^2 - 6\left(\frac{x}{l}\right)\right\}\left(\frac{1}{l}\right)$$

$$H_1'' = \left\{12\left(\frac{x}{l}\right) - 6)\right\}\left(\frac{1}{l^2}\right)$$

Let

$$K_{(11)1} = \int_0^l EIH_1''H_1''dx$$

$$K_{(11)1} = \int_0^l EI\left(\left\{12\left(\frac{x}{l}\right) - 6)\right\}\left(\frac{1}{l^2}\right)\right)^2 dx$$

$$K_{(11)1} = EI\left(\frac{36}{l^4}\right)\int_0^l \left(\frac{4x^2}{l^2} + 1 - \frac{4x}{l}\right)dx$$

$$K_{(11)1} = EI\left(\frac{36}{l^4}\right)\left[\frac{4x^3}{3l^2} + x - \frac{4x^2}{2l}\right]_0^l = EI\left(\frac{36}{l^4}\right)\left[\frac{4l}{3} - l\right]$$

$$K_{(11)1} = \frac{12EI}{l^3}\ldots(1)$$

$$K_{(11)2} = \frac{\Omega^2 A_i}{2}\int_0^l H_1'H_1'dx$$

$$K_{(11)2} = \frac{\Omega^2 A_i}{2}\frac{36}{l^2}\int_0^l \left[\left\{6\left(\frac{x}{l}\right)^2 - 6\left(\frac{x}{l}\right)\right\}\left(\frac{1}{l}\right)\right]^2 dx$$

$$K_{(11)2} = \frac{18\Omega^2 A_i}{l^2}\int_0^l \left[\left(\frac{x}{l}\right)^4 + \left(\frac{x}{l}\right)^2 - 2\left(\frac{x}{l}\right)^3\right]dx$$

$$K_{(11)2} = \frac{18\Omega^2 A_i}{l^2}\left[\frac{x^5}{5l^4} + \frac{x^3}{3l^2} - 2\frac{x^4}{4l^3}\right]_0^l = \frac{18\Omega^2 A_i}{l^2}\left[\frac{6l + 10l - 15l}{30}\right]$$

$$K_{(11)2} = \frac{\Omega^2 A_i}{2}\left[\frac{6}{5l}\right]\ldots(2)$$

$$K_{(11)3} = -\frac{m_i\Omega^2}{2}\int_0^l \left(x^2 + 2xx_i\right) H_1' H_1' dx$$

$$K_{(11)3} = -\frac{m_i\Omega^2}{2}\int_0^l \left(x^2 + 2xx_i\right) \left[\left\{6\left(\frac{x}{l}\right)^2 - 6\left(\frac{x}{l}\right)\right\}\left(\frac{1}{l}\right)\right]^2 dx$$

$$K_{(11)3} = -\frac{36m_i\Omega^2}{2l^2}\int_0^l \left[\frac{x^6}{l^4} + \frac{2x^5x_i}{l^4} + \frac{x^4}{l^2} + \frac{2x^3x_i}{l^2} - \frac{2x^5}{l^3} - \frac{4x^4x_i}{l^3}\right] dx$$

$$= -\frac{36m_i\Omega^2}{2l^2}\left[\frac{x^7}{7l^4} + \frac{2x_ix^6}{6l^4} + \frac{x^5}{5l^2} + \frac{2x_ix^4}{4l^2} - \frac{2x^6}{6l^3} - \frac{4x_ix^5}{5l^3}\right]_0^l$$

$$K_{(11)3} = -m_i\Omega^2\left[\frac{6l}{35} + \frac{3x_i}{5}\right] \ldots (3)$$

$$K_{11} = K_{(11)1} + K_{(11)2} + K_{(11)3}$$

Substitute Eqs. 1, 2 and 3 in above to get,

$$K_{11} = \frac{12EI}{l^3} + \frac{\Omega^2 A_i}{2}\left[\frac{6}{5l}\right] - m_i\Omega^2\left[\frac{3}{5}x_i + \frac{6l}{35}\right]$$

2. i = 1, j = 2

$$H_1 = 2\left(\frac{x}{l}\right)^3 - 3\left(\frac{x}{l}\right)^2 + 1$$

$$H_1' = \left\{6\left(\frac{x}{l}\right)^2 - 6\left(\frac{x}{l}\right)\right\}\left(\frac{1}{l}\right)$$

$$H_1'' = \left\{12\left(\frac{x}{l}\right) - 6)\right\}\left(\frac{1}{l^2}\right)$$

$$H_2 = \left[\left(\frac{x}{l}\right)^3 - 2\left(\frac{x}{l}\right)^2 + \frac{x}{l}\right] l$$

$$H_2' = 6\left(3\left(\frac{x}{l}\right)^2 - 4\left(3\left(\frac{x}{l}\right) + 1\right)\right)$$

$$H_2'' = \frac{1}{l}\left[6\frac{x}{l} - 4\right]$$

Let

$$K_{(12)1} = \int_0^l EIH_1''H_2''dx$$

$$K_{(12)1} = \int_0^l EI\left[\left\{12\left(\frac{x}{l}\right) - 6)\right\}\left(\frac{1}{l^2}\right)\right]\left[\frac{1}{l}\left[6\frac{x}{l} - 4\right]\right]dx$$

$$K_{(12)1} = EI\frac{6}{l^3}\int_0^l \left[12\left(\frac{x}{l}\right)^2 - 14\left(\frac{x}{l}\right) + 4\right]dx$$

$$K_{(12)1} = EI\frac{6}{l^3}\left[12\frac{x^3}{3l^2} - 14\frac{x^2}{2l} + 4x\right]_0^l$$

$$K_{(12)1} = \frac{6EI}{l^2}\ldots(1)$$

$$K_{(12)2} = \frac{\Omega^2 A_i}{2}\int_0^l H_1'H_2'dx$$

$$K_{(12)2} = \frac{\Omega^2 A_i}{2}\int_0^l \left[\left\{6\left(\frac{x}{l}\right)^2 - 6\left(\frac{x}{l}\right)\right\}\left(\frac{1}{l}\right)\right]\left[6\left(3\left(\frac{x}{l}\right)^2 - 4\left(\frac{x}{l}\right) + 1\right)\right]dx$$

$$K_{(12)2} = \frac{\Omega^2 A_i}{2}\frac{6}{l}\int_0^l \left[3\left(\frac{x}{l}\right)^4 - 7\left(\frac{x}{l}\right)^3 + 5\left(\frac{x}{l}\right)^2 - \left(\frac{x}{l}\right)\right]dx$$

$$K_{(12)2} = \frac{\Omega^2 A_i}{2}\frac{6}{l}\left[3\left(\frac{x^5}{5l^4}\right) - 7\left(\frac{x^4}{4l^3}\right) + 5\left(\frac{x^3}{3l^2}\right) - \left(\frac{x^2}{2l}\right)\right]_0^l$$

$$K_{(12)2} = \frac{6\Omega^2 A_i}{2}\left[\frac{72 - 210 + 200 - 60}{120}\right]$$

$$K_{(12)2} = \frac{\Omega^2 A_i}{2}\left[\frac{1}{10}\right]\ldots(2)$$

$$K_{(12)3} = -\frac{m_i\Omega^2}{2}\int_0^l \left(x^2 + 2xx_i\right)H_1'H_2'dx$$

$$K_{(12)3} = -\frac{m_i\Omega^2}{2}\int_0^l \left(x^2 + 2xx_i\right)\left[\left\{6\left(\frac{x}{l}\right)^2 - 6\left(\frac{x}{l}\right)\right\}\left(\frac{1}{l}\right)\right]\left[6\left(3\left(\frac{x}{l}\right)^2 - 4\left(\frac{x}{l}\right) + 1\right)\right]dx$$

$$K_{(12)3} = -\frac{m_i\Omega^2}{2}\frac{6}{l}\int_0^l [\, 3\left(\frac{x}{l}\right)^4 x^2 - 7\left(\frac{x}{l}\right)^3 x^2 + 5\left(\frac{x}{l}\right)^2 x^2 - \left(\frac{x}{l}\right)x^2 + 6xx_i\left(\frac{x}{l}\right)^4$$

$$-14xx_i\left(\frac{x}{l}\right)^3 + 10xx_i\left(\frac{x}{l}\right)^2 - 2xx_i\left(\frac{x}{l}\right)\,]\,dx$$

$$K_{(12)3} = -\frac{6m_i\Omega^2}{2l}\left[\frac{3x^7}{7l^4} - \frac{7x^6}{6l^3} + \frac{5x^5}{5l^2} - \frac{x^4}{4l^2} + \frac{2x_ix^6}{6l^4} - \frac{14x_ix^5}{5l^3} + \frac{10x_ix^4}{4l^2} - \frac{2x_ix^3}{3l}\right]_0^l$$

$$K_{(12)3} = -m_i\Omega^2\left[\frac{l^2}{28} + \frac{lx_i}{10}\right]\ldots(3)$$

$$K_{12} = K_{(12)1} + K_{(12)2} + K_{(12)3}$$

Substitute Eqs. 1, 2 and 3 in above to get,

$$K_{12} = \frac{6EI}{l^2} + \frac{\Omega^2 A_i}{2}\left[\frac{1}{10}\right] - m_i\Omega^2\left[\frac{l^2}{28} + \frac{lx_i}{10}\right]$$

3. i = 1, j = 3

$$H_1 = 2\left(\frac{x}{l}\right)^3 - 3\left(\frac{x}{l}\right)^2 + 1$$

$$H_1' = \left\{6\left(\frac{x}{l}\right)^2 - 6\left(\frac{x}{l}\right)\right\}\left(\frac{1}{l}\right)$$

$$H_1'' = \left\{12\left(\frac{x}{l}\right) - 6)\right\}\left(\frac{1}{l^2}\right)$$

$$H_3 = -2\left(\frac{x}{l}\right)^3 + 3\left(\frac{x}{l}\right)^2$$

$$H_3' = 6\left(\frac{x}{l^2}\right)\left[-\left(\frac{x}{l}\right) + 1\right]$$

$$H_3'' = -12\left(\frac{x}{l^3}\right) + \frac{6}{l^2}$$

Let

$$K_{(13)1} = \int_0^l EIH_1''H_3''dx$$

$$K_{(13)1} = \int_0^l EI\left[\left\{12\left(\frac{x}{l}\right) - 6)\right\}\left(\frac{1}{l^2}\right)\right]\left[-12\left(\frac{x}{l^3}\right) + \frac{6}{l^2}\right]dx$$

$$K_{(13)1} = EI\frac{36}{l^4}\int_0^l \left[-4\left(\frac{x}{l}\right)^2 + 4\left(\frac{x}{l}\right) - 1\right]dx$$

$$K_{(13)1} = EI\frac{36}{l^4}\left[-4\frac{x^3}{3l^2} + 4\frac{x^2}{2l} - x\right]_0^l$$

$$K_{(13)1} = \frac{-12EI}{l^3}\ldots(1)$$

$$K_{(13)2} = \frac{\Omega^2 A_i}{2}\int_0^l H_1' H_3' dx$$

$$K_{(13)2} = \frac{\Omega^2 A_i}{2}\int_0^l \left[\left\{6\left(\frac{x}{l}\right)^2 - 6\left(\frac{x}{l}\right)\right\}\left(\frac{1}{l}\right)\right]\left[6\left(\frac{x}{l^2}\right)\left[-\left(\frac{x}{l}\right)+1\right]\right]dx$$

$$K_{(13)2} = \frac{\Omega^2 A_i}{2}\frac{36}{l^2}\int_0^l \left[-\left(\frac{x}{l}\right)^4 + \left(\frac{x}{l}\right)^3 + \left(\frac{x}{l}\right)^3 - \left(\frac{x}{l}\right)^2\right]dx$$

$$K_{(13)2} = \frac{\Omega^2 A_i}{2}\frac{36}{l^2}\left[-\frac{x^5}{5l^4} + 2\frac{x^4}{4l^3} - \frac{x^3}{3l^2}\right]_0^l$$

$$K_{(13)2} = \frac{\Omega^2 A_i}{2}\frac{36}{l}\left[\frac{-6+15-10}{30}\right]$$

$$K_{(13)2} = \frac{\Omega^2 A_i}{2}\left[\frac{-6}{5l}\right]\ldots(2)$$

$$K_{(13)3} = -\frac{m_i\Omega^2}{2}\int_0^l \left(x^2 + 2xx_i\right) H_1' H_3' dx$$

$$K_{(13)3} = -\frac{m_i\Omega^2}{2}\int_0^l \left(x^2 + 2xx_i\right)\left[\left\{6\left(\frac{x}{l}\right)^2 - 6\left(\frac{x}{l}\right)\right\}\left(\frac{1}{l}\right)\right]\left[6\left(\frac{x}{l^2}\right)\left[-\left(\frac{x}{l}\right)+1\right]\right]dx$$

$$K_{(13)3} = -\frac{m_i\Omega^2}{2}\frac{36}{l^2}\int_0^l \left[-\frac{x^6}{l^4} + \frac{2x^5}{l^3} - \frac{x^4}{l^2} - \frac{2x_ix^5}{l^4} + \frac{4x_ix^4}{l^3} - \frac{2x_ix^3}{l^2}\right]dx$$

$$K_{(13)3} = -\frac{m_i\Omega^2}{2}\frac{36}{l^2}\left[-\frac{x^7}{7l^4} + \frac{2x^6}{6l^3} - \frac{x^5}{5l^2} - \frac{2x_ix^6}{6l^4} + \frac{4x_ix^5}{5l^3} - \frac{2x_ix^4}{4l^2}\right]_0^l$$

$$K_{(13)3} = -m_i\Omega^2\left[\frac{-6l}{35} + \frac{-3x_i}{5}\right]\ldots(3)$$

$$K_{13} = K_{(13)1} + K_{(13)2} + K_{(13)3}$$

Substitute Eqs. 1, 2 and 3 in above to get,

$$K_{13} = \frac{-12EI}{l^3} + \frac{\Omega^2 A_i}{2}\left[\frac{-6}{5l}\right] - m_i\Omega^2\left[\frac{-6l}{35} + \frac{-3x_i}{5}\right]$$

4. i = 1, j = 4

$$H_1 = 2\left(\frac{x}{l}\right)^3 - 3\left(\frac{x}{l}\right)^2 + 1$$

$$H_1' = \left\{6\left(\frac{x}{l}\right)^2 - 6\left(\frac{x}{l}\right)\right\}\left(\frac{1}{l}\right)$$

$$H_1'' = \left\{12\left(\frac{x}{l}\right) - 6)\right\}\left(\frac{1}{l^2}\right)$$

$$H_4 = \left[\left(\frac{x}{l}\right)^3 - \left(\frac{x}{l}\right)^2\right]l$$

$$H_4' = \left[3\left(\frac{x}{l}\right)^2 - 2\frac{x}{l}\right]$$

$$H_4'' = \frac{1}{l}\left[\frac{6x}{l} - 2\right]$$

Let

$$K_{(14)1} = \int_0^l EIH_1''H_4''dx$$

$$K_{(14)1} = \int_0^l EI\left[\left\{12\left(\frac{x}{l}\right) - 6)\right\}\left(\frac{1}{l^2}\right)\right]\left[\frac{1}{l}\left[\frac{6x}{l} - 2\right]\right]dx$$

$$K_{(14)1} = EI\frac{6}{l^3}\int_0^l \left[\left(12\frac{x}{l}\right)^2 - 10\left(\frac{x}{l}\right) + 2\right]dx$$

$$K_{(14)1} = EI\frac{6}{l^3}\left[\frac{12}{l^2}\left(\frac{x^3}{3}\right) - \frac{10}{l}\left(\frac{x^2}{2}\right) + 2x\right]_0^l$$

$$K_{(14)1} = \frac{6EI}{l^2}\ldots(1)$$

$$K_{(14)2} = \frac{\Omega^2 A_i}{2}\int_0^l H_1'H_4'dx$$

$$K_{(14)2} = \frac{\Omega^2 A_i}{2} \int_0^l \left[\left\{6\left(\frac{x}{l}\right)^2 - 6\left(\frac{x}{l}\right)\right\}\left(\frac{1}{l}\right)\right]\left[\left[3\left(\frac{x}{l}\right)^2 - 2\frac{x}{l}\right]\right] dx$$

$$K_{(14)2} = \frac{\Omega^2 A_i}{2} \frac{6}{l} \int_0^l \left[3\left(\frac{x}{l}\right)^4 - 5\left(\frac{x}{l}\right)^3 + 2\left(\frac{x}{l}\right)^2\right] dx$$

$$K_{(14)2} = \frac{\Omega^2 A_i}{2} \frac{6}{l} \left[\frac{3x^5}{5l^4} - \frac{5x^4}{4l^3} + \frac{2x^3}{3l^2}\right]_0^l$$

$$K_{(14)2} = \frac{\Omega^2 A_i}{2} \left[\frac{36 - 75 + 40}{10}\right]$$

$$K_{(14)2} = \frac{\Omega^2 A_i}{2} \left[\frac{1}{10}\right] \ldots (2)$$

$$K_{(14)3} = -\frac{m_i \Omega^2}{2} \int_0^l \left(x^2 + 2xx_i\right) H_1' H_4' dx$$

$$K_{(14)3} = -\frac{m_i \Omega^2}{2} \int_0^l \left(x^2 + 2xx_i\right) \left[\left\{6\left(\frac{x}{l}\right)^2 - 6\left(\frac{x}{l}\right)\right\}\left(\frac{1}{l}\right)\right]\left[\left[3\left(\frac{x}{l}\right)^2 - 2\frac{x}{l}\right]\right] dx$$

$$K_{(14)3} = -\frac{m_i \Omega^2}{2} \frac{6}{l} \int_0^l \left[\frac{3x^6}{l^4} - \frac{5x^5}{l^3} + \frac{2x^4}{l^2} + \frac{6x_i x^5}{l^4} - \frac{10x_i x^4}{l^3} + \frac{4x_i x^3}{l^2}\right] dx$$

$$K_{(14)3} = -\frac{m_i \Omega^2}{2} \frac{6}{l} \left[\frac{3x^7}{7l^4} - \frac{5x^6}{6l^3} + \frac{2x^5}{5l^2} + \frac{6x_i x^6}{6l^4} - \frac{10x_i x^5}{5l^3} + \frac{4x_i x^4}{4l^2}\right]_0^l$$

$$K_{(14)3} = -m_i \Omega^2 \left[\frac{-l^2}{70}\right] \ldots (3)$$

$$K_{14} = K_{(14)1} + K_{(14)2} + K_{(14)3}$$

Substitute Eqs. 1, 2 and 3 in above to get,

$$K_{14} = \frac{6EI}{l^2} + \frac{\Omega^2 A_i}{2}\left[\frac{1}{10}\right] - m_i \Omega^2 \left[\frac{-l^2}{70}\right]$$

5. i = 2, j = 2

$$H_2 = \left[\left(\frac{x}{l}\right)^3 - 2\left(\frac{x}{l}\right)^2 + \frac{x}{l}\right] l$$

$$H_2' = \left(3\left(\frac{x}{l}\right)^2 - 4\left(\frac{x}{l}\right) + 1\right)$$

$$H_2'' = \frac{1}{l}\left[6\frac{x}{l} - 4\right]$$

Let

$$K_{(22)1} = \int_0^l EIH_2''H_2''dx$$

$$K_{(22)1} = \int_0^l EI\left(\frac{1}{l}\left[6\frac{x}{l} - 4\right]\right)^2 dx$$

$$K_{(22)1} = EI\left(\frac{1}{l^2}\right)\int_0^l \left(36\left(\frac{x}{l}\right)^2 - 48\left(\frac{x}{l}\right) + 16\right)dx$$

$$K_{(22)1} = EI\left(\frac{1}{l^2}\right)\left[\frac{36x^3}{3l^2} - \frac{48x^2}{2l} + 16x\right]_0^l$$

$$K_{(22)1} = \frac{4EI}{l}\ldots(1)$$

$$K_{(22)2} = \frac{\Omega^2 A_i}{2}\int_0^l H_2'H_2'dx$$

$$K_{(22)2} = \frac{\Omega^2 A_i}{2}\int_0^l \left[\left(3\left(\frac{x}{l}\right)^2\right) - 4\left(\left(\frac{x}{l}\right) + 1\right)\right]^2 dx$$

$$K_{(22)2} = \frac{\Omega^2 A_i}{2}\int_0^l \left[9\left(\frac{x}{l}\right)^4 - 24\left(\frac{x}{l}\right)^3 + 22\left(\frac{x}{l}\right)^2 - 8\left(\frac{x}{l}\right) + 1\right]dx$$

$$K_{(22)2} = \frac{\Omega^2 A_i}{2}\left[\left(\frac{9x^5}{5l^4}\right) - \left(\frac{24x^4}{4l^3}\right) + \left(\frac{22x^3}{3l^2}\right) - \left(\frac{8x^2}{2l}\right) + x\right]_0^l$$

$$K_{(22)2} = \frac{\Omega^2 A_i}{2}l\left[\frac{27 - 90 + 110 - 60 + 15}{15}\right]$$

$$K_{(22)2} = \frac{\Omega^2 A_i}{2}\left[\frac{2l}{15}\right]\ldots(2)$$

$$K_{(22)3} = -\frac{m_i\Omega^2}{2}\int_0^l \left(x^2 + 2xx_i\right)H_2'H_2'dx$$

$$K_{(22)3} = -\frac{m_i\Omega^2}{2}\int_0^l \left(x^2 + 2xx_i\right)\left[\left(3\left(\frac{x}{l}\right)^2 - 4(\left(\frac{x}{l}\right) + 1\right)\right]^2 dx$$

$$K_{(22)3} = -\frac{m_i\Omega^2}{2}\int_0^l \left[\frac{9x^6}{l^4} - \frac{24x^5}{l^3} - \frac{22x^4}{l^2} - \frac{8x^3}{l} + x^2 + \frac{18x^5x_i}{l^4} - \frac{48x_ix^4}{l^3} + \frac{44x^3x_i}{l^2} - \frac{16x^2x_i}{l} + 2xx_i\right]dx$$

$$K_{(22)3} = -\frac{m_i\Omega^2}{2}\left[\frac{9x^7}{7l^4} - \frac{24x^6}{6l^3} + \frac{22x^5}{5l^2} - \frac{8x^4}{4l} + \frac{x^3}{3} + \frac{18x_ix^6}{6l^4} - \frac{48x_ix^5}{5l^3} + \frac{44x_ix^4}{4l^2} - \frac{16x_ix^3}{3l} + \frac{2x_ix^2}{2}\right]_0^l$$

$$K_{(22)3} = -m_i\Omega^2\left[\frac{l^3}{105} + \frac{x_il^2}{30}\right]\ldots(3)$$

$$K_{22} = K_{(22)1} + K_{(22)2} + K_{(22)3}$$

Substitute Eqs. 1, 2 and 3 in above to get,

$$K_{22} = \frac{4EI}{l} + \frac{\Omega^2A_i}{2}\left[\frac{2l}{15}\right] - m_i\Omega^2\left[\frac{l^3}{105} + \frac{x_il^2}{30}\right]$$

6. i = 2, j = 3

$$H_2 = \left[\left(\frac{x}{l}\right)^3 - 2\left(\frac{x}{l}\right)^2 + \frac{x}{l}\right]l$$

$$H_2' = \left(3\left(\frac{x}{l}\right)^2 - 4\left(\frac{x}{l}\right) + 1\right)$$

$$H_2'' = \frac{1}{l}\left[6\frac{x}{l} - 4\right]$$

$$H_3 = -2\left(\frac{x}{l}\right)^3 + 3\left(\frac{x}{l}\right)^2$$

$$H_3' = 6\left(\frac{x}{l^2}\right)\left[-\left(\frac{x}{l}\right) + 1\right]$$

$$H_3'' = -12\left(\frac{x}{l^3}\right) + \frac{6}{l^2}$$

Let

$$K_{(23)1} = \int_0^l EIH_2''H_3''dx$$

$$K_{(23)1} = \int_0^l EI\left[\frac{1}{l}\left[6\frac{x}{l} - 4\right]\right]\left[-12\left(\frac{x}{l^3}\right) + \frac{6}{l^2}\right]dx$$

$$K_{(23)1} = \int_0^l EI\frac{6}{l^3}\left[-12\left(\frac{x}{l}\right)^2 + 14\left(\frac{x}{l}\right) - 4\right]dx$$

$$K_{(23)1} = EI\frac{6}{l^3}\left[-12\frac{x^3}{3l^2} + 14\frac{x^2}{2l} - 4x\right]_0^l$$

$$K_{(23)1} = \frac{-6EI}{l^2}\ldots(1)$$

$$K_{(23)2} = \frac{\Omega^2 A_i}{2}\int_0^l H_2'H_3'dx$$

$$K_{(23)2} = \frac{\Omega^2 A_i}{2}\int_0^l \left[\left(3\left(\frac{x}{l}\right)^2 - 4\left(\frac{x}{l}\right) + 1\right)\right]\left[6\left(\frac{x}{l^2}\right)\left[-\left(\frac{x}{l}\right) + 1\right]\right]dx$$

$$K_{(23)2} = \frac{\Omega^2 A_i}{2}\frac{6}{l}\int_0^l \left[-3\left(\frac{x}{l}\right)^4 + 7\left(\frac{x}{l}\right)^3 - 5\left(\frac{x}{l}\right)^2 + \left(\frac{x}{l}\right)\right]dx$$

$$K_{(23)2} = \frac{\Omega^2 A_i}{2}\frac{6}{l}\left[-3\frac{x^5}{5l^4} + 7\frac{x^4}{4l^3} - 5\frac{x^3}{3l^2} + \frac{x^2}{2l}\right]_0^l$$

$$K_{(23)2} = \frac{\Omega^2 A_i}{2}\left[\frac{-72 + 210 - 200 + 60}{20}\right]$$

$$K_{(23)2} = \frac{\Omega^2 A_i}{2}\left[\frac{-1}{10}\right]\ldots(2)$$

$$K_{(23)3} = -\frac{m_i\Omega^2}{2}\int_0^l \left(x^2 + 2xx_i\right)H_2'H_3'dx$$

$$K_{(23)3} = -\frac{m_i\Omega^2}{2}\int_0^l \left(x^2 + 2xx_i\right)\left[\left(3\left(\frac{x}{l}\right)^2 - 4\left(\frac{x}{l}\right) + 1\right)\right]\left[6\left(\frac{x}{l^2}\right)\left[-\left(\frac{x}{l}\right) + 1\right]\right]dx$$

$$K_{(23)3} = -\frac{m_i\Omega^2}{2}\frac{6}{l}\int_0^l \left[-3\frac{x^6}{l^4} + \frac{7x^5}{l^3} - \frac{5x^4}{l^2} + \frac{x^3}{l} - \frac{6x_ix^5}{l^4} + \frac{14x_ix^4}{l^3} - \frac{10x_ix^3}{l^2} + \frac{2x_ix^2}{l}\right]dx$$

$$K_{(23)3} = -\frac{m_i\Omega^2}{2}\frac{6}{l}\left[\frac{-3}{l^4}\left(\frac{x^7}{7}\right) + \frac{7}{l^3}\left(\frac{x^6}{6}\right) - \frac{5}{l^2}\left(\frac{x^5}{5}\right) + \frac{1}{l}\left(\frac{x^4}{4}\right) - \frac{6x_i}{l^4}\left(\frac{x^6}{6}\right)\right.$$
$$\left. + \frac{14x_i}{l^3}\left(\frac{x^5}{5}\right) - \frac{10x_i}{l^2}\left(\frac{x^4}{4}\right) + \frac{2x_i}{l}\left(\frac{x^3}{3}\right)\right]_0^l$$

$$K_{(23)3} = -m_i\Omega^2\left[\frac{-l^2}{28} + \frac{-lx_i}{10}\right]\ldots(3)$$

$$K_{23} = K_{(23)1} + K_{(23)2} + K_{(23)3}$$

Substitute Eqs. 1, 2 and 3 in above to get,

$$K_{23} = \frac{-6EI}{l^2} + \frac{\Omega^2 A_i}{2}\left[\frac{-1}{10}\right] - m_i\Omega^2\left[\frac{-l^2}{28} + \frac{-lx_i}{10}\right]$$

7. i = 2, j = 4

$$H_2 = \left[\left(\frac{x}{l}\right)^3 - 2\left(\frac{x}{l}\right)^2 + \frac{x}{l}\right]l$$

$$H_2' = \left(3\left(\frac{x}{l}\right)^2 - 4\left(\frac{x}{l}\right) + 1\right)$$

$$H_2'' = \frac{1}{l}\left[6\frac{x}{l} - 4\right]$$

$$H_4 = \left[\left(\frac{x}{l}\right)^3 - \left(\frac{x}{l}\right)^2\right]l$$

$$H_4' = \left[3\left(\frac{x}{l}\right)^2 - 2\frac{x}{l}\right]$$

$$H_4'' = \frac{1}{l}\left[\frac{6x}{l} - 2\right]$$

Let

$$K_{(24)1} = \int_0^l EIH_2''H_4''dx$$

$$K_{(24)1} = \int_0^l EI\left[\frac{1}{l}\left[6\frac{x}{l} - 4\right]\right]\left[\frac{1}{l}\left[\frac{6x}{l} - 2\right]\right]dx$$

$$K_{(24)1} = \int_0^l EI\frac{1}{l^2}\left[36\left(\frac{x}{l}\right)^2 - 12\left(\frac{x}{l}\right) - 24\left(\frac{x}{l}\right) + 8\right]dx$$

$$K_{(24)1} = EI\frac{1}{l^2}\left[\frac{36}{l^2}\left(\frac{x^3}{3}\right) - \frac{12}{l}\left(\frac{x^2}{2}\right) - \frac{24}{l}\left(\frac{x^2}{2}\right) + 8x\right]_0^l$$

$$K_{(24)1} = \frac{2EI}{l} \ldots (1)$$

$$K_{(24)2} = \frac{\Omega^2 A_i}{2}\int_0^l H_2' H_4' dx$$

$$K_{(24)2} = \frac{\Omega^2 A_i}{2}\int_0^l \left[\left(3\left(\frac{x}{l}\right)^2 - 4\left(\frac{x}{l}\right) + 1\right)\right]\left[\left[3\left(\frac{x}{l}\right)^2 - 2\frac{x}{l}\right]\right]dx$$

$$K_{(24)2} = \frac{\Omega^2 A_i}{2}\int_0^l \left[9\left(\frac{x}{l}\right)^4 - 18\left(\frac{x}{l}\right)^3 + 11\left(\frac{x}{l}\right)^2 - 2\left(\frac{x}{l}\right)\right]dx$$

$$K_{(24)2} = \frac{\Omega^2 A_i}{2}\left[\frac{9x^5}{5l^4} - \frac{18x^4}{4l^3} + \frac{11x^3}{3l^2} - \frac{2x^2}{2l}\right]_0^l$$

$$K_{(24)2} = \frac{\Omega^2 A_i}{2} l\left[\frac{108 - 270 + 220 - 60}{60}\right]$$

$$K_{(24)2} = \frac{\Omega^2 A_i}{2}\left[\frac{-l}{30}\right] \ldots (2)$$

$$K_{(24)3} = -\frac{m_i \Omega^2}{2}\int_0^l \left(x^2 + 2xx_i\right) H_2' H_4' dx$$

$$K_{(24)3} = -\frac{m_i \Omega^2}{2}\int_0^l \left(x^2 + 2xx_i\right)\left[\left(3\left(\frac{x}{l}\right)^2 - 4\left(\frac{x}{l}\right) + 1\right)\right]\left[\left[3\left(\frac{x}{l}\right)^2 - 2\frac{x}{l}\right]\right]dx$$

$$K_{(24)3} = -\frac{m_i \Omega^2}{2}\int_0^l \left[\frac{9x^6}{l^4} - \frac{18x^5}{l^3} + \frac{11x^4}{l^2} - \frac{2x^3}{l} + \frac{18x_i x^5}{l^4} - \frac{36x_i x^4}{l^3} + \frac{22x_i x^3}{l^2} - \frac{4x_i x^2}{l}\right]dx$$

$$K_{(24)3} = -\frac{m_i \Omega^2}{2}\left[\frac{9x^7}{7l^4} - \frac{18x^6}{6l^3} + \frac{11x^5}{5l^2} - \frac{2x^4}{4l} + \frac{18x_i x^6}{6l^4} - \frac{36x_i x^5}{5l^3} + \frac{22x_i x^4}{4l^2} - \frac{4x_i x^3}{3l}\right]_0^l$$

$$K_{(24)3} = -m_i \Omega^2\left[\frac{-l^3}{140} - \frac{l^2 x_i}{60}\right] \ldots (3)$$

$$K_{24} = K_{(24)1} + K_{(24)2} + K_{(24)3}$$

Substitute Eqs. 1, 2 and 3 in above to get,

$$K_{24} = \frac{2EI}{l} + \frac{\Omega^2 A_i}{2}\left[\frac{-l}{30}\right] - m_i\Omega^2\left[\frac{-l^3}{140} - \frac{l^2 x_i}{60}\right]$$

8. i = 3, j = 3

$$H_3 = -2\left(\frac{x}{l}\right)^3 + 3\left(\frac{x}{l}\right)^2$$

$$H_3' = 6\left(\frac{x}{l^2}\right)\left[-\left(\frac{x}{l}\right) + 1\right]$$

$$H_3'' = -12\left(\frac{x}{l^3}\right) + \frac{6}{l^2}$$

Let

$$K_{(33)1} = \int_0^l EIH_3''H_3''dx$$

$$K_{(33)1} = \int_0^l EI\left(-12\left(\frac{x}{l^3}\right) + \frac{6}{l^2}\right)^2 dx$$

$$K_{(33)1} = EI\left(\frac{36}{l^4}\right)\int_0^l \left(\frac{4x^2}{l^2} + 1 - \frac{4x}{l}\right)dx$$

$$K_{(33)1} = EI\left(\frac{36}{l^4}\right)\left[\frac{4x^3}{3l^2} + x - \frac{4x^2}{2l}\right]_0^l$$

$$K_{(33)1} = EI\left(\frac{36}{l^4}\right)\left[\frac{4l}{3} - l\right]$$

$$K_{(33)1} = \frac{12EI}{l^3} \ldots (1)$$

$$K_{(33)2} = \frac{\Omega^2 A_i}{2}\int_0^l H_3'H_3'dx$$

$$K_{(33)2} = \frac{\Omega^2 A_i}{2}\int_0^l \left[6\left(\frac{x}{l^2}\right)\left[-\left(\frac{x}{l}\right) + 1\right]\right]^2 dx$$

$$K_{(33)2} = \frac{\Omega^2 A_i}{2}\frac{36}{l^2}\int_0^l \left[\left(\frac{x}{l}\right)^4 - 2\left(\frac{x}{l}\right)^3 + \left(\frac{x}{l}\right)^2\right]dx$$

$$K_{(33)2} = \frac{36\Omega^2 A_i}{2l^2}\left[\frac{x^5}{5l^4} + \frac{x^3}{3l^2} - 2\frac{x^4}{4l^3}\right]_0^l$$

$$K_{(33)2} = \frac{18\Omega^2 A_i}{l}\left[\frac{6 - 15 + 10}{30}\right]$$

$$K_{(33)2} = \frac{\Omega^2 A_i}{2}\left[\frac{6}{5l}\right] \ldots (2)$$

$$K_{(33)3} = -\frac{m_i\Omega^2}{2}\int_0^l \left(x^2 + 2xx_i\right) H_3' H_3' dx$$

$$K_{(33)3} = -\frac{m_i\Omega^2}{2}\int_0^l \left(x^2 + 2xx_i\right)\left[6\left(\frac{x}{l^2}\right)\left[-\left(\frac{x}{l}\right) + 1\right]\right]^2 dx$$

$$K_{(33)3} = -\frac{36m_i\Omega^2}{2l^2}\int_0^l \left[\frac{x^6}{l^4} + \frac{2x^5x_i}{l^4} + \frac{x^4}{l^2} + \frac{2x^3x_i}{l^2} - \frac{2x^5}{l^3} - \frac{4x^4x_i}{l^3}\right] dx$$

$$K_{(33)3} = -\frac{36m_i\Omega^2}{2l^2}\left[\frac{x^7}{7l^4} + \frac{2x_ix^6}{6l^4} + \frac{x^5}{5l^2} + \frac{2x_ix^4}{4l^2} - \frac{2x^6}{6l^3} - \frac{4x_ix^5}{5l^3}\right]_0^l$$

$$K_{(33)3} = -m_i\Omega^2\left[\frac{6l}{35} + \frac{3x_i}{5}\right] \ldots (3)$$

$$K_{33} = K_{(33)1} + K_{(33)2} + K_{(33)3}$$

Substitute Eqs. 1, 2 and 3 in above to get,

$$K_{33} = \frac{12EI}{l^3} + \frac{\Omega^2 A_i}{2}\left[\frac{6}{5l}\right] - m_i\Omega^2\left[\frac{3}{5}x_i + \frac{6l}{35}\right]$$

9. i = 3, j = 4

$$H_3 = -2\left(\frac{x}{l}\right)^3 + 3\left(\frac{x}{l}\right)^2$$

$$H_3' = 6\left(\frac{x}{l^2}\right)\left[-\left(\frac{x}{l}\right) + 1\right]$$

$$H_3'' = -12\left(\frac{x}{l^3}\right) + \frac{6}{l^2}$$

$$H_4 = \left[\left(\frac{x}{l}\right)^3 - \left(\frac{x}{l}\right)^2\right] l$$

$$H_4' = \left[3\left(\frac{x}{l}\right)^2 - 2\frac{x}{l}\right]$$

$$H_4'' = \frac{1}{l}\left[\frac{6x}{l} - 2\right]$$

Let

$$K_{(34)1} = \int_0^l EIH_3''H_4''dx$$

$$K_{(34)1} = \int_0^l EI\left[-12\left(\frac{x}{l^3}\right) + \frac{6}{l^2}\right]\left[\frac{1}{l}\left[\frac{6x}{l} - 2\right]\right]dx$$

$$K_{(34)1} = EI\frac{-6}{l^3}\int_0^l \left[12\left(\frac{x}{l}\right)^2 - 4\left(\frac{x}{l}\right) - 6\left(\frac{x}{l}\right) + 2\right]dx$$

$$K_{(34)1} = EI\frac{-6}{l^3}\left[\frac{12}{l^2}\left(\frac{x^3}{3}\right) - \frac{10}{l}\left(\frac{x^2}{2}\right) + 2x\right]_0^l$$

$$K_{(34)1} = \frac{-6EI}{l^2}\ldots(1)$$

$$K_{(34)2} = \frac{\Omega^2 A_i}{2}\int_0^l H_3'H_4'dx$$

$$K_{(34)2} = \frac{\Omega^2 A_i}{2}\int_0^l \left[6\left(\frac{x}{l^2}\right)\left[-\left(\frac{x}{l}\right) + 1\right]\right]\left[\left[3\left(\frac{x}{l}\right)^2 - 2\frac{x}{l}\right]\right]dx$$

$$K_{(34)2} = \frac{\Omega^2 A_i}{2}\frac{-6}{l}\int_0^l \left[3\left(\frac{x}{l}\right)^4 - 5\left(\frac{x}{l}\right)^3 + 2\left(\frac{x}{l}\right)^2\right]dx$$

$$K_{(34)2} = \frac{\Omega^2 A_i}{2}\frac{-6}{l}\left[\frac{3x^5}{5l^4} - \frac{5x^4}{4l^3} + \frac{2x^3}{3l^2}\right]_0^l$$

$$K_{(34)2} = \frac{-\Omega^2 A_i}{2}\left[\frac{36 - 75 + 40}{10}\right]$$

$$K_{(34)2} = \frac{\Omega^2 A_i}{2}\left[\frac{-1}{10}\right]\ldots(2)$$

$$K_{(34)3} = -\frac{m_i\Omega^2}{2}\int_0^l \left(x^2 + 2xx_i\right)H_3'H_4'dx$$

$$K_{(34)3} = -\frac{m_i\Omega^2}{2}\int_0^l \left(x^2 + 2xx_i\right)\left[6\left(\frac{x}{l^2}\right)\left[-\left(\frac{x}{l}\right) + 1\right]\right]\left[\left[3\left(\frac{x}{l}\right)^2 - 2\frac{x}{l}\right]\right]dx$$

$$K_{(34)3} = -\frac{m_i\Omega^2}{2}\frac{-6}{l}\int_0^l \left[\frac{3x^6}{l^4} - \frac{5x^5}{l^3} + \frac{2x^4}{l^2} + \frac{6x_ix^5}{l^4} - \frac{10x_ix^4}{l^3} + \frac{4x_ix^3}{l^2}\right] dx$$

$$K_{(34)3} = -\frac{m_i\Omega^2}{2}\frac{-6}{l}\left[\frac{3x^7}{7l^4} - \frac{5x^6}{6l^3} + \frac{2x^5}{5l^2} + \frac{6x_ix^6}{6l^4} - \frac{10x_ix^5}{5l^3} + \frac{4x_ix^4}{4l^2}\right]_0^l$$

$$K_{(34)3} = -m_i\Omega^2\left[\frac{l^2}{70}\right]\ldots(3)$$

$$K_{34} = K_{(34)1} + K_{(34)2} + K_{(34)3}$$

Substitute Eqs. 1, 2 and 3 in above to get,

$$K_{34} = \frac{-6EI}{l^2} + \frac{\Omega^2 A_i}{2}\left[\frac{-1}{10}\right] - m_i\Omega^2\left[\frac{l^2}{70}\right]$$

10. i = 4, j = 4

$$H_4 = \left[\left(\frac{x}{l}\right)^3 - \left(\frac{x}{l}\right)^2\right] l$$

$$H_4' = \left[3\left(\frac{x}{l}\right)^2 - 2\frac{x}{l}\right]$$

$$H_4'' = \frac{1}{l}\left[\frac{6x}{l} - 2\right]$$

Let

$$K_{(44)1} = \int_0^l EIH_4''H_4''dx$$

$$K_{(44)1} = \int_0^l EI\left(\frac{1}{l}\left[\frac{6x}{l} - 2\right]\right)^2 dx$$

$$K_{(44)1} = EI\left(\frac{1}{l^2}\right)\int_0^l \left(36\left(\frac{x}{l}\right)^2 - 24\left(\frac{x}{l}\right) + 4\right) dx$$

$$K_{(44)1} = EI\left(\frac{1}{l^2}\right)\left[\frac{36x^3}{3l^2} + x - \frac{24x^2}{2l} + 4x\right]_0^l$$

$$K_{(44)1} = \frac{4EI}{l}\ldots(1)$$

$$K_{(44)2} = \frac{\Omega^2 A_i}{2} \int_0^l H_4' H_4' dx$$

$$K_{(44)2} = \frac{\Omega^2 A_i}{2} \int_0^l \left[\left[3\left(\frac{x}{l}\right)^2 - 2\frac{x}{l}\right]\right]^2 dx$$

$$K_{(44)2} = \frac{\Omega^2 A_i}{2} \int_0^l \left[9\left(\frac{x}{l}\right)^4 - 12\left(\frac{x}{l}\right)^3 + 4\left(\frac{x}{l}\right)^2\right] dx$$

$$K_{(44)2} = \frac{\Omega^2 A_i}{2} \left[\left(\frac{9x^5}{5l^4}\right) - \left(\frac{12x^4}{4l^3}\right) + \left(\frac{4x^3}{3l^2}\right)\right]_0^l$$

$$K_{(44)2} = \frac{\Omega^2 A_i}{2} l \left[\frac{27 - 45 + 20}{15}\right]$$

$$K_{(44)2} = \frac{\Omega^2 A_i}{2} \left[\frac{2l}{15}\right] \ldots (2)$$

$$K_{(44)3} = -\frac{m_i \Omega^2}{2} \int_0^l \left(x^2 + 2xx_i\right) H_4' H_4' dx$$

$$K_{(44)3} = -\frac{m_i \Omega^2}{2} \int_0^l \left(x^2 + 2xx_i\right) \left[\left[3\left(\frac{x}{l}\right)^2 - 2\frac{x}{l}\right]\right]^2 dx$$

$$K_{(44)3} = -\frac{m_i \Omega^2}{2} \left[\frac{9x^6}{l^4} + \frac{4x^4}{l^2} + \frac{18x_i x^5}{l^4} - \frac{24x_i x^4}{l^3} + \frac{8x_i x^3}{l^2}\right] dx$$

$$K_{(44)3} = -\frac{m_i \Omega^2}{2} \left[\frac{9x^7}{7l^4} - \frac{12x^6}{6l^3} + \frac{4x^5}{5l^2} + \frac{18x_i x^6}{6l^4} - \frac{24x_i x^5}{5l^3} + \frac{8x_i x^4}{4l^2}\right]_0^l$$

$$K_{(44)3} = -m_i \Omega^2 \left[\frac{3l^3}{70} + \frac{x_i l^2}{10}\right] \ldots (3)$$

$$K_{44} = K_{(44)1} + K_{(44)2} + K_{(44)3}$$

Substitute Eqs. 1, 2 and 3 in above to get,

$$K_{44} = \frac{4EI}{l} + \frac{\Omega^2 A_i}{2} \left[\frac{2l}{15}\right] - m_i \Omega^2 \left[\frac{3l^3}{70} + \frac{x_i l^2}{10}\right]$$

We know that,

$$K_{12} = K_{21}, K_{13} = K_{31}, K_{14} = K_{41}, K_{23} = K_{32}, K_{24} = K_{42} \;\; and \;\; K_{34} = K_{43}$$

The stiffness matrix is given by,

$$k_{ij} = EI \begin{bmatrix} \frac{12}{l^3} & \frac{6}{l^2} & \frac{-12}{l^3} & \frac{6}{l^2} \\ \frac{6}{l^2} & \frac{4}{l} & \frac{-6}{l^2} & \frac{2}{l} \\ \frac{-12}{l^3} & \frac{-6}{l^2} & \frac{12}{l^3} & \frac{-6}{l^2} \\ \frac{6}{l^2} & \frac{2}{l} & \frac{-6}{l^2} & \frac{4}{l} \end{bmatrix} + \frac{\Omega^2 A_i}{2} \begin{bmatrix} \frac{6}{5l} & \frac{1}{10} & \frac{-6}{5l} & \frac{1}{10} \\ \frac{1}{10} & \frac{2l}{15} & \frac{-1}{10} & \frac{-l}{30} \\ \frac{-6}{5l} & \frac{-1}{10} & \frac{6}{5l} & \frac{-1}{10} \\ \frac{1}{10} & \frac{-l}{30} & \frac{-1}{10} & \frac{2l}{15} \end{bmatrix}$$

$$-m_i\Omega^2 \begin{bmatrix} \frac{3}{5}x_i + \frac{6l}{35} & \frac{l}{10}x_i + \frac{l^2}{28} & \frac{-3}{5}x_i - \frac{6l}{35} & \frac{-l^2}{70} \\ \frac{l}{10}x_i + \frac{l^2}{28} & \frac{l^2}{30}x_i + \frac{l^3}{105} & \frac{-l}{10}x_i - \frac{l^2}{28} & \frac{-l^2}{60}x_i - \frac{l^3}{140} \\ \frac{-3}{5}x_i - \frac{6l}{35} & \frac{-l}{10}x_i - \frac{l^2}{28} & \frac{3}{5}x_i + \frac{6l}{35} & \frac{l^2}{70} \\ \frac{-l^2}{70} & \frac{-l^2}{60}x_i - \frac{l^3}{140} & \frac{l^2}{70} & \frac{l^2}{30}x_i + \frac{l^3}{105} \end{bmatrix}$$

Appendix B
MATLAB Code

Code for Calculating Rotating and Non-rotating Natural Frequencies of Cantilevered Uniform Beam

```
% FEA to calculate Rotating and Non-Rotating Natural Frequencies of
% cantilevered uniform beam
% Code Written By Kalaiselvam.K
% Project Assistant of Dr.Ranjan Ganguli
% Rotorcraft and Mechanics lab
% Indian Institute of Science
clear all;
clc;
R=10;        % Radius of Blade in m
m=5;         % Mass per Unit Length of Blade in kg/m
EI=100000;   % Flexural Rigidity of Blade in N-m^2
Nel=50;      % Number of Elements
Dof=2;       % Degrees of Freedom per Node
l=(R/Nel);   % Length of Each Element in m
lambda = input('Enter value of lambda(Non-Dimensional Rotating speed)\n');
omega=lambda*sqrt((EI)/(m*(R^4)));% Rotating Speed in radian/sec
format loose
syms x;
H_1=((2*(x/l)^3)-(3*(x/l)^2)+1);
H_2=((((x/l)^3)-(2*(x/l)^2)+(x/l))*l);% Shape function's
H_3=(-(2*(x/l)^3)+(3*(x/l)^2));
H_4=((((x/l)^3)-((x/l)^2))*l);
H=[H_1 H_2 H_3 H_4].';
 H=vpa(H,6);
for i=1:(Dof*2)
    for j=1:(Dof*2)
M(i,j)=(m*int((H(i,1)*H(j,1)),0,l));% Element Mass Matrix
    end
end
M= vpa(M,6);
% Element Stiffness Matrix for simplification we divided the
% equation into 4 parts namely K_1,K_2,K_3 and K_4
for i=1:(Dof*2)
    for j=1:(Dof*2)
```


R. Ganguli, *Finite Element Analysis of Rotating Beams*,
Foundations of Engineering Mechanics, DOI 10.1007/978-981-10-1902-9

```
K_1(i,j)=(EI*int((diff(H(i,1),x,2)*(diff(H(j,1),x,2))),0,1));
    end
end
 K_1=vpa(K_1,6);
for i=1:(Dof*2)
    for j=1:(Dof*2)
K_2(i,j)=int((diff(H(i,1),x,1)*(diff(H(j,1),x,1))),0,1);
    end
end
 K_2=vpa(K_2,6);
for i=1:(Dof*2)
    for j=1:(Dof*2)
K_3(i,j)=int(2*x*(diff(H(i,1),x,1)*(diff(H(j,1),x,1))),0,1);
    end
end
K_3=vpa(K_3,6);
for i=1:(Dof*2)
    for j=1:(Dof*2)
K_4(i,j)=int((x^2)*(diff(H(i,1),x,1)*(diff(H(j,1),x,1))),0,1);
    end
end
 K_4=vpa(K_4,6);
 % Global Mass Matrix :- Assembly of Element Mass Matrices
GM = zeros(2*(Nel+1),2*(Nel+1));
for n=1:Nel
for i=(2*(n-1)+1):(2*(n+1))
    for j=(2*(n-1)+1):(2*(n+1))
GM(i,j)=GM(i,j)+M((i-(2*(n-1))),((j-(2*(n-1)))));
    end
end
end
GM=vpa(GM,6);
% Global Stiffness Matrix:- Assembly of Element Stiffness Matrices for K_1
GK1 = zeros(2*(Nel+1),2*(Nel+1));
for n=1:Nel
for i=(2*(n-1)+1):(2*(n+1))
    for j=(2*(n-1)+1):(2*(n+1))
GK1(i,j)=GK1(i,j)+K_1((i-(2*(n-1))),((j-(2*(n-1)))));
    end
end
end
GK1=vpa(GK1,6);
% Calculating where X is the position of element from Fixed End
for n=1:Nel
    for j=n:Nel+1
        X(j)=(j-1)*l;
    end
end
X=vpa(X,6);
% Calculating A value for each element
A(i)=0;
for i=1:Nel
    for j=i:Nel
        A(i)= A(i)+((X(j+1))^2)-((X(j))^2);

    end
end
A=vpa(A,6);
% Global Stiffness Matrix:- Assembly of Element Stiffness Matrices for K_2
GK2 = zeros(2*(Nel+1),2*(Nel+1));
for n=1:Nel
for i=(2*(n-1)+1):(2*(n+1))
    for j=(2*(n-1)+1):(2*(n+1))
```

```
GK2(i,j)=GK2(i,j)+(m*(omega^2))*(A(n)/2)
*(K_2((i-(2*(n-1))),((j-(2*(n-1))))));
    end
end
end
GK2=vpa(GK2,6);
% Global Stiffness Matrix:- Assembly of Element Stiffness Matrices for K_3
GK3 = zeros(2*(Nel+1),2*(Nel+1));
for n=1:Nel
for i=(2*(n-1)+1):(2*(n+1))
    for j=(2*(n-1)+1):(2*(n+1))
GK3(i,j)=GK3(i,j)+(m*(omega^2))*(X(n)/2)
*(K_3((i-(2*(n-1))),((j-(2*(n-1))))));
    end
end
end
GK3=vpa(GK3,6);
% Global Stiffness Matrix:- Assembly of Element Stiffness Matrices for K_4
GK4 = zeros(2*(Nel+1),2*(Nel+1));
for n=1:Nel
for i=(2*(n-1)+1):(2*(n+1))
    for j=(2*(n-1)+1):(2*(n+1))
GK4(i,j)=GK4(i,j)+(m*(omega^2)/2)*(K_4((i-(2*(n-1))),((j-(2*(n-1))))));
    end
end
end
GK4=vpa(GK4,6);
GK=GK1+GK2-GK3-GK4; % Global Stiffness Matrix
GK=vpa(GK,6);
% Global Stiffness Matrix: Applying Boundary Conditions for a Cantilever Beam
GK(1,:)=[];
GK(:,1) = [];
GK(1,:)=[];
GK(:,1) = [];
GK_B=vpa(GK,4);
% Global Mass Matrix: Applying Boundary Conditions for a Cantilever Beam
GM(1,:)=[];
GM(:,1) = [];
GM(1,:)=[];
GM(:,1) = [];
GM_B=vpa(GM,4);
% double converts the symbolic object to a numeric object
 GM_B = double(GM_B);
 GK_B = double(GK_B);
 % Calculating Eigen values(D) and Eigen vectors(V) of Reduced Global matrices
[V,D]=eig(GK_B,GM_B);
% Natural Frequencies from the Reduced Global Matrices in rad/s
for i = 1:(2*Nel)
    Frequency(i) = sqrt(D(i,i));
end
Frequency=vpa(Frequency,6)
% Natural Frequencies of rotating blade in non-dimensional form
for i=1:(2*Nel)
 Non_dim_nat_freq(i) = Frequency(i)/sqrt((EI)/(m*(R^4)));
end
Non_dim_nat_freq=vpa(Non_dim_nat_freq,6)
```

OUTPUT: 1

```
Enter value of lambda(Non-Dimensional Rotating speed)
0

Frequency =

[4.9724, 31.1615, 87.2531, 170.981, 282.645, 422.225, 589.725, 785.149, 1008.5,
1259.79, 1539.04, 1846.25, 2181.46, 2544.69, 2935.98, 3355.37, 3802.93, 4278.72,
4782.83, 5315.34, 5876.36, 6466.03, 7084.49, 7731.9, 8408.44, 9114.32, 9849.77,
10615.0, 11410.4, 12236.1, 13092.6, 13980.1, 14899.0, 15849.8, 16832.7, 17848.3,
18896.9, 19979.0, 21094.8, 22244.8, 23428.9, 24647.2, 25899.2, 27183.5, 28497.5,
29835.8, 31186.3, 32521.9, 33772.8, 34757.0, 39181.4, 40309.6, 41775.3, 43411.3,
45155.3, 46983.2, 48885.0, 50856.5, 52896.1, 55004.0, 57180.9, 59428.4, 61748.0,
64141.7, 66611.6, 69160.1, 71789.4, 74501.8, 77299.7, 80185.5, 83161.2, 86228.9,
89390.3, 92646.8, 95999.3, 99448.2, 102993.0, 106632.0, 110363.0, 114182.0,
118083.0,122059.0, 126098.0, 130188.0, 134313.0, 138453.0, 142583.0, 146676.0,
150698.0,154613.0, 158377.0, 161944.0, 165266.0, 168289.0, 170963.0, 173235.0,
175059.0,176394.0, 177209.0, 211700.0]

Non_dim_nat_freq =

[3.51602, 22.0345, 61.6972, 120.902, 199.86, 298.558, 416.999, 555.184,
713.119, 890.809, 1088.26, 1305.5, 1542.52, 1799.37, 2076.05, 2372.61,
2689.08, 3025.51, 3381.97, 3758.51, 4155.22, 4572.18, 5009.49, 5467.28,
5945.67, 6444.8, 6964.84, 7505.96, 8068.36, 8652.25, 9257.85, 9885.42,
10535.2, 11207.5, 11902.5, 12620.7,13362.1, 14127.3, 14916.3, 15729.4,
16566.7, 17428.2, 18313.5, 19221.6, 20150.8, 21097.1, 22052.1, 22996.5,
23880.9, 24576.9, 27705.4, 28503.2, 29539.6, 30696.4, 31929.6, 33222.2,
34566.9, 35960.9, 37403.2, 38893.7, 40433.0, 42022.2, 43662.4, 45355.0,
47101.5, 48903.6, 50762.7, 52680.7, 54659.2, 56699.7, 58803.8, 60973.0,
63208.5, 65511.2, 67881.8, 70320.5, 72827.0, 75400.4, 78038.7, 80739.1,
83497.6, 86308.6, 89164.8, 92057.1, 94973.9, 97901.1, 100822.0,
103716.0, 106560.0, 109328.0, 111989.0, 114512.0, 116860.0, 118999.0,
120889.0, 122496.0, 123786.0, 124730.0, 125305.0, 149694.0]
```

OUTPUT: 2

```
Enter value of lambda(Non-Dimensional Rotating speed)
12

Frequency =

[18.6254, 53.1788, 112.592, 198.746, 311.887, 452.417, 620.565, 816.454,
1040.16, 1291.72, 1571.17, 1878.56, 2213.91, 2577.26, 2968.65, 3388.13,
3835.76, 4311.61, 4815.77, 5348.32, 5909.39, 6499.09, 7117.57, 7765.0,
8441.56, 9147.45, 9882.91, 10648.2, 11443.5, 12269.3, 13125.7, 14013.2,
14932.1, 15882.9, 16865.8, 17881.3, 18929.9, 20012.0, 21127.8, 22277.7,
23461.8, 24680.0, 25931.9, 27216.2, 28530.2, 29868.4, 31218.9, 32554.5,
33805.5, 34792.4, 39213.4, 40340.3, 41806.2, 43442.3, 45186.4, 47014.3,
48916.0, 50887.4, 52926.9, 55034.7, 57211.6, 59458.9, 61778.3, 64171.9,
66641.8, 69190.1, 71819.2, 74531.5, 77329.4, 80215.0, 83190.6, 86258.1,
89419.4, 92675.8, 96028.2, 99477.0, 103022.0, 106661.0, 110392.0,
114211.0, 118112.0, 122087.0, 126126.0, 130217.0, 134342.0, 138481.0,
142612.0, 146704.0, 150727.0, 154641.0, 158405.0, 161973.0, 165294.0,
168318.0, 170992.0, 173264.0, 175089.0, 176424.0, 177240.0, 211701.0]
```

```
Non_dim_nat_freq =

[13.1702, 37.6031, 79.6145, 140.535, 220.537, 319.907, 438.806, 577.32, 735.501,
913.382, 1110.99, 1328.34, 1565.47, 1822.4, 2099.15, 2395.77, 2712.29, 3048.77,
3405.26, 3781.83, 4178.57, 4595.55, 5032.88, 5490.68, 5969.08, 6468.22, 6988.27,
7529.4, 8091.79, 8675.68, 9281.28, 9908.83, 10558.6, 11230.9, 11925.9, 12644.0,
13385.5, 14150.6, 14939.6, 15752.7, 16590.0, 17451.4, 18336.6, 19244.8, 20173.9,
21120.2, 22075.1, 23019.5, 23904.1, 24601.9, 27728.1, 28524.9, 29561.5, 30718.4,
31951.6, 33244.1, 34588.8, 35982.8, 37425.0, 38915.4, 40454.7, 42043.8, 43683.9,
45376.4, 47122.8, 48924.8, 50783.9, 52701.8, 54680.1, 56720.5, 58824.6, 60993.7,
63229.1, 65531.7, 67902.2, 70340.9, 72847.3, 75420.6, 78058.8, 80759.2, 83517.6,
86328.6, 89184.8, 92077.1, 94993.9, 97921.0, 100842.0, 103736.0, 106580.0,
109348.0, 112009.0, 114532.0, 116881.0, 119019.0, 120909.0, 122516.0, 123806.0,
124750.0, 125327.0, 149695.0]
```

Appendix C
Governing Equation for Rotating Timoshenko Beam

Consider the Timoshenko beam element shown in Fig. C.1.

The origin of this element is at a distance of r_i from the axis of rotation. The x-axis coincides with the centroidal axis of the beam. The y-axis is perpendicular and points away from the plane of the paper. The z-axis is parallel to the axis of rotation and coincides with the axis of rotation of the beam.

The uniform strain $\varepsilon(x)$ of the cross-section due to the centrifugal force $T(x)$ is given by,

$$\varepsilon(x) = \frac{T(x)}{EA} \tag{C.1}$$

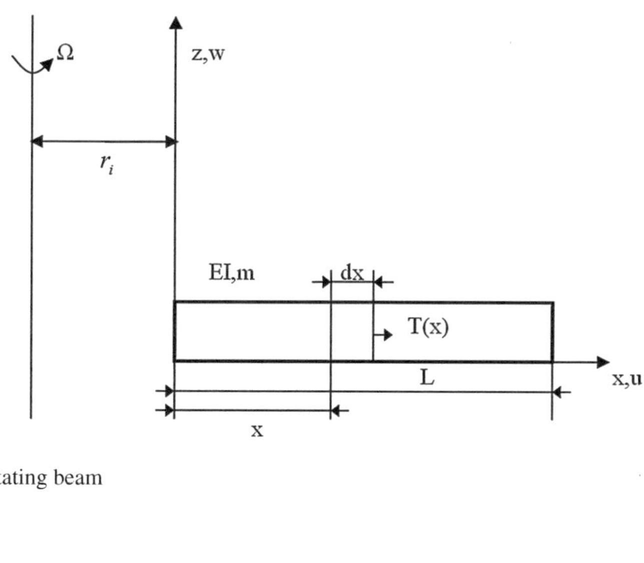

Fig. C.1 Rotating beam

R. Ganguli, *Finite Element Analysis of Rotating Beams*,
Foundations of Engineering Mechanics, DOI 10.1007/978-981-10-1902-9

Here, x is the distance of the cross-section from the origin. Also,

$$T(x) = \int_{x}^{L} \rho A\Omega^2 (r_i + x)\, dx = \rho A\Omega^2 \left\{ r_i (L - x) + \frac{(L^2 - x^2)}{2} \right\} \tag{C.2}$$

The strain $\varepsilon_0(x)$ can be related to the axial displacement $u_0(x)$ by,

$$u_0^{'}(x) = \varepsilon_0(x) = \frac{T(x)}{EA} \tag{C.3}$$

In Timoshenko beam theory, shear strain γ is given as,

$$\gamma = w^{'} - \theta \tag{C.4}$$

where $w(x)$ is the flexural displacement of the beam neutral axis.

The element dy deforms due to the combined axial and flexural displacements. At the left edge of the element, a point at a distance of $\bar{\eta}$ away from the neutral axis in the z-direction has the co-ordinates: $\{(x + u_0 - \bar{\eta}\theta)\,, 0, (\bar{\eta} + w)\}$. At the right edge, the corresponding point has the coordinates:

$$\left[\left\{x - u_0 - \bar{\eta}\theta + \left(1 + u_0^{'} - \bar{\eta}\theta^{'}\right) dx\right\}, 0, \left(\bar{\eta} + w + w^{'} dx\right)\right]$$

So, the strain of the element at a distance $\bar{\eta}$ from the neutral axis is,

$$\varepsilon(x, n) = \left[\left(1 + u_0^{'} - \bar{\eta}\theta^{'}\right)^2 + \left(w^{'}\right)^2\right] - 1 \cong u_0^{'} - \bar{\eta}\theta' + \frac{(w')^2}{2} \tag{C.5}$$

The strain energy due to flexure can be written as

$$U_f = \int\int\int_v \frac{E\varepsilon^2}{2} dV = \frac{E}{2}\int_A \int_0^L \left\{u_0^{'} - \bar{\eta}\theta' + \frac{(w')^2}{2}\right\}^2 dxdA \tag{C.6}$$

This can also be written as

$$U_f = \frac{EA}{2}\int_0^L (u_0^{'})^2 dx + \frac{EI}{2}\int_0^L (\theta')^2 dx + EA\int_0^L u_0^{'}(w')^2 dx \tag{C.7}$$

Since

$A = \int_A dA$ and $I = \int_A (\bar{\eta})^2 dA$

Using Eqs. C.2 and C.3, Eq. C.7 can be written as

$$U_f = C_1 + \frac{1}{2}\left[\int_0^L EI(\theta')^2 dx + \int_0^L T(w')^2 dx\right] \tag{C.8}$$

where C_1 is a constant.

The strain energy due to shear is

$$U_s = \int\int\int_v \frac{G\gamma^2}{2} dV = \frac{1}{2}\int_A\int_0^L G\gamma^2 dAdx = \frac{1}{2}kAG\int_0^L \gamma^2 dx \tag{C.9}$$

substituting for γ yields,

$$U_s = \frac{1}{2}\int_0^L kAG\left(w' - \theta\right)^2 dx \tag{C.10}$$

Now, the total strain energy of the beam is given by

$$U = U_f + U_s = \frac{1}{2}\left[\int_0^L \left\{EI(\theta')^2 + T(w')^2 + kAG\left(w' - \theta\right)^2\right\} dx\right] + C_1 \tag{C.11}$$

Next, we need to calculate the kinetic energy of the Timoshenko beam. Consider point at a distance of $\bar{\eta}$ from the neutral axis. The velocities at this point along the x, y and z directions are

$$V_x = -\bar{\eta}\dot{\theta}, \quad V_y = -\Omega\left(x + u_0 - \bar{\eta}\theta\right), \quad V_z = \dot{w} \tag{C.12}$$

The kinetic energy of the rotating timoshenko beam is

$$J = \frac{1}{2}\int_A\int_0^L \left(V_x^2 + V_y^2 + V_z^2\right)\rho dAdx \tag{C.13}$$

substituting Eq. C.12 into Eq. C.13 gives

$$J = \frac{1}{2}\int_A\int_0^L \rho\Omega^2(x + u_0)^2 dxdA + \frac{1}{2}\int_A\int_0^L \rho\Omega^2\bar{\eta}^2\theta^2 dxdA$$
$$-\frac{1}{2}\int_A\int_0^L 2\rho\Omega^2\bar{\eta}\theta(x + u_0)dxdA + \frac{1}{2}\int_A\int_0^L \rho\bar{\eta}^2\dot{\theta}^2 dxdA + \frac{1}{2}\int_A\int_0^L \rho\dot{w}^2 dxdA \tag{C.14}$$

close observation of the above equation reveals that the first integral is constant and the third one is zero. This yields a simplex expression for the kinetic energy,

$$J = C_2 + \frac{1}{2}\left[\int_0^L \rho I(\Omega^2\theta^2 + \dot{\theta}^2)dx + \int_0^L \rho A\dot{w}^2 dx\right] \tag{C.15}$$

where C_2 is a constant. The lagrangian $L = J - U$ can now be written as,

$$L = \frac{1}{2}\int_0^L \{\rho I(\Omega^2\theta^2 + \dot{\theta}^2) + \rho A\dot{w}^2\}dx$$

$$-\frac{1}{2}\int_0^L \left\{EI(\theta')^2 + T(w')^2 + kAG\left(w' - \theta\right)^2\right\} dx + C_2 - C_1 \tag{C.16}$$

According to Hamilton's principle,

$$\delta\int_{t_1}^{t_2} Ldt = 0 \tag{C.17}$$

which leads to,

$$\int_{t_1}^{t_2}\int_0^L \{\rho I(\Omega^2\theta\delta\theta + \dot{\theta}\delta\dot{\theta}) + \rho\dot{w}A\delta\dot{w}\}dxdt$$

$$-\int_{t_1}^{t_2}\int_0^L \{EI\theta'\delta\theta' + Tw'\delta w' + kAG\left(w' - \theta\right)\left(\delta w' - \delta\theta\right)\}dxdt = 0 \tag{C.18}$$

Integrating by parts yields,

$$\int_{t_1}^{t_2}\int_0^L \{(Tw')' - \rho A\ddot{w} + kAG(w'' - \theta')\}\delta w dt$$

$$+\int_{t_1}^{t_2}\int_0^L \{EI\theta'' + kAG(w' - \theta) - \rho I\ddot{\theta} + \rho I\Omega^2\theta\}\delta\theta dt + \int_{t_1}^{t_2}\left[-EI\theta'\delta\theta\right]_0^L dt$$

$$+\int_{t_1}^{t_2}[-\{Tw' + kAG(w' - \theta)\}\delta w]_0^L dt + \int_0^L [\rho I\dot{\theta}\delta\theta + \rho A\dot{w}\delta w]_{t_1}^{t_2} dx = 0 \tag{C.19}$$

Now, δw and $\delta\theta$ are completely arbitrary. The above equation therefore yields two equations.

$$\left(Tw'\right)' - \rho A\ddot{w} + kAG\left(w'' - \theta'\right) = 0 \tag{C.20}$$

and

$$EI\theta'' + kAG\left(w' - \theta\right) - \rho I\ddot{\theta} + \rho I\Omega^2\theta = 0 \tag{C.21}$$

The bending moment and shear force are also obtained as,

$$M = -EI\theta' \tag{C.22}$$

and

$$Q = -Tw' - kAG\left(w' - \theta\right) = -Tw' - M' - \rho I\ddot{\theta} + \rho I\Omega^2\theta. \tag{C.23}$$

MIX
Papier aus verantwortungsvollen Quellen
Paper from responsible sources
FSC® C105338

If you have any concerns about our products,
you can contact us on
ProductSafety@springernature.com

In case Publisher is established outside the EU,
the EU authorized representative is:
Springer Nature Customer Service Center GmbH
Europaplatz 3, 69115 Heidelberg, Germany

Printed by Libri Plureos GmbH
in Hamburg, Germany